Series:

Genome Designing of Crops

Series editor: Chittaranjan Kole, Kolkata, West Bengal, India

1. Biofortification for Nutrient-Rich Crops
   Monika Garg, Saloni Sharma and Apoorv Tiwari (Editors), 2025
2. Omics and System Biology Approaches for Delivering Better Cereals
   Dinesh Kumar Saini and Chittaranjan Kole (Editors), 2025

Genome Designing of Crops
Series editor: Chittaranjan Kole

Biofortification for Nutrient-Rich Crops

Editors

Monika Garg, Saloni Sharma and Apoorv Tiwari

National Agri-Food Biotechnology Institute (NABI)
Punjab, India

CRC Press
Taylor & Francis Group
Boca Raton London New York

CRC Press is an imprint of the
Taylor & Francis Group, an **informa** business

First edition published 2025
by CRC Press
2385 NW Executive Center Drive, Suite 320, Boca Raton FL 33431

and by CRC Press
4 Park Square, Milton Park, Abingdon, Oxon, OX14 4RN

CRC Press is an imprint of Taylor & Francis Group, LLC

Library of Congress Cataloging-in-Publication Data (applied for)

ISBN: 978-1-032-66727-0 (hbk)
ISBN: 978-1-032-69062-9 (pbk)
ISBN: 978-1-032-69063-6 (ebk)

DOI: 10.1201/10.1201/9781032690636

Typeset in Times New Roman
by Prime Publishing Services

Dedicated to

Prof. M.S. Swaminathan
(7 August 1925–28 September 2023)

"A world scientist of rare distinction"

Mankombu Sambasivan Swaminathan: A visionary scientist, Father of India's historic "Green Revolution", staunch advocate of farmers' welfare. His groundbreaking work in agriculture transformed the lives of millions and ensured food security for our nation.

Preface

Strength of the Book

This comprehensive volume harnesses the expertise of globally renowned experts in biofortification, delving meticulously into pivotal nutrients across dedicated chapters. These luminaries' collective wisdom and practical insight infuse the text with unparalleled depth and authority. The inclusion of illustrative figures aids in summarizing complex concepts, accompanied by an exploration of cutting-edge innovations. The text's accessible language makes advanced ideas understandable, embracing a wide readership. As research in this field continues to evolve, with new projects emerging and others concluding, this book is up-to-date, encapsulating the latest work and strategies.

All this Book Covers

This comprehensive book on biofortification provides a scientifically rich and engaging exploration of strategies, from pioneering innovations to meticulously refined methods. It tackles global nutrient deficiencies head-on and their implications for agriculture. What makes this book stand out is its in-depth coverage of various aspects of biofortification and nutrients. It delves into food and nutritional security, placing a strong emphasis on the vital role of micronutrients and vitamins in human health. It offers insights into historical and contemporary approaches, spanning diverse breeding methods, biofortification efficacy, bioavailability, and strategic development.

The book boasts a broad scope, encompassing fundamental biofortification principles, global efforts, challenges, and integrating cutting-edge 'Omic' technologies with traditional methods. It explores well-studied micronutrients and vitamin biofortification. Furthermore, it addresses anti-nutrients such as phytic acid, lectins, and glucosinolates. Readers will also appreciate its exploration of scientific research methodologies and meta-analyses, providing a critical assessment of nutrient effectiveness. In addition, it highlights emerging elements like phenolics, starch, and dietary fiber, particularly relevant in the context of addressing obesity concerns in both developed and developing countries.

With 19 chapters, this book serves as a comprehensive and captivating resource in the field of biofortification. It combines scientific rigor with engaging reading, making it an excellent choice for researchers and students seeking to delve deeper into this critical area of study.

Go through the Book in 15 minutes

Exploring the book's figures within individual chapters reveals a scientific narrative of progress in the field. These figures intricately compile pertinent information, while the abstracts offer concise yet concrete insights, eliminating redundancy in statements. Dive into these visual and textual aids to unravel the captivating journey of biofortification's evolution.

HarvestPlus and Beyond It

HarvestPlus, a visionary nonprofit founded in 2004, leads the biofortification frontier. It works across CGIAR, enriching 12 vital crops including—rice, wheat, maize, beans, cassava, sweet potatoes, and pearl millet—with vitamin A, iron, or zinc. Over 400 biofortified varieties have blossomed across 40+ nations, nourishing countless lives.

Beyond HarvestPlus, global institutes and industry leaders like Syngenta, BASF SE, DuPont, Bayer AG, Nestle, Okanagan Specialty Fruits Inc, ArborGen Inc, Arcadia Biosciences, Corteva, MAURO Seed Co., Mahyco, AgBioForum, and Evogene Ltd. take up the mantle of biofortification. They expand nutrients, such as vitamins, phytochemicals, prebiotics, dietary fiber and lipids, optimize amino acids, and tame anti-nutrients in more crops. This symphony of innovation cultivates a banquet of unprecedented nourishment where science and sustenance intertwine.

Contributing Authors

Distinguished scientists from the United States, South Africa, Switzerland, Israel, and India, all active practitioners in their respective fields, have generously shared their extensive knowledge and invaluable insights within this volume, making substantial contributions to their respective disciplines.

The Editing Process and Acknowledgements

The editors, Monika Garg, Saloni Sharma, and Apoorv Tiwari, performed an in-depth analysis of the chapters' text, figures, and graphs and then modified those aspects per requirements. In addition, Satveer Kaur, a Research Scholar at NABI, offered her assistance throughout the editorial process for this book.

Our Audience

Our primary readership encompasses academic institutions and libraries. Nonetheless, this compendium extends its reach to a diverse array of stakeholders in agriculture, nutrition, and health, including researchers, academics, extension workers, policymakers, students, and all involved in the value chain. Within its pages lies a mission – the battle against micronutrient deficiency and hidden hunger, achieved through education and the widespread adoption of biofortification. It serves as a valuable tool for educating professionals in agriculture, health, nutrition, and community development on cultivating and utilizing biofortified crop varieties.

Dr. Monika Garg
Dr. Saloni Sharma
Dr. Apoorv Tiwari

List of Contributors

1. **Michael Baum** International Centre for Agricultural Research in the Dry Areas, Morocco
2. **Uttam Kumar:** Borlaug Institute for South Asia Ludhiana, Punjab, India
3. **Shiv Kumar** ICARDA, NASC Complex, New Delhi, India
4. **Neeraja C.N.** ICAR-Indian Institute of Rice Research, Rajendranagar, Hyderabad, India
5. **Dinesh Pandey** G.B. Pant University of Ag. & Tech. Pantnagar, India
6. **Rohit Kumar** Clemson University, South Carolina, USA
7. **Maryke Labuschagne** University of the Free State, Bloemfontein, South Africa
8. **Teresa Fitzpatrick** University of Geneva, Geneva, Switzerland.
9. **Catherine Simpson** Texas Tech University, Lubbock, Texas, USA
10. **Firoz Hossain** ICAR-Indian Agricultural Research Institute, New Delhi, India
11. **Upendra Kumar** CCS Haryana Agricultural University, Hisar, India
12. **Dr. Joy K. Roy** National Agri-Food Biotechnology Institute, Mohali, India.
13. **Dr. Moshe Goldsmith** Weizmann Institute of Science, Rehovot, Israel.
14. **Wusirika Ramakrishna** Central University of Punjab, Punjab, India
15. **Gurpreet Kaur** Punjab Agricultural University, Punjab, India
16. **Manjeet Kaur Sangha** Punjab Agricultural University, Punjab, India
17. **Parveen Chuuneja** Punjab Agricultural University, Ludhiana, Punjab, India

16\. **Apoorv Tiwari** National Agri-Food Biotechnology Institute, Punjab, India

19\. **Monika Garg** National Agri-Food Biotechnology Institute, Punjab, India

About the Editors

Dr. Monika Garg: She is a scientist at the premier Indian institute, NABI, Mohali for the last 13 years. She gained her PhD and post-doctoral experience from Tottori University, Japan. She had also served ICARDA, CGIAR, Syria, as Associate Scientist. She has more than 70 publications in reputed journals with an h-index of 21 and Editor/Reviewer for various journals. She is working on the biofortification of wheat to improve its nutritional quality. She is involved in the exploration of biofortified wheat as a commercial crop in India and understanding its effect on lifestyle disorders. She is also working on the pre-breeding platform for improved heat tolerance and sustainable production. Her special interest lies in anthocyanin biofortified colored wheat.

Dr. Saloni Sharma: Currently, she is working as an ICMR-postdoctoral fellow at NABI, to validate wheatgrass health benefits. Simultaneously, due to interest and experience in plants genetics, genomics, and physiology, she is also working to understand the colored wheat crop genetics, genomics, and stress-responsive physiology and studying wheat biofortification. She completed her PhD degree from PU, Chandigarh, India while performing her research work at NABI itself. Before PhD she had 4-year research experience at Agriculture University, Palampur (H.P.).

Dr. Apoorv Tiwari: Dr. Apoorv Tiwari holds an M.Sc. from CSJM University, Kanpur, and M.Tech. and Ph.D. in Bioinformatics from SHUATS, Prayagraj, India. Currently a Project Scientist at the National Agri-Food Biotechnology Institute in Mohali, Punjab, he conducted research on the genomics of colored wheat. Previously, as a Senior Researcher at GBPUA&T, Pantnagar, Uttarakhand, he focused on GWAS, identifying potential SNP markers for seed protein content in finger millet. With over 50 publications in renowned journals and editorial roles, Dr. Tiwari's expertise encompasses crop genomics, proteomics, transcriptomics, GWAS, QTL mapping, statistical methodologies, and computer-aided drug designing. His contributions have earned him recognition with the Young Scientist and Governor's awards.

Contents

CHAPTER

1

Biofortification: Concept and Methodologies

Michael Baum[a*[0000-0002-8248-6088]], Vandita Tiwari[b], Maninder Meenu[b], Kiran Khandare[c], Bhawna Sheoron[b], Satveer Kaur[b] and Mona Yadav[b]

[a] International Centre for Agricultural Research in the Dry Areas (ICARDA)

[b] Department of Agri-Biotechnology, National Agri-Food Biotechnology Institute, Mohali, Punjab - 140306, India

[c] Chemical Engineering Division, Center of Innovative and Applied Bioprocessing (CIAB), Mohali, Punjab - 140306, India

1. Why is Biofortification Research Necessary?

Biofortification research is essential for several socioeconomic development and human health reasons. Humans require approximately 40 nutrients in adequate quantities to maintain healthy and productive lives (Table 1). These nutrients can be broadly classified into two categories: essential and non-essential. While essential nutrients must be obtained from food since the human body cannot synthesize them, non-essential nutrients, which the body can produce, do not require direct food consumption. Essential macronutrients, such as sodium, potassium, calcium, magnesium, phosphorous, chlorine, and sulphur, are needed in medium to large amounts. While essential micronutrients, including iron, zinc, copper, manganese, iodine, selenium, molybdenum, cobalt, and nickel, as well as various vitamins like A, D, E, K, and water-soluble vitamins from the B series and vitamin C are required in small quantities but both are equally important for the body (Prashanth et al., 2015). These nutrients play a critical role in human health and impact physical and mental growth.

When the intake of energy and nutrients becomes imbalanced or insufficient, it results in a condition defined as malnutrition by the World Health Organization (WHO). Malnutrition can manifest in various forms and affects individuals across

*Corresponding author: m.baum@cgiar.org

Table 1: Tabular illustration of micro- and macronutrients required for human health (Adopted from Garg et al., 2018, Frontiers in Nutrition)

Micronutrients		**Macronutrients**		
Micro-minerals	**Vitamins**	**Amino Acids (Essential)**	**Fatty Acids (Essential)**	**Macrominerals**
Fe	A (Retinol)	Histidine	Linoleic acid	K
Zn	D (Calciferol)	Isoleucine	Linolenic acid	Ca
Cu	E (x-Tocopherol)	Leucine		Mg
Mn	K (Phylloquinone)	Lysine		S
I	C (Ascorbic acid)	Methionine		P
Se	B1 (Thiamin)	Phenylalanine		Na
Mo	B2 (Riboflavin)	Threonine		Cl
Co	B3 (Niacin)	Tryptophan		
Ni	B5 (Pantothenic acid)	Valine		
	B6 (Pyridoxine)			
	B7 (Biotin)			
	B9 (Folic acid)			
	B12 (Cobalamin)			

the globe, posing a significant challenge to global health (Thompson et al., 2017). Malnutrition can lead to stunted growth, cognitive impairments, and heightened infection susceptibility.

In developing countries, particularly among populations adhering to non-vegetarian dietary practices, agriculture is the primary source of nourishment. Modern agriculture has successfully met the energy demands of underprivileged communities in emerging nations and supplied essential macroelements like carbon, hydrogen, nitrogen, phosphorus, sulphur, potassium, magnesium, sodium, and chlorine. However, specific segments of the population, such as the rural poor, heavily rely on locally produced staple foods due to limited resources and access to markets (Hotz, 2013). These staple crops can provide carbohydrates, fats, and fibre. Unfortunately, these staple foods often lack sufficient nutrients like proteins and essential macro and micronutrients (Table 1) to support optimal health.

Furthermore, a micronutrient-rich diet-based approach can offer additional benefits alongside inorganic supplements during abnormal instances of diseases or infections involving increased nutrient excretion, such as diarrhoea and dysentery. This strategy helps to maintain adequate levels of essential nutrients within the body in normal instances and can contribute to preventing diseases.

1.1. Meaning of Biofortification

'Biofortification', also known as 'biological fortification', is an agricultural strategy that boosts the nutrient content of crops by increasing the density of vitamins and minerals in staple foods, enhancing their bioavailability for the human population (Bouis et al., 2011). The key goal of biofortification is to enrich staple crops with essential micronutrients, enhancing their value for human health. This approach offers a sustainable solution to address widespread micronutrient deficiencies, particularly in regions with limited dietary diversity and supplement access. By enhancing the presence of vital nutrients in commonly consumed foods, biofortification can potentially mitigate malnutrition and its related health impacts.

Biofortification integrates modern scientific advancements with traditional agricultural practices, demonstrating a holistic approach. It holds promise for enhancing global food security, improving population nutrition, fostering human health, and reducing diet-related health issues. Unlike ongoing supplementation, a one-time investment in plant breeding results in nutrient rich crops, making a lasting global impact on nutrition. This cost-effective method enhances the nutritional value of staple crops and diminishes reliance on resource-intensive interventions, fostering sustainability.

1.2. Advantages and Limitations of Biofortification

Advantages

(i) *Reaching Malnourished Rural Populations:* Biofortification benefits underserved rural communities with limited access to fortified foods and supplements. Malnutrition can be tackled at its source by cultivating modern crop varieties enriched with vital nutrients, often developed by agricultural research centers, and distributed by NGOs and government agencies (Bouis et al., 2011).

(ii) *Cost Effectiveness:* Although biofortified staple foods may not match the nutrient levels of supplements or industrially fortified foods, they can significantly alleviate micronutrient deficiencies. This cost-effective strategy offers a sustainable way to elevate millions from malnourishment to better nutrition, especially when targeting those with mild deficiencies.

(iii) *Sustainability:* Once established, the system becomes sustainable. Initial development costs may be high, but recurrent expenses for maintaining and monitoring crop nutrient traits are modest compared to the benefits, making the strategy long-lasting.

Limitations

(i) *Varying Impact Across Life Stages:* Biofortified staple foods contribute to lifelong micronutrient stores, yet their impact varies among age groups due to consumption patterns, existing deficiencies, and specific nutrient needs.

(ii) *Time Dimension:* Adopting biofortified crops across developing nations takes about a decade. Confirming effectiveness through deficiency reductions is essential, placing biofortification alongside supplementation and fortification for micronutrient malnutrition reduction.

(iii) *Productivity, Bioavailability, and Diversity:* Micronutrient deficiency worsens due to bioavailability issues, nutrient inhibitors like phytate, and insufficient vitamin C or meat consumption (Lockyer et al., 2018). Balancing yield and micronutrient levels pose challenges (Calayugan et al., 2020). Genetic uniformity and limited breeding alleles impact diversity and crop productivity.

(iv) *Limited Evidence and Classification:* WHO assessed biofortification in 2016, noting scarce evidence and incomplete reviews; classified it as a "Category 3 Intervention", so more data is needed. Micronutrient rise is observed, but health outcomes lack proof. Further research is imperative to comprehend biofortification's full potential and public health impact for broader acceptance.

(v) *Significant investment:* To substantially increase micronutrient in staple crops such as wheat e.g., Zn and Fe it requires a significant investment over years. A balanced and diverse diet with staple crops, legumes, and vegetables is appropriate to avoid micronutrient malnutrition in most of the instances, but in many parts of the world they are not available and dependence on staple cereal specially in case of environmental stresses and famines is the only solution.

2. Methodologies Employed for Biofortification

Biofortification employs three primary strategies: agronomic, breeding-based, and genetic. Since its initial inception, several advancements have been made. The following sections provide a detailed discussion of the methodologies used for plant biofortification.

2.1. Agronomic Biofortification

Agronomic biofortification refers to the practice of enhancing the nutrient content of crops by manipulating their growing conditions, primarily through agricultural practices and soil management. Typically, it involves the usage of mineral fertilizers and increasing their solubilization in the soil, as well as their mobilization into edible plant parts. The goal is to improve the nutritional quality of the harvested produce and subsequently address micronutrient deficiencies in populations that rely heavily on these crops as a staple food source. Macrominerals such as nitrogen, phosphorus, and potassium are essential in achieving increased agricultural yields and are present in most fertilizers (Erisman et al., 2008). Iron, zinc, copper, cobalt, manganese, iodine, molybdenum, selenium, and nickel are microminerals typically absorbed by the soil and present in different amounts in the edible portions of specific plants. Numerous studies have shown that agronomic biofortification of these micronutrient fertilizers (Fe, Zn, Se, and iodine) efficiently increases these nutrients in the edible parts of plants (Jaiswal et al., 2022; Cakmak and Kutman, 2018). Thus, we can reduce human micronutrient deficiency by improving the soil's micronutrient status by applying micronutrient-rich fertilizers (Cakmak and Kutman, 2018). As elucidated by Bouis and Saltzman (2017), the success of agronomic biofortification hinges on the tailoring of approaches to the specific agroecological contexts and dietary habits of target populations. Customizing interventions based on regional conditions ensures the optimal impact of agronomic practices on nutritional enhancement.

Techniques Used for Agronomic Biofortification

Fertilization: Adding specific nutrients to the soil through fertilizers to enhance their availability to plants. For example, fertilization with N will increase protein levels in plants, and adding zinc or iron fertilizers to the soil can increase crop micro-nutrient content. These fertilizers should be applied regularly in every crop season. The use of mineral fertilizers includes the success of zinc (Zn) fertilization in Turkey and thiamin (Se) fertilization of crops in Finland (compiled in Garg et al., 2018).

Soil Management: Improving soil health and structure to enhance nutrient uptake by plants. It includes crop rotation, intercropping, and organic matter incorporation.

pH Adjustment: Modifying the pH of the soil to optimize nutrient absorption. Certain nutrients become more available to plants at specific pH levels. For example, iron uptake by the roots is better at lower pH.

Irrigation Management: Ensuring proper irrigation practices to prevent nutrient leaching and to promote efficient nutrient uptake. This strategy is region specific and depends on irrigation practice. Iodine fertilization in irrigation water in China is one such example (Garg et al., 2018).

Biofortified Fertilizers: Developing fertilizers that are specifically designed to deliver targeted nutrients to plants. For example, the emerging field of utilizing nanoparticles to provide desired nutrients to edible plants, also known as nano-biofortification, can be an important strategy. To date, several publications have shown positive effect of nanoparticle including copper, zinc, selenium, and iron employed in nano-biofortification (Saritha et al., 2022). There are several examples nano-Fe treatment in soybean and wheat, nano-Zn in maize, wheat and rice, and nano-Se in tea (Kapoor et al., 2022).

Enhancing Plant-Microbe Interactions: Beneficial soil microbes can assist in nutrient uptake by forming symbiotic relationships with plants. Living microbes like bacteria and fungi that can colonize the plant roots or leaves improve the soil's solubilization, mobilization, or fixation of nutrients by chelation, acidification, exchange reactions, and the production of organic acids (Ku et al., 2019). Phyto-availability of mineral elements can be increased using soil microorganisms from the genera *Bacillus*, *Pseudomonas*, *Rhizobium*, *Azotobacter*, etc. (Rengel et al., 1999). Utilizing microorganisms for biofortification is 'greener' and can help in specific nutrients like vitamin B12, which can only be produced by bacteria and archaea, not plants (Ku et al., 2019).

Challenges

Improving crop nutrients through agronomic biofortification has some challenges. Soils vary, so methods must match and balancing of nutrients is essential. Training farmers, getting supplies, and keeping them in the process is essential. It increases the cost of farm input. People's tastes and habits affect success. Growing at a bigger scale needs policy support and farm-level monitoring. It is also about not harming the environment, maintaining the natural soil microbiome, and respecting traditions. The major problem is continuous monitoring and assistance for an extended time (Kaur et al., 2020b). Solving these problems together can help agronomic biofortification fight malnutrition and improve food.

2.2. Plant Breeding-based Biofortification

Biofortification by traditional breeding is one of the most acknowledged and economical and environment-friendly techniques to release and adopt crop cultivars (Figure 1) compared to genetic transformation that involves huge research investments. Conventional breeding is only possible if the trait of interest has sufficient genotypic variability (Hirschi, 2009; Fu, 2015). This is typically done based on the nutritional needs of the target population and the prevalence of nutrient deficiencies in the region. Identifying germplasm with high nutrient content or plant varieties with naturally high levels of the target nutrient can be done by screening existing germplasm collections or field trials. Once germplasm with high nutrient content has been identified, crossbreeding using controlled pollination can be used to combine desirable traits and develop new plant varieties with even higher levels of the target nutrient. Once the new plant varieties have been developed, they are screened to confirm that they have the desired traits, such as high levels of the target nutrient, good agronomic performance, and resistance to pests and diseases (Bado et al., 2015). Farmers can then grow the crop, and consumers can purchase it (Kuckuck et al., 2020; Cooper et al., 2014). Using molecular markers for selection could speed up the development of new biofortified crop varieties (Bado et al., 2015; Crossa et al. 2017).

On the other hand, there are instances when only a tiny amount of genetic variation is present in the gene pool. One way to get around this is to perform a cross with a more remote relative, allowing the trait to be gradually introduced into commercial cultivars. Alternatively, mutagenesis can directly integrate new characteristics into commercial cultivars (Robertsen et al., 2019). Many international organizations have launched initiatives to increase crops' nutritional value through breeding. These initiatives aim to develop nutrient-dense crops that can provide better nutrition to people in regions where malnutrition is a significant issue. One such initiative is the "Biofortification Program", which is being led by the HarvestPlus organization. The program focuses on breeding crops rich in vitamins and minerals mainly iron, zinc, and vitamin A.

Through crop breeding, biofortification has been planned for many crops. The utilization of wild relatives of lentils has led to a significant increase in Zn and Fe contents in the pre-breeding materials (Figure 1) that can be used to generate high yielding regionally adapted cultivars. The Bangladesh Rice Research Institute (BRRI) released the world's first zinc-enriched rice varieties in 2013 (Ohly et al., 2019). Other countries such as Ethiopia, India, and Nepal have also released zinc, pro-vitamin A, and iron biofortified staple crop varieties, including wheat, rice, pearl millet, cassava, and sweet potato. Different colored wheat varieties, such as black, blue, and purple, have been utilized in breeding efforts across multiple countries. Breeding efforts that use the genetic diversity of colored wheat varieties can improve wheat-based foods' nutritional and functional properties and provide new options for consumers interested in healthy and flavorful foods (Garg et al., 2018b, c, d).

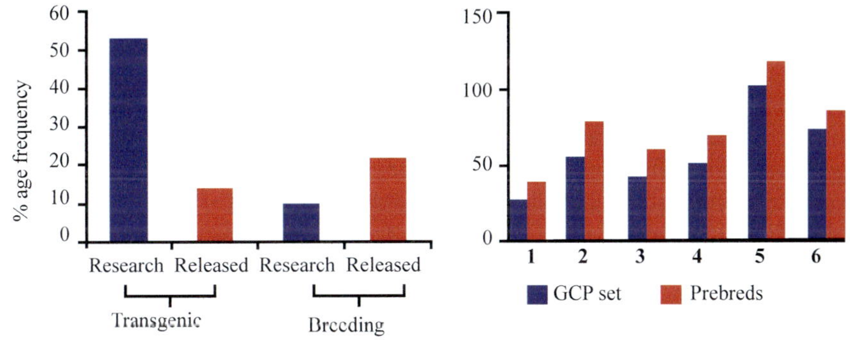

Figure 1: (a) Despite transgenic data, breeding is most successful for releasing biofortified crops (Adopted from Garg et al., 2018, *Fron. Nutr.*). (b) Wild relative derived lentils prebreds have higher Fe and Zn than conventional breeding (GCP) (HarvestPlus program).

Overall, conventional breeding is the traditional approach to plant breeding, which involves selecting and crossing plants with desirable traits over several generations to create new varieties with improved characteristics. This approach has been used for centuries and has successfully developed many of the crop varieties grown today.

Challenges

Biofortifying crops with nutrients through breeding has some associated challenges. Crafting improved crop varieties demands prolonged dedication, as the breeding process is a time-consuming journey spanning several years of development and meticulous testing. However, the aspiration isn't solely nutritional enhancement. Vital traits such as yield, disease resistance, and flavor must also be upheld to ensure the crops' overall efficacy (Ginkel and Cherfas, 2023). Wide adoption of these crops is also a challenge. Collaborations between plant breeders, agronomists, and nutritionists are crucial to ensure that biofortified crops are nutritionally enriched and agronomically viable.

2.3. Transgenic-based Biofortification

Transgenic breeding, also called genetic engineering, involves inserting desired genetic material into crops to provide them with specific nutrients.

The genetic material may originate from either the same or a different species. It can be introduced using various techniques, including gene guns, agrobacterium-mediated transformation, and CRISPR/Cas9.

In the last 15 years, the development of novel DNA sequencing techniques has enabled the identification of significant areas within the plant genome, leading to a substantial improvement in both speed and precision (Hickey et al., 2019; De Steur et al., 2015). The first step in transgenic biofortification is to identify the nutrient whose content is to be increased in the crop based on the nutritional needs of the target population and the prevalence of nutrient deficiencies in the region. The next step is to identify a gene responsible for producing the target nutrient. Once the gene has been identified, it is inserted into the genome of the crop, and the edited plant cells are cultured to produce new plants. These plants are then screened to confirm that the gene has been successfully inserted into the genome. Crop biofortification has harnessed a plethora of genes. The transgenic approach boasts a notable advantage: a valuable gene, once identified, can be repurposed for various crops. Key genes, such as phytoene synthase, carotene desaturase, nicotinamide synthase, and ferritin have found utility across multiple events and crop types (Figure 2).

The next step is to test the nutritional content of the transgenic crop to confirm that the target nutrient has been increased. Before the transgenic crop can be released for commercial use, it must be evaluated for safety and efficacy. It involves testing the crop for potential adverse effects on human health or the environment. If the transgenic crop is safe and effective, it can be released for commercial use. Farmers can then grow the crop, and consumers can purchase it (Vats et al., 2019; Zhu et al., 2007). Many crops have already undergone genetic modification to increase their micronutrient levels, such as high lysine maize, high unsaturated fatty acid soybean, iron-rich cassava with high levels of pro-vitamin A, and Golden rice with high levels of pro-vitamin A (Garg et al., 2018a). Reports are available for biofortified cereals, legumes, vegetables, oilseeds, fruits, and fodder crops.

In conclusion, transgenic biofortification offers several benefits for crops, including improved nutritional content and faster development.

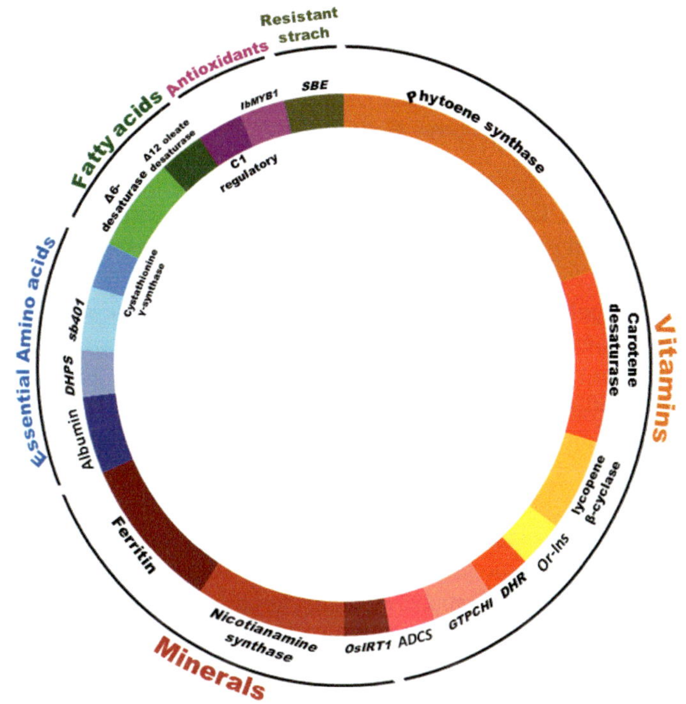

Figure 2: Transgenic biofortification enables versatile use of identified genes across multiple crops (Adopted from Garg et al., 2018, *Fron. Nutr.*)

Challenges

Creating transgenic biofortified crops requires significant time, effort, and funding during the research and development phase. Biofortification also results in genetic uniformity due to using the same genetic package to introduce high micronutrient levels. Another significant drawback of this approach is its limited adoption among the general population (Ofori et al., 2022). Another constraint arises from the variation in laws and regulations employed by different nations to approve and commercialize transgenic crops. Unfortunately, the prevailing political and economic conditions do not demonstrate a favorable attitude towards adopting this technology. In addition, it should be noted that the aforementioned regulatory procedures incur significant financial costs and need a substantial amount of time to complete. Let us consider the case of Bt Brinjal (Buiatti et al., 2013); the initial product development was undertaken by a seed firm based in India. Regrettably, the modified crop product distribution in India was hindered due to concerns voiced by anti-GMO activists. Consequently, a moratorium was established, necessitating additional studies before its potential release. But recent CRISPR/Cas9 technology offers some relief over genetic modification, as some governments are accepting SDN-1 (mutation without addition of foreign DNA) and SDN-2 (mutation with addition of one to small sequences) events out of the preview of regulatory procedures required for genetic transformation.

2.4. CRISPR/Cas Genome Editing

While transgenic and genome editing technologies share close ties, they exhibit distinct differences. Transgenic technology involves the introduction of foreign DNA into the crop genome, resulting in random integration within the genome and the emergence of crops with novel traits. In contrast, genome editing focuses on precise manipulation at specific genome locations. This manipulation includes creating knockouts, replacements, or altering the gene expression of existing target genes. The aim is to enhance desired traits in a predetermined and accurate manner. Meganucleases (MeN), transcription activator-like effector nucleases (TALEN), Zinc-Finger Nucleases (ZFN), and clustered regularly interspaced short palindromic repeats (CRISPR) associated protein (Cas) are vital tools. They induce targeted breaks in the host genome, prompting cellular repair via nonhomologous end joining (NHEJ) or homologous recombination (HR). NHEJ creates minor deletions to disable genes, while HR, with a repair template, introduces precise DNA alterations, including insertions or substitutions. Notably, CRISPR-Cas technology, favored for its advantages over MeNs, ZFNs, and TALENs, dominates current genome editing. It adeptly mitigates drawbacks and offers streamlined editing processes. Lately, novel techniques in genome editing have emerged, encompassing epigenome modifiers, base editing, and prime editing. Due to these distinctions, genome editing presents several advantages. It operates at a faster pace, demonstrates greater accuracy, and often eliminates the need for introducing exogenous DNA to improve crops. This efficiency contributes to genome editing's potential acceptance and more favorable reception in the agricultural sector for rectifying nutrient deficiencies, ensuring food security, and cultivating readily adaptable crop varieties. Genome editing, particularly the CRISPR-Cas approach, holds immense potential in developing biofortified crops that can effectively counter nutrient deficiencies and elevate food value (Nagamine et al., 2022). This revolutionary technology has facilitated the creation of vitamin-enriched crops such as vitamin D-enhanced tomatoes by silencing the Sl7-DR2 enzyme (Li et al., 2022), vitamin A-enriched rice and bananas through CRISPR/Cas9-based techniques (Dong et al., 2020; Kaur et al., 2020a). Moreover, genome editing has been instrumental in boosting nutrient accumulation in crops, addressing malnutrition concerns, as demonstrated by the reduction of phytic acid and concurrent increase in iron and zinc levels in wheat (Ibrahim et al., 2022) and enhancing nutrient usage efficiency in rice (Lu et al., 2017). These biofortified crops stand as promising solutions to global health and nutrition challenges, positioning genome editing as a transformative tool in agricultural and nutritional advancements.

Challenges

Genome editing faces key challenges. Constructing CRISPR/Cas vectors is complex. Off-target mutations are a major drawback, harder to predict with incomplete genome sequences. Cas nucleases need specific PAM sequences, limiting editing regions, although new Cas enzymes are easing this. Base editing is precise but limited in changes. Prime editing allows versatile alterations but lacks efficiency. HDR-mediated insertion helps but is infrequent due to donor template need. Efficiently delivering editing components and regenerating crops are hurdles. Direct (e.g., PEG, bombardment) and indirect (e.g., Agrobacterium) methods exist, but variations and specific needs can complicate. Viral and non-viral vectors aid delivery, though viral vectors restrict large fragments. Non-viral options like nanoparticles offer promise. Regulations for gene-edited crops are developing, impacting applications. Biosafety rules differ by nation, while consumer acceptance depends on proven benefits. Overall, genome editing needs improved vectors, reduced off-target risks, broader editing scope, higher efficiency, optimized delivery, better crop regeneration, clear regulations, and public trust to reach its full potential.

3. Conclusion

The biofortification of staple food crops is an economical and sustainable method used to alleviate the issue of malnutrition. Conventional plant breeding, agronomic methods, and transgenic methods conduct the biofortification of crops. Globally, zinc, iron, and vitamin A biofortified crops such as wheat, rice maize, millet, banana, beans, lentils and cassava are released. However, these biofortified crop varieties faced significant challenges, such as the cost of the final commodity in the market and consumers' acceptability towards newly released varieties. It ultimately hindered the goal of providing nutritional security to an exponentially increasing global population.

References

Bado, S., B.P. Forster, S. Nielen, A.M. Ali, P.J. Lagoda, B.J. Till, and M. Laimer. 2015. Plant mutation breeding: Current progress and future assessment. *Plant Breed. Rev.*, 39: 23–88.

Bouis, H.E., and A. Saltzman. 2017. Improving nutrition through biofortification: A review of evidence from HarvestPlus, 2003 through 2016. *Glob. Food Sec.*, 12: 49–58.

Bouis, H.E., C. Hotz, B. McClafferty, J.V. Meenakshi, and W.H. Pfeiffer. 2011. Biofortification: A new tool to reduce micronutrient malnutrition. *Food Nutr. Bull.*, 32(Suppl 1): S31–S40.

Buiatti, M.P. Christou, and G. Pastore. 2013. The application of GMOs in agriculture and in food production for a better nutrition: Two different scientific points of view. *Genes Nutr.*, 8: 255–270.

Cakmak, I., and U.Á. Kutman. 2018. Agronomic biofortification of cereals with zinc: A review. *Eur. J. Soil Sci.*, 69: 172–180.

Calayugan, M.I.C., B.M. Swamy, C.T. Nha, A.D. Palanog, P.S. Biswas, G.I. Descalsota-Empleo, and M.A. Inabangan-Asilo. 2021. Zinc-biofortified rice: A sustainable food-based product for fighting zinc malnutrition. *Rice Improvement: Physiological, Molecular Breeding and Genetic Perspectives*, 449–470. Cham: Springer International Publishing.

Cooper, M., C.D. Messina, D. Podlich, L.R. Totir, A. Baumgarten, N.J. Hausmann, and G. Graham. 2014. Predicting the future of plant breeding: Complementing empirical evaluation with genetic prediction. *Crop Pasture Sci.*, 65: 311–336.

Crossa, J., P. Pérez-Rodríguez, J. Cuevas, O. Montesinos-López, D. Jarquín, G. De Los Campos, and R.K. Varshney. 2017. Genomic selection in plant breeding: Methods, models, and perspectives. *Trends Plant Sci.*, 22: 961–975.

De Steur, H., D. Blancquaert, S. Strobbe, W. Lambert, X. Gellynck, and D. Van Der Straeten. 2015. Status and market potential of transgenic biofortified crops. *Nat. Biotechnol.*, 33: 25–29.

Dong, O.X., S. Yu, R. Jain, N. Zhang, P.Q. Duong, C. Butler, Y. Li, A. Lipzen, J.A. Martin, and K.W. Barry. 2020. Marker-free carotenoid-enriched rice generated through targeted gene insertion using CRISPR-Cas9. *Nat. Commun.*, 11: 1–10.

Erisman, J.W., M.A. Sutton, J. Galloway, Z. Klimont, and W. Winiwarter. 2008. How a century of ammonia synthesis changed the world. *Nat. Geosci.*, 1: 636–639.

Fu, Y.B. 2015. Understanding crop genetic diversity under modern plant breeding. *Theor. Appl. Genet.*, 128: 2131–2142.

Garg, M., N. Sharma, S. Sharma, P. Kapoor, A. Kumar, V. Chunduri, and P. Arora. 2018a. Biofortified crops generated by breeding, agronomy, and transgenic approaches are improving lives of millions of people around the world. *Front. Nutr.*, 12.

Garg, M. 2018b. NABIMG-11-Black (BW/2* PBW621) (IC0620916; INGR17003), a wheat (*Triticum aestivum*) germplasm with black grain colour (purple pericarp + blue aleuron). *IJPGR*, 31: 334–335.

Garg, M. 2018c. NABIMG-10-Purple (BW/2* PBW621) (IC0620915; INGR17002), a wheat (*Triticum aestivum*) germplasm with purple grain (*pericarp*) colour. *IJPGR*, 31: 333–334.

Garg, M. 2018d. NABIMG-9-Blue; BW/2*/PBW621 (IC0620914; INGR17001), a wheat (*Triticum aestivum*) germplasm with blue grain (*aleurone*) colour. *IJPGR*, 31: 332–333.

Hickey, L.T., A.N. Hafeez, H. Robinson, S.A. Jackson, S.C. Leal-Bertioli, M. Tester, and B.B. Wulff. 2019. Breeding crops to feed 10 billion. *Nat. Biotechnol.*, 37: 744–754.

Hirschi, K.D. 2009. Nutrient biofortification of food crops. *Annu. Rev. Nutr.*, 29: 401–421.

Hotz, C. 2013. *In:* Benjamin Caballero (Ed.), *Encyclopedia of Human Nutrition*, 3rd Edn. Academic Press, Waltham.

Ibrahim, S., B. Saleem, N. Rehman, S.A. Zafar, M.K. Naeem, and M.R. Khan. 2022. CRISPR/Cas9 mediated disruption of Inositol Pentakisphosphate 2-Kinase 1 (TaIPK1) reduces phytic acid and improves iron and zinc accumulation in wheat grains. *J. Adv. Res.*, 37: 33–41.

Jaiswal, D.K., R. Krishna, G.K. Chouhan, A.P. de Araujo Pereira, A.B. Ade, S. Prakash, and Verma, J.P. 2022. Bio-fortification of minerals in crops: Current scenario and future prospects for sustainable agriculture and human health. *Plant Growth Regul.*, 98: 5–22.

Kapoor, P., R.K. Dhaka, P. Sihag, S. Mehla, V. Sagwal, Y. Singh, and U. Kumar. 2022. Nanotechnology-enabled biofortification strategies for micronutrients enrichment of food crops: Current understanding and future scope. *NanoImpact.*, 26: 100407.

Kaur, N., A. Alok, P. Kumar, N. Kaur, P. Awasthi, S. Chaturvedi, P. Pandey, A. Pandey, A.K. Pandey, and S. Tiwari. 2020a. CRISPR/Cas9 directed editing of lycopene epsilon-cyclase modulates metabolic flux for β-carotene biosynthesis in banana fruit. *Metab. Eng.*, 59: 76–86.

Kaur, T., K.L Rana, D. Kour, I. Sheikh, N. Yadav, V. Kumar, and A.K. Saxena. 2020b. Microbe-mediated biofortification for micronutrients: Present status and future challenges. *New and Future Developments in Microbial Biotechnology and Bioengineering*, 1–17, Elsevier.

Ku, Y.S., H.M. Rehman, and H.M. Lam. 2019. Possible roles of rhizospheric and endophytic microbes to provide a safe and affordable means of crop biofortification. *Agronomy*, 9: 764.

Kuckuck, H., G. Kobabe, and G. Wenzel. 2020. *Fundamentals of Plant Breeding.* Walter de Gruyter GmbH and Co KG.

Li, J., A. Scarano, N.M. Gonzalez, F. D'Orso, Y. Yue, K. Nemeth, G. Saalbach, L. Hill, C. de Oliveira Martins, and R. Moran. 2022. Biofortified tomatoes provide a new route to vitamin D sufficiency. *Nat. Plants*, 1–6.

Lockyer, S., A. White, and J.L. Buttriss. 2018. Biofortified crops for tackling micronutrient deficiencies – What impact are these having in developing countries and could they be of relevance within Europe. *Nutr. Bull.*, 43: 319–357.

Lu, H., S. Liu, S. Xu, W. Chen, X. Zhou, Y. Tan, J. Huang, and Q. Shu. 2017. CRISPR-S: A active interference element for a rapid and inexpensive selection of genome-edited, transgene-free rice plants. *Plant Biotechnol. J.*, 15: 1371.

Nagamine, A., M. Takayama, and H. Ezura. 2022. Genetic improvement of tomato using gene editing technologies. *J. Hortic. Sci. Biotechnol.*, 1–9.

Ofori, K.F., S. Antoniello, M.M. English, and A.N. Aryee. 2022. Improving nutrition through biofortification: A systematic review. *Front. Nutr.*, 9: 1043655.

Ohly, H., M.R. Broadley, E.J.M. Joy, M.J. Khan, H. McArdle, M. Zaman, and N. Lowe. 2019. The BiZiFED project: Biofortified zinc flour to eliminate deficiency in Pakistan. *Nutr. Bull.*, 44: 60–64.

Prashanth, L., K.K. Kattapagari, R.T. Chitturi, V.R.R. Baddam, and L.K. Prasad. 2015. A review on role of essential trace elements in health and disease. *J. Dr. NTR Univ. Health Sci.*, 4: 75–85.

Rengel, Z., G.D. Batten, and D.D. Crowley. 1999. Agronomic approaches for improving the micronutrient density in edible portions of field crops. *Field Crops Res.*, 60: 27–40.

Robertsen, C.D., R.L. Hjortshøj, and L.L. Janss. 2019. Genomic selection in cereal breeding. *Agronomy*, 9: 95.

Saritha, G.N.G., T. Anju, and A. Kumar. 2022. Nanotechnology – Big impact: How nanotechnology is changing the future of agriculture. *J. Agric. Food Res.*, 100457.

Thompson, D.S., N. Younger-Coleman, P. Lyew-Ayee, L.G. Greene, M.S. Boyne, and T.E. Forrester. 2017. Socioeconomic factors associated with severe acute malnutrition in Jamaica. *PloS One*, 12: e0173101.

Van Ginkel, M., and J. Cherfas. 2023. What is wrong with biofortification? *Glob. Food Sec.*, 37: 100689.

Vats, S., S. Kumawat, V. Kumar, G.B. Patil, T. Joshi, H. Sonah, and R. Deshmukh. 2019. Genome editing in plants: Exploration of technological advancements and challenges. *Cells*, 8: 1386.

Zhu, C., S. Naqvi, S. Gomez-Galera, A.M. Pelacho, T. Capell, and P. Christou. 2007. Transgenic strategies for the nutritional enhancement of plants. *Trends Plant Sci.*, 12: 548–555.

CHAPTER

2

Deciding the Targets: Crops and Nutrients

Uttam Kumar[a*][0000-0002-6618-3810], **Anita Kumari**[b,c][0000-0002-3942-7234], **Harish Kumar**[d][0000-0003-0127-2430], **Jaspreet Kaur**[c][0000-0002-4219-9703] **and Mona Yadav**[b][0009-0006-3033-1697]

[a] Astralyan Agro Pvt. Ltd, Shamli, Uttar Pradesh - 247773
[b] Department of Agri-Biotechnology, National Agri-Food Biotechnology Institute, Mohali, Punjab - 140306, India
[c] University Institute of Engineering and Technology, Panjab University, Chandigarh, India
[d] Indian Institute of Science Education and Research, Mohali, Punjab - 140306

1. Introduction

Human growth and development depend on the right proportions of food and nutrients. Diet quality is a critical link between food security and nutrition because the human body cannot synthesize micronutrients; thus, a well-balanced diet is necessary. While required in small amounts, micronutrients play essential roles in human growth and development.

Malnutrition encompasses deficiencies, excesses, or imbalances in an individual's energy and/or nutrient intake. With the rapid increase in population and declining food availability, the number of nutritionally deficient or imbalanced individuals is also rising proportionally. Consuming calorie-rich but nutrient-deficient food leads to micronutrient deficiency, often called hidden hunger (FAO, 2022). Malnutrition has become a global issue affecting millions worldwide, particularly in developing countries (Gonmei and Toteja, 2018; Venkatesh et al., 2021; Chaudhary et al., 2022). Globally, an estimated 767.9 million people, accounting for 9.8% of the global population, suffer from undernourishment (FAO, 2022). Approximately 670 million people are also projected to experience malnutrition by 2030.

*Corresponding author: kumaruttam@hotmail.com

Due to micronutrient deficiencies, newborns may experience low birth weight, higher mortality rates, and impaired mental development. Children may suffer from stunted growth, frequent infections, reduced mental and learning capacities, and higher mortality rates due to malnutrition. Adolescents can face stunted growth, reduced mental capacity, fatigue, and increased infection vulnerability. Adults may experience reduced productivity, poor socioeconomic status, and higher rates of chronic diseases (Bailey et al., 2015). Pregnant women may face increased mortality and perinatal complications due to micronutrient deficiencies (Gernand et al., 2016). Elderly individuals may experience higher morbidity and mortality rates (Lassi et al., 2017). The nutritional quality of the food consumed by an individual determines the total amount of nutrient intake. Therefore, a balanced diet is essential to fulfil daily nutritional requirements. A balanced diet consists of various foods containing adequate nutrients required for good health. It is unlikely that all the necessary nutrients can be obtained from a single source. To meet human nutritional needs, it is essential to include multiple sources. Another strategy is to consume foods containing multiple nutrients, allowing the fulfilment of all required nutrients through a few varieties of food. For example, mineral malnutrition can be addressed by consuming more fish and animal products, taking mineral supplements, adding minerals to food, or enhancing the body's mineral absorption from crops. Interventions such as fortifying food with selected vitamins and minerals have been implemented to overcome such conditions. According to the World Health Organization (WHO), fortification is the deliberate practice of increasing the content of one or more micronutrients (i.e., vitamins and minerals) in a food to enhance its nutritional quality and provide public health benefits with minimal health risks. However, strategies like increasing diet diversification, mineral supplementation, and food fortification have not always been effective. This has led to a push for the biofortification of crops using mineral fertilizers and breeding varieties with better mineral absorption abilities (White and Broadley, 2009). It is the process of enhancing the nutritional content of food crops rather than supplementing food with nutrients (Campos-Bowers and Wittenmyer, 2007; De Moura et al., 2014).

Biofortification is a better alternative to address this global challenge by increasing the nutritional values of conventional crops in terms of essential micronutrients such as vitamins and minerals. By improving the nutritional values of staple crops, biofortification offers a sustainable and cost-effective solution to the widespread problem of malnutrition in many regions of the world. Biofortification of staple crops in nutritionally deprived areas can alleviate health issues and raise living standards by reducing health-related expenses. Biofortification can benefit the poor and wealthy populations. Biofortification allows additional nutrients to be incorporated into commonly used staple crops. Therefore, it is important to identify which crop should be fortified and which compound should be added to that crop to enhance its nutritional value. Biofortification allows additional nutrients to be incorporated into staple crops, but several questions must be addressed before investing in it.

- How can we determine which nutrients should be increased in the food?
- What is the bioavailability of the selected nutrient?
- Which source should be prioritized for the selected nutrient?
- Is that source sufficient to meet the requirements of the entire population?
- If not, what alternatives should be considered?
- What is the most effective method to increase nutrient content in the crop?

2. Selection of the Target Crop

Selecting a target crop for biofortification requires a comprehensive strategy considering several factors. The factors include nutrient content, dietary habits, nutrient bioavailability, agronomic considerations, cultural acceptability, cost, and sustainability (Figure 1).

- ***Nutrient Content:*** One of the primary considerations is evaluating the baseline nutrient content of different crops. The objective is to identify crops that are naturally deficient in the target nutrient, making the impact of biofortification more significant. For instance, if a region faces widespread vitamin A deficiency, choosing crops with inherently low vitamin A levels becomes paramount.
- ***Dietary Contribution:*** To ensure that biofortified foods make a substantial contribution to the diets of the target population, priority should be given to staple foods that are widely consumed. These foods constitute a significant portion of daily caloric intake for the majority. Thus, enhancing the nutrient content of these staples can lead to considerable health improvements (Saltzman et al., 2017).
- ***Cultural Acceptance:*** Cultural preferences and culinary traditions significantly influence dietary choices. Selecting crops that align with local food habits and preferences is crucial for ensuring the uptake and consumption of biofortified foods. When foods resonate with cultural norms, their acceptance is more likely, leading to a greater impact (Hotz and McClafferty, 2007).
- ***Accessibility:*** Ensuring the selected crop is accessible to the target population is essential. It's about choosing readily available and affordable crops to the communities that need them the most. Accessibility is a key factor in addressing malnutrition at the grassroots level.
- ***Agroecological Suitability:*** Different crops thrive under distinct agroecological conditions. A crop's adaptability to the local climate, soil type, and growing season is essential for its successful cultivation. Opting for crops naturally aligning with the region's environmental factors promotes higher productivity.
- ***Genetic Manipulation:*** In cases where genetic engineering is part of the biofortification process, choosing crops with well-established genetic transformation protocols is vital. This ensures that the necessary modifications can be introduced effectively into the selected crop's genome.
- ***Bioavailability:*** The bioavailability of the target nutrient in the selected crop is critical for its effectiveness. Certain crops may contain nutrients, but the

human body might not easily absorb them. Opting for crops that offer high bioavailability ensures that the enhanced nutrient content translates into tangible health benefits.

- ***Market Demand:*** Assessing consumer demand and market potential is crucial for sustainability. A significant demand for biofortified foods can drive commercial viability and create a self-sustaining cycle of production and consumption.
- ***Feasibility:*** Biofortification should be seamlessly integrated into existing agricultural practices. Choosing crops that can be easily integrated into local farming systems ensures that the intervention is adopted and scaled up effectively.
- ***Nutrient Interaction:*** Consideration should be given to potential nutrient interactions. Opting for crops where the biofortification of the target nutrient does not interfere with the absorption of other essential nutrients ensures a balanced overall impact on health (Garg et al., 2018).
- ***Environmental Impact:*** The environmental impact of cultivating biofortified crops should be evaluated. It includes assessing resource use, ecological considerations, and potential effects on the surrounding ecosystem.

By thoroughly evaluating these multifaceted considerations, the selection of the target crop for biofortification can be tailored to the target population's specific needs, preferences, and characteristics. Making an informed decision regarding the crop not only maximizes the intervention's impact but also contributes to long-term improvements in public health and nutrition.

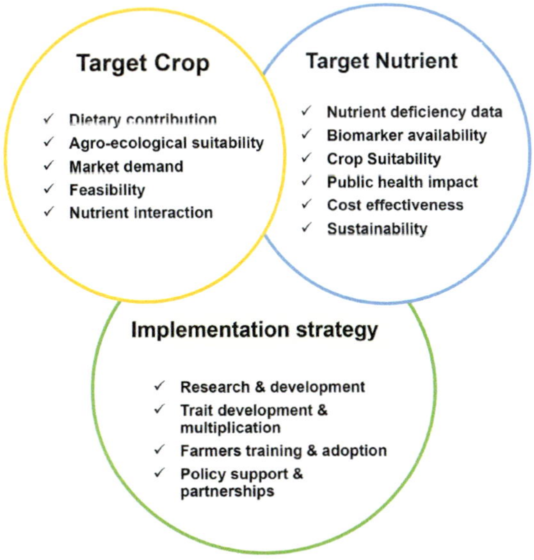

Figure 1: Selection criteria and implementation strategies for crop and nutrients in biofortification.

3. Selection of the Target Nutrient

Selecting the target nutrient for biofortification is a critical process that involves strategic considerations to ensure its effectiveness (Siwela et al., 2020; Ofori et al., 2022). Here's how these factors guide the decision-making (Figure 1):

- ***Nutrient Deficiency Data:*** Analyzing existing deficiency data helps identify the most lacking nutrients, directing focus towards addressing prevalent nutritional gaps (WHO/FAO, 2006).
- ***Health Impact:*** Prioritizing nutrients with significant health benefits ensures the biofortification efforts lead to tangible improvements in public health (Bouis et al., 2011).
- ***Biomarker Availability:*** Opting for nutrients with measurable biomarkers enables effective monitoring of the program's success and impact.
- ***Cultural Acceptance:*** Considering dietary habits and cultural preferences ensures the chosen nutrient aligns with local consumption patterns, enhancing acceptance.
- ***Bioavailability:*** Selecting nutrients easily absorbed by the body maximizes the translated nutritional benefits for consumers.
- ***Crop Suitability:*** Assessing crops compatible with nutrient accumulation, factoring genetic traits and metabolic pathways, aids effective biofortification.
- ***Environmental Factors:*** Factoring in regional conditions like soil quality and climate impacts nutrient content and bioavailability in fortified crops.
- ***Genetic Manipulation:*** If genetic engineering is employed, opting for nutrients amenable to stable genetic integration ensures sustained enhancement.
- ***Sustainability:*** Evaluating long-term viability includes weighing production costs, resource demands, and potential ecological effects (Saltzman et al., 2017).
- ***Cost-Effectiveness:*** Balancing benefits and expenses ensures impactful biofortification strategies without compromising affordability.
- ***Public Health Impact:*** Emphasizing nutrients that address widespread deficiencies and elevate overall nutritional status contributes significantly to public health.

Ultimately, a comprehensive assessment considering nutrient deficiencies, health effects, cultural appropriateness, and feasibility guides the choice of the target nutrient. This complex decision necessitates collaboration between nutritionists, agronomists, geneticists, and public health experts.

4. Potential Crops that can be Biofortified

- ***Rice:*** Rice plays a vital role in regions where it serves as a staple food, addressing prevalent nutrient deficiencies. Biofortification of rice is mainly targeted for vitamin A, and sometimes zinc and iron. Improved nutrition within farming communities can increase labor productivity and economic advantages. One well-known biofortified rice variety is "Golden Rice", named for its golden or yellow appearance due to its increased beta-carotene content. Golden Rice was developed using genetic modification to introduce genes responsible for beta-carotene production. Researchers and breeders also work on conventionally bred varieties with elevated beta-carotene levels. These non-GMO biofortified rice varieties are created through traditional breeding techniques involving selecting and crossbreeding rice strains naturally rich in beta-carotene.

- ***Wheat:*** Biofortified wheat with higher levels of zinc and iron can significantly contribute to addressing mineral deficiencies. Since many people consume wheat directly or indirectly, wheat biofortification is a cost-effective strategy for improving nutrition. Many biofortified wheat varieties, such as WB 02, PBW-Zn 1, HUW711, BHU-25, BHU-31, Rajendra Gehun 02, etc. have been developed through conventional breeding techniques (Wani et al., 2022). The first biofortified wheat variety released in 2016 in India was Zinc Shakti (Chitra). It involves selecting and crossbreeding wheat strains with naturally higher zinc and iron content. In some cases, genetic modification techniques are used to introduce genes responsible for increased zinc and iron content. So far, no genetically modified wheat variety has been released.
- ***Maize (Corn):*** Biofortified maize is a variety of maize with elevated levels of essential nutrients like vitamin A, iron, and zinc. The primary goal is to combat nutritional deficiencies in regions heavily reliant on maize as a staple food. Varieties like "Golden Maize" are enriched with beta-carotene. Some biofortified maize also contains higher levels of iron and zinc. These varieties are created through conventional breeding or genetic modification techniques. Overall, biofortified maize represents a promising solution to improve the nutritional well-being of communities worldwide.
- ***Sweet Potatoes:*** Orange-fleshed sweet potatoes are rich in vitamin A, providing a natural solution to combat vitamin A deficiency. This biofortified crop is precious in regions where sweet potatoes are a dietary staple.
- ***Cassava:*** Biofortified cassava with elevated vitamin A and iron levels addresses key nutrient gaps in diets heavily reliant on this drought-tolerant root crop. It is consumed in various forms, including boiled, chips, and as an ingredient in certain traditional dishes. It is essential in various industrial applications, including starch, flour, and bioethanol production. India is one of the largest cassava producers globally. Cassava holds significant cultural and economic importance in Kerala and Tamil Nadu regions.
- ***Legumes (Beans, Lentils, Peas):*** Biofortifying legumes with higher levels of iron, zinc, and protein can significantly improve the nutritional content of diets, particularly for vegetarian populations or areas with limited access to animal-sourced foods.
- ***Sorghum and Millet:*** These biofortified crops can offer essential nutrients like iron and zinc, helping to combat anemia and other mineral deficiencies.
- ***Leafy Greens (Spinach, Kale, etc.):*** Biofortification of leafy greens can elevate their nutritional value, providing a rich source of vitamins and minerals. It is

essential for overall health, including bone health (due to calcium content) and immune function (due to vitamins and minerals).

- ***Oilseeds (Soybeans, Sunflowers, etc.):*** Biofortifying oilseeds can lead to oils with enhanced nutritional profiles, contributing to overall dietary diversity and health.
- ***Barley:*** Biofortified barley can become a more nutrient-dense grain, providing a broader spectrum of essential nutrients like iron and zinc. Barley serves various purposes, including food, fodder, and malting for the brewing industry.
- ***Fruits (Bananas, Mangoes, etc.):*** Biofortifying fruits can improve vitamin and mineral content, enhancing the nutritional quality of diets. Bananas are quite accessible fruits affordable to nearly all income groups. It is calories rich, valued for its natural energy-boosting properties and a popular choice for athletes and individuals seeking a quick energy source. It is a significant vitamin C, B6, manganese, potassium, magnesium, and dietary fiber source. Vitamin A-rich bananas can play a crucial role in combating vitamin A deficiency (Nabuuma et al., 2022).

  Similarly, if successfully developed and adopted, biofortified vitamin A-rich mangos could substantially influence global nutrition, given the widespread consumption of mangos in many countries.
- ***Quinoa:*** Elevating the nutrient content of quinoa, an already highly nutritious grain, can further contribute to addressing malnutrition and providing essential nutrients for growth, development, and overall health.
- ***Pulses (Chickpeas, Lentils, etc.):*** Biofortified pulses can significantly contribute to protein and mineral intake, especially in vegetarian or low-meat diets. It is crucial for meeting protein and nutrient needs.
- ***Tomatoes:*** Biofortified tomatoes with higher levels of vitamin C, vitamin A, and antioxidants can enhance their nutritional value, providing a rich source of essential vitamins and protective compounds.
- ***Potatoes:*** Biofortifying potatoes can improve their nutritional content, making them a more substantial source of essential nutrients like vitamins A, C, iron, and potassium. Biofortified potatoes are part of a global effort to improve nutrition and combat malnutrition (Singh et al., 2022). Various initiatives and research projects are underway to develop and promote biofortified potato varieties in different parts of the world. For instance, organizations like the International Potato Center (CIP) have been working on biofortification programs to enhance the nutritional content of potatoes.

5. Implementation of Biofortification Programs

For the successful implementation of the biofortification program, the following steps are used (Figure 1):

- ***Research & Development:*** The continuous betterment of crops is an essential undertaking motivated by unpredictable shifts in weather patterns and the evolving demands of consumers. The initial phase of executing a biofortification initiative involves the identification of crops that can be feasibly enhanced through biofortification techniques and the precise nutrients that require augmentation. This process entails researching to ascertain crop varieties that inherently accumulate elevated levels of specific nutrients or creating genetically modified varieties via breeding methods or biotechnology (Douthwaite et al., 2023).

- ***Trait Development and Multiplication:*** Once the target crops and nutrients are identified, the search for a method to achieve the trait begins. Researchers work to develop new crop varieties with improved nutritional profiles. Depending on the desired traits and available technology, this can involve traditional breeding methods, marker-assisted selection, or genetic engineering techniques (Garg et al., 2018). After developing improved quality of seed, the next step is seed multiplication. Seed multiplication can be carried out through collaboration between public and private sector entities and typically follows the subsequent steps: The newly released varieties are subject to licensing agreements with seed companies or community-based multipliers. The multiplication of planting material takes place, and subsequently, the seed is packaged and transported to various distribution points.
- ***Field Trials:*** Before the dissemination of biofortified crop varieties to agricultural practitioners, comprehensive field trials should be carried out to assess their efficacy, agronomic attributes, and nutritional composition. This data generation will help convince farmers and consumers regarding the quality of the product (Chaudhary et al., 2022).
- ***Farmer Training and Adoption:*** Obtaining farmers' acceptance is essential because 40% of the global population relies on their own agricultural output or limited foreign resources. The primary concern of farmers is to produce high-yielding crops with better nutrition profile (Johns and Eyzaguirre, 2007). Farmers are familiar with consumer preferences, so they will not cultivate commodities that they do not believe will sell. For example, in a study conducted in Uganda on the adoption of orange-fleshed sweet potatoes among farmers, it was determined that yield, taste, and price were the main factors that determined its adoption (Talsma, 2017). Popularize the new varieties by working with existing farmer organizations at the village level, training farmers, demonstrating new techniques, and identifying the lead farmers as early adopters who can motivate and influence others, also, by increasing the availability of seed and grain via farmer-to-farmer distribution and public and private partners.
- ***Policy Support and Partnerships:*** The effectiveness of biofortification depends on the involvement of all pertinent sectors, including international centers, Ministry of Agriculture, Ministry of Health, national agricultural research institutes, private sector and/or state seed companies, NGOs/CBOs, national universities, continental and regional research teams and regional economic communities. These sectors provide information regarding crop

development, seed multiplication, extension services, awareness creation, technical assistance, policy support, messaging and training materials, food product development, and inclusion into community health services (Olson et al., 2021; Douthwaite et al., 2023). Strong community involvement increases understanding and improves relationships for better biofortification acceptance going forward.

6. Role of Policy and Regulatory Frameworks in Biofortification Programs

Policy and regulatory frameworks are crucial in promoting and sustaining biofortification programs. They provide the necessary support and structure to facilitate these programs' adoption, implementation, and long-term success. Some of the roles of policy and regulatory frameworks in promoting and sustaining biofortification programs are setting targets and standards, integrating into national nutrition strategies, research and development support, and protecting breeders' rights and encouraging private sector involvement in biofortification. Policy frameworks establish regulations and guidelines for producing, quality control, and distributing biofortified seeds. Policy frameworks help in the market development, consumer awareness programs, and serve as a means for advocacy endeavors to promote the consciousness among policymakers and stakeholders regarding the significance of biofortification (Mejia et al., 2017).

7. Biofortification Priority Index

HarvestPlus has played a significant role in advancing biofortification strategy. The Biofortification Priority Index (BPI), a tool developed by HarvestPlus, helps identify and prioritize crops for biofortification based on key factors such as nutrient content, harvest index, nutrient deficiency prevalence, cost of implementation, and potential nutrient intake. The formula for calculating the BPI guides decision-making by evaluating the impact and feasibility of biofortification strategies for different crops. The BPI provides a systematic approach to determine which crops should be targeted for biofortification efforts, ensuring that resources are allocated effectively to combat malnutrition and improve public health. This index uses country-level data for under consideration crop that is usually a staple. It calculates the geometric mean of three sub-indices: production and consumption indices for biofortifiable staples and a micronutrient deficiency index for the micronutrient with which the crop can be biofortified (Asare-Marfo et al., 2013; Birol and Bouis, 2023). The formula for calculating the BPI can be expressed as follows:

$$\text{BPI} = (\sqrt{}\text{Production Index} \times \text{Consumption Index}) \times \sqrt{}\ \text{Micronutrient Deficiency Index}$$

8. Challenges Faced during the Implementation of Biofortification of Target Crops

The successful implementation of biofortification programs on targeted crops may encounter challenges that require attention and resolution (Siwela et al., 2020; Ofori et al., 2022; Bouis et al., 2011). The following are prevalent obstacles:

- Generating biofortified crop varieties with enhanced nutritional properties can be complex and time-consuming. It requires scientific expertise, breeding techniques, and adequate funding. Overcoming technical challenges and ensuring the availability of a diverse range of biofortified crops suitable for different regions and agroecological conditions is crucial.
- Establishing robust seed systems for biofortified crops is crucial but can be challenging in remote areas for their effective distribution and availability to farmers.
- Long-term financial sustainability and resource allocation are essential to ensure program continuity.
- With the rapid development of genome-editing tools in the past decade, the generation of GM crops is still a concern in many countries for human health and environmental safety considerations.
- Government policies for biofortified need review and update with a provision of incentives for biofortified crop cultivation. It may require subsidies, tax benefits, or grants targeted explicitly at these crops. Farmers may hesitate to grow biofortified crops if they don't have guaranteed market access or if there are no price premiums for these crops. Creating and promoting markets for biofortified products can incentivize farmers to grow them.

9. Conclusion

There are currently huge pressures on the global food system, with the growing global population expected to increase from 7.6 billion in 2017 to 8.6 billion by 2030 and 9.8 billion by 2050 (United Nations, 2017). Although many biofortified crops have been developed, there is still a need of know how to select specific crops with specific nutritional content for biofortification programs. Scientists, policymakers, regulators and journalists must coordinate and discuss how to improve global food security using genome-editing tools, and at the same time alleviate public concerns about selecting the target crop and their biofortification.

References

Asare-Marfo, D., E. Birol, C. Gonzalez, M. Moursi, S. Perez, J. Schwarz, and M. Zeller.

2013. *Prioritizing Countries for Biofortification Interventions Using Country-Level Data*. International Food Policy Research Institute (IFPRI).

Bailey, R.L., K.P. West Jr., and R.E. Black. 2015. The epidemiology of global micronutrient deficiencies. *Ann. Nutr. Metab*., 66: 22–33.

Birol, E., and H.E. Bouis. 2023. Role of socio-economic research in developing, delivering, and scaling new crop varieties: The case of staple crop biofortification. *Front. Plant Sci.*, 14: 1099496.

Bouis, H.E., C. Hotz, B. McClafferty, J.V. Meenakshi, and W.H. Pfeiffer. 2011. Biofortification: A new tool to reduce micronutrient malnutrition. *Food Nutr. Bull.*, 32: S31–S40.

Campos-Bowers, M.H., and B.F. Wittenmyer. 2007. Biofortification in China: Policy and practice. *Health Res. Policy Syst.*, 5: 1–7.

Chaudhary, M., A. Mandal, S. Muduli, and A. Deepasree. 2022. Agronomic biofortification of food crops: A sustainable way to boost nutritional security. *In:* V. Meena, H.P. Parewa, and S.K. Meena (Eds.), *Revisiting Plant Biostimulants*. IntechOpen, United Kingdom.

Chaudhary, V., K.N. Saraswathy, and R. Sarwal. 2022. Dietary diversity as a sustainable approach towards micronutrient deficiencies in India. *Indian J. Med. Res.*, 156: 31–45.

De Moura, F.F., A.C. Palmer, J.L. Finkelstein, J.D. Haas, L.E. Murray-Kolb, M.J. Wenger, E. Birol, E. Boy, and J.P. Peña-Rosas. 2014. Are biofortified staple food crops improving vitamin A and iron status in women and children? New evidence from efficacy trials. *Adv. Nutr.*, 5: 568–570.

Douthwaite, B., N. Johnson, and A. Wyatt. 2023. Using outcome trajectory evaluation to assess HarvestPlus' contribution to the development of national biofortification breeding programs. *Eur J. Dev. Res.*, 35: 426–451.

FAO, IFAD, UNICEF, WFP, WHO. 2022. The State of Food Security and Nutrition in the World 2022. Repurposing food and agricultural policies to make healthy diets more affordable. Data are available on FAOSTAT (https://www.fao.org/faostat/en/#data/FS)

Garg, M., N. Sharma, S. Sharma, P. Kapoor, A. Kumar, V. Chunduri, and P. Arora. 2018. Biofortified crops generated by breeding, agronomy, and transgenic approaches are improving lives of millions of people around the world. *Front. Nutr.*, 12.

Gernand, A.D., K.J. Schulze, C.P. Stewart, K.P. West Jr., and P. Christian. 2016. Micronutrient deficiencies in pregnancy worldwide: Health effects and prevention. *Nat. Rev. Endocrinol.*, 12: 274–289.

Gonmei, Z. and G.S. Toteja. 2018. Micronutrient status of Indian population. *Indian J. Med. Res.*, 148: 511–521.

HarvestPlus. (n.d.). Crops. https://www.harvestplus.org/what-we-do/crops.

Hotz, C., and B. McClafferty. 2007. From harvest to health: Challenges for developing biofortified staple foods and determining their impact on micronutrient status. *Food Nutr. Bull.*, 28: S271–S279.

Johns, T. and P.B. Eyzaguirre. 2007. Biofortification, biodiversity and diet: A search for complementary applications against poverty and malnutrition. *Food Policy*, 32: 1–24.

Lassi, Z., A. Moin, and Z. Bhutta. 2017. Nutrition in middle childhood and adolescence. *In:* D.A.P. Bundy, de N. Silva, S. Horton, D.T. Jamison, and G.C. Patton. (Eds.), *Child and Adolescent Health and Development*. The International Bank for Reconstruction and Development/The World Bank, Washington (DC).

Mejia, L.A., O. Dary, and H. Boukerdenna. 2017. Global regulatory framework for production and marketing of crops biofortified with vitamins and minerals. *Ann N.Y. Acad. Sci.*, 1390: 47–58.

Nabuuma, D., W. Ocimati, A. Simbare, and B. Ekesa. 2022. Comparative analysis of the sensory acceptability of introduced Pro-Vitamin A-rich bananas in Eastern Africa. *Fruits*, 77(6): 10. ISSN: 0248-1294 https://doi.org/10.17660/th2022/023

Ofori, K.F., S. Antoniello, M.M. English, and A.N. Aryee. 2022. Improving nutrition through biofortification: A systematic review. *Front. Nutr.*, 9: 1043655.

Olson, R., B. Gavin-Smith, C. Ferraboschi, and K. Kraemer. 2021. Food fortification: The advantages, disadvantages, and lessons from sight-and-life programs. *Nutrients*, 13: 1118.

Saltzman, A., E. Birol, A. Oparinde, M.S. Andersson, D. Asare-Marfo, M.T. Diressie, C. Gonzalez, K. Lividini, M. Moursi, and M. Zeller. 2017. Availability, production, and consumption of crops biofortified by plant breeding: Current evidence and future potential. *Ann. N.Y. Acad. Sci.*, 1390: 104–114.

Singh, B., U. Goutam, S. Kukreja, S. Siddappa, S. Sood, J. Sharma, and V. Bhardwaj. 2022. Biofortification strategies to improve iron concentrations in potato tubers: Lessons and future opportunities. *Potato Res.*, 65: 51–64.

Siwela, M., K. Pillay, L. Govender, S. Lottering, F.N. Mudau, A.T. Modi, and T. Mabhaudhi. 2020. Biofortified crops for combating hidden hunger in South Africa: Availability, acceptability, micronutrient retention, and bioavailability. *Foods*, 9: 815.

Talsma, E.F., A. Melse-Boonstra, and I.D. Brouwer. 2017. Acceptance and adoption of biofortified crops in low- and middle-income countries: A systematic review. *Nutr. Rev.*, 75: 798–829.

Venkatesh, U., A. Sharma, V.A. Ananthan, P. Subbiah, and R. Durga. 2021. Micronutrient's deficiency in India: A systematic review and meta-analysis. *J. Nutr. Sci.*, 10: e110.

Wani, S.H., Gaikwad, K., Razzaq, A., Samantara, K., Kumar, M., and V. Govindan. 2022. Improving zinc and iron biofortification in wheat through genomics approaches. *Mol. Biol. Rep.*, 49: 8007–8023.

White, P.J., and M.R. Broadley. 2009. Biofortification of crops with seven mineral elements often lacking in human diets: Iron, zinc, copper, calcium, magnesium, selenium, and iodine. *New Phytol.*, 182.

WHO/FAO 2006. *Guidelines on Food Fortification with Micronutrients*. World Health Organization and Food and Agriculture Organization of the United Nations.

CHAPTER 3

Global Scenario of Biofortification

Shiv Kumar[a*][0000-0001-8407-3562], Akanksha Singh[b], H.K. Dikshit[b], Gyan Prakash Mishra[b], M. Aski[b], Ruchi Bansal[c], and Priyanka Gupta[d]

[a] ICARDA, NASC Complex, New Delhi - 110012
[b] Division of Genetics, ICAR-IARI, New Delhi - 110012
[c] Division of Plant Physiology, ICAR-IARI, New Delhi - 110012
[d] University Laval, Quebec, Canada

1. Introduction

The global population is expected to reach over 10 billion by 2050, necessitating up to 70% increase in the global food output. At present, one-third of the world's population, mainly in Asia and Africa, suffers from micronutrient deficiencies (Kumar et al., 2022). Micronutrient deficiency (MND), also referred as hidden hunger, affects pregnant women and young children below 5-year age the most. Lack of optimal nutrition in women's diet during pregnancy and lactation may result in impaired physical and mental development (Marshall et al., 2021). Nevertheless, the most notable and widespread deficiency is of iron (Fe), zinc (Zn), vitamin A, and vitamin B9 (folate). Fe is required for metabolic processes like electron transport and DNA synthesis. Its deficiency (below 40 μg/day) results in anemia, weakness, and fatigue, impaired mental development and learning capacity, and increased risk during childbirth. According to Skalny et al. (2020), zinc exhibits potent antiviral and antibacterial properties within the human body, contributing to the enhancement of immune function and the body's ability to combat a range of viruses, including the coronavirus. Vitamin A has a crucial role in maintaining good vision, facilitating bone formation, and promoting cell division in animals. According to Conaway et al. (2013), the compound β-Cryptoxanthin has been found to promote the process of osteoblastic bone formation. Carotenoids possess potent anti-carcinogenic capabilities and serve as

*Corresponding author: SK.Agrawal@cgiar.org

protective agents against oxidative damage to cellular organelles. Deficiency of vitamin A/carotenoid impairs growth and causes eye damage leading to blindness and increased risk of infections such as diarrheal disease. Similarly, vitamin B9 (folate) is required to produce red and white blood cells in bone marrow and to transform carbohydrates into energy during pregnancy, infancy and adolescence and its deficiency can lead to cause megaloblastic anemia. Therefore, elevating the state of human health by providing multiple nutrients through enhanced agricultural production has been identified as a major challenge. This requires genetic and agronomic innovations to enhance food production augmented with higher nutritive value.

2. Genetic Biofortification

Enhancing the concentration of micronutrients in foods using traditional plant breeding and/or genetic engineering represents a very efficient strategy, devoid of additional expenses at the consumer level. Increasing the concentration of micronutrients in foods using traditional plant breeding and/or genetic engineering, stands out as the most efficient strategy, without imposing any additional costs on consumer end. A global programme, HarvestPlus, was implemented to reduce micronutrient malnutrition among the poor by breeding staple food crops (beans, rice, wheat, maize, pearl millets, sweet potato, cassava) that are rich in micronutrients (Fe, Zn, and vitamin A) through biofortification. Genetic biofortification is a process of increasing the density of minerals and vitamins in crop varieties through conventional and molecular tools of crop improvement.

Global breeding programs have mainstreamed conventional methods of biofortification for crop productivity and micronutrient enrichment with stable performance across environments. Like the transfer of any other traits, conventional breeding scheme for biofortification involves selection recombination selection evaluation pipeline. It starts with the identification of desired germplasm through extensive screening, study on the mode of inheritance of the traits, hybridization with appropriate agronomic background, selection and fixing of the best recombinants, and their testing at a target population of environments. To increase selection accuracy and intensity, genomic-assisted selection has been integrated. Biofortified varieties emerging through this process are widely accepted by farmers and consumers as it does not involve regulatory hurdles and is regarded as the sustainable route for providing more nutritious crops to consumers. Crop wild relatives are reservoirs of substantial micronutrient variation. This existing variability has been harnessed by breeders to develop varieties with improved micronutrient content. In recent years, CRISPR-Cas genome editing technology has also been utilized for improving nutritional properties in food crops (Nagmina and Ezura, 2022).

Spectacular progress has been made in the biofortification of one or the other micronutrients in food crops under the umbrella of HarvestPlus. This collaboration has resulted in the development of >330 biofortified varieties of seven target crops in 40 countries through conventional breeding (Table 1, Figure 1) (https://www.harvestplus.org). These included 65 and 11 iron-biofortified varieties of beans and pearl millet; 14, 12, and 11 zinc-enriched varieties of wheat, rice, and maize; and 18, 64, and 128 varieties of cassava, maize, and sweet potato biofortified with vitamin A, respectively. Recently, Bangladesh and the Philippines approved genetically modified golden rice with vitamin A for commercial cultivation. Biofortified crops grown in 35 countries have benefited 40 million people who cannot afford diet diversification (Mishra et al., 2022). These varieties have significantly improved target populations micronutrient status.

3. Natural Genetic Variation for Micronutrients in Target Crops

Genetic diversity for target micronutrients is key to mainstreaming biofortification in crop plants. Assessing micronutrient diversity and identification of potential donor parents has been integrated globally in biofortified programs. Among cereals, pearl millet, rice, wheat, and maize have been targeted for biofortification with Fe, Zn, and Vitamin A. Pearl millet exhibits relatively high concentrations of both iron (Fe) ranging from 24.2 to 48.8 ppm and zinc (Zn) ranging from 19.8 to 43.4 ppm (Pucher et al., 2014). In contrast, polished rice shows lower levels of Fe (0.8–12.3 ppm) and Zn (8.2–40.8 ppm) (Bollinedi et al., 2020), while wheat have moderate range for Fe (up to 88 ppm) and Zn (14–190 ppm) (Khokhar et al., 2020). However, maize, enriched with β-carotene, demonstrates wide variability from 0.01 to 16.50 μg/g (Vignesh et al., 2012).

Among legumes, common bean, lentil, mungbean, and cowpea have been subjected to biofortification due to significant genetic variability. The CIAT found wide diversity in common bean (Fe 34–96; Zn 21–59) by involving 1072 cultivated and 119 wild accessions (Islam et al., 2002; Singh et al., 2017). Among the released lentil varieties, Barimasur-4, Barimasur-5, Barimasur-6, Barimasur-7 Barimasur-8, Bariasur-9 (75–86 ppm Fe and 51–63 ppm Zn content) in Bangladesh, Alemaya (82 ppm Fe and 62 ppm Zn) in Ethiopia, Pusa Vaibhav, Pusa Ageti Masoor, and IPL220 in India, Sisir, Shital, Khajurah-1, Khajurah-2, ILL 7723, and Shekhar (82–98 ppm Fe) in Nepal and Idlib-2 and Idlib-3 (72 ppm Fe) in Syria have been deployed for large-scale cultivation. Cowpea is a major source of Fe and Zn in West African countries and its diversity has been utilized to improve its Fe and Zn content (Table 1).

4. Genetic Engineering for Biofortification

Biofortification using transgenic is another option when crop genetic resources have limited micronutrient variation. Bioavailability, micronutrient enhancement, and antinutritional factor reduction can be achieved by targeting the related

Table 1: Biofortified varieties released at the global level

Vitamin/Micronutrient	Country	Biofortified Varieties Released (ppm)
Vitamin A Cassava	Sierra Leone	Slicass 12 (8.8)
	Nigeria	UMUCASS 44 (10.8), UMUCASS 37 (7.4), UMUCASS 36 (6.9), UMUCASS 45 (11.4), UMUCASS 46 (10.9), UMUCASS 38 (7.8), UMUCASS 52 (16.8), UMUCASS 53 (15.8), UMUCASS 54 (15.6)
	Ghana	Nyonkuagbeli (9.6), Tetteh bankye (6.9), Kpornuagbeli (7.9), CRI-Manu (10.4), CRI-OpemfoBankye (7.8), CRI-Lizzy (7.7), CRI-MmofraBankye (12.7), CRI-Kent (7.8), CRI-Peprah (12.7), CRI-Manu (10.7),
	DR Congo	Kindisa (7.6), Vimpi (10.4), Lumonu (9.5), Mukoleshi (7.9)
	Cameroon	I070593 (11.4)
	Brazil	BRS Jari (7.7), BRS Gema de Ovo (3.4), BRS Dourada (4.5)
Vitamin A Sweet Potato	Zambia	Chiwoko (110), Olympia (49), Chumfwa (342.1), Kokota
	Uganda	Ejumula (91), Kakamega (41), Kabode (97), Gerald (72), Joweria (95), Vita (75), NASPOT 7 (108), NASPOT 8 (144)
	Tanzania	Mayai (110), Ejumula (144), Kakamega (38), Kiegea (18), Kabode (97)
	South Africa	Impilo (51), Khano, Purple Sunset, Bophelo, Serolane, Isondlo
	Mozambique	Ejumula (53.8), Persistente, Cordner, Amelia (50), Bela (83.9), Cecilia (60.1), Delvia (55.4), Erica (101.6), Esther (47.2), Ininda (53.1), Irene (60.6), Jane (55.9), Lourdes (99.4), Melinda (57.1), Namanga (83.9), Sumaia (77), Tio Joe (103.2), Gaba Gaba, Caelan, Ivone (70.6), Victoria (47.7), Alisha (73.3), Lawrence (62.5)
	Malawi	Ana Akwanire (55), Kadyaubwerere (89), Kaphulira (32), Mathuthu (29), Zondeni (90), Chipika (124.7)

(Contd.)

Table 1: *(Contd.)*

Vitamin/Micronutrient	Country	Biofortified Varieties Released (ppm)
	Madagascar	Mendrika, Bara (46), Gloria (78), Zambezi (38), P162, J664, J314, Delvia, Jane, Lourdes, Irene, Erica
	Keyna	K-117, Kenspot-3 (14), Kenspot-4 (40), Kenspot-5 (55), Vita (315), K/KA2004/215 (10), Kabode (97), NASPOT 10 O, Irene (332.3)
	Rwanda	Cacearpedo, Kakamega, Terimbere, Ndamirabana, Gihingamukungu, 2000-040
	Nigeria	King J, UMUSPO4, Solo Gold
	Ghana	CRI-Bohye (243.1), CRI-Apomuden (55), SARI-Nan (383.7)
	Ethiopia	Kulfo (210.6), Tulla (146), Kero, Birtukane, Ukr/Eju-10 (124), Ukr/Eju-13
	Cote de Ivore	Irene, Bela Bela, TIB-440060, TIB
	Burundi	Cacearpedo, 97062
	Burkina Faso	Tiebele-2, Caromex, CIP199062.1, Bagre, Jewel, Nanyoumondo-1, Nanyoumondo-2, Heere
	Angola	Camuto, Cenoura, Helena, Zapallo, Nemanete, Huambachero, Mayai
	Peru	Mejorada, Arne, Benjamin, Abigail, Isabel, Sumy
	Panama	IDIAP C9017 (52), IDIAP C0317 (58)
	Nicaragua	INTA Nutritivo
	Guatemala	ICTA Dorado-BC (133), ICTA Pacifico-BC (77)
	Colombia	Agrosavia Aurora (214)
	Brazil	Beauregard (115), BRS Amelia (115), SCS-367 Favorita, BRS Nuti (149)
	Timor-Leste	Hohrae-3
	South Korea	Haepymi, Dahomi

	Indonesia	Beta-1, Beta-2
	India	Sree Kanaka (100), Indira Madhur (35), Kamala Sundari, Bhu Sona (140), Bhu Kanti, Bhu Ja
	China	Yanshu 25 (36), Zheshu 13, Guangshu 87, Xiangxiang
	Bangladesh	BARI-SP-12, BARI-SP-13, BARI-SP-14, BARI-SP-15
Vitamin A Banana/ Plantain	DR Congo	Apantu (110.6), Bira (106.3), Pelipita (17.4), To'o (77.6), Lahi (105.1), Muracho (154.6), Pisang Papan (108.6)
	Burundi	Apantu (100.6), To'o (77.6), Muracho (154.6) Bira (106.3), Pelipita (17.4), Lai (105.1), Pisang Papan (108.1)
Vitamin A Maize	Zimbabwe	ZS242A (9.8), ZS244A (12), ZS246A (8.7), ZS248A (7.8), ZS500A (14.9)
	Zambia	GV665A (7.8), GV662A (4.7), GV664A (7.2), GV671A (11.4), GV672A (10.2), GV673A (10.9), GV6023A (13), GV6029A 914.9), GV6027A (9.5), GV6017A (10.8), GV6025A (11.8)
	Tanzania	Meru VAH 517 (14.9), Meru VAH 519 (7.8)
	Rwanda	RAHA03 (5.57), RAHA04 (7.28), RAHA02 (4.11)
	Nigeria	Sammaz 38 (5.9), Sammaz 39 (6.7), Ife Hyb 3 (8), Ife Hyb 4 (7.8), Sammaz 43 (7), Sammaz 44 (7.8), Sammaz 49 (8.2), Sammaz 52 (8.6), SAMMAZ 59 (16.3), SAMMAZ 60 (15.53)
	Mali	Nafama (7), Abebe (8), Duba (9), Kodialan (9), Dakan (7)
	Malawi	MH39A (12), MH40A (10), MH42A (14.9), MH43A (9.3), MH44A (11), MH46A, MH45A, MH47A, MH48A, MH49A
	Ghana	CSIR-CRI Honampa (7), CSIR-CRI Odomfo (6), CSIR-CRI Owanwa (6), Aho dzin (9.3), Dzifoo (11.3), Ahof (9.2), CRI-Nkwagye (14.8), CRI-Abebe (14), CRI-Ewool (13.2), ZAMS666A (14.2), CRI-HarvestPlus (13.2)
	Ethiopia	BHA5211 (11), BHA5212 (11)
	DR Congo	Sam 4 Vita (8.8), Muibaki 3 (10), Muibaki 2 (7), Muibaki 1 (6), GV664A (7.2), GV665A (7.8), LY1001-14 (9.2), PVA SYN 10-F2, PVA SYN 18-F2

(*Contd.*)

Table 1: (*Contd.*)

Vitamin/Micronutrient	Country	Biofortified Varieties Released (ppm)
	Cameroon	PVA SYN 6 F2 (7), PVA SYN 13 F2 (9)
	Brazil	BRS 4104 (8)
Zn Rice	Nicaragua	INTA Las Minas (22)
	El Salvador	CENTA A-Nutremas (22.8)
	Colombia	FedearrozBIOZn 035 (26)
	Bolivia	CIAT BIO-44 +Zinc (22)
	Indonesia	INPARI IR Nutri Zinc (25), Inpara 11 Siam HiZInc (33.9), Inpara 12 Mayas (29.8)
	India	DRR Dhan 49 (25.2)
	Bangladesh	BRRI Dhan62 (20), BRRI Dhan64 (24), BRRI Dhan72 (23), BRRI Dhan74 (24.2), BU Aromatic Hybrid Dhan-1 (22),Binadhan 20 (27.5), BU Aromatic Dhan-2 (22), BRRI Dhan84 (27.6), BRRI dhan100 (25), BRRI dhan102 (25.5)
Fe Pearl Millet	Niger	Chakti (65)
	India	Dhanshakti (72), Shakti 1201 (Dhoni) (74), HHB 299 (73), AHB 1200 Fe (77), HHB 311 (62), RHB 234 (62), AHB 1269 Fe (73), RHB 233 (64), Moti Shakti (GHB 1225) (74), Phule Mahashakti (DHBH 1211) (78)
Zn Wheat	Mexico	Nohely F2018 (31)
	Brazil	BRS 331 (37.3)
	Bolivia	INIAF Okinawa (32)
	Pakistan	Zincol-2016 (37), Akhbar-2019 (37), Nawab 21 (37)
	Nepal	Zinc Gahun-1 (38), Himgange (54), Panchakoshi (39.4), Zinc Gahun-2 (39.4), Zinc wheat-3 (48), Borlaug100 (31)

	India	BHU-3 (30.5), Zn-Shakti (34.2), BHU-1 (34.8), BHU-5 (29.5), WB-02 (31), HPBW01 (31), BHU-25 (31), BHU-31 (39.5), HUW 711 (31)
	Bangladesh	BARI-Gom33 (33)
Zn Maize	Nicaragua	Fortinica (34.9), INTA-Nutremas (35)
	Honduras	DICTA B02 (34.50), DICTA B03 (35.1)
	Guatemala	ICTA HB-18ACP+Zn (31), ICTA B-15ACP+Zn (30), Fortaleza 17 (32)
	El Salvador	CENTA Porrillo 2020 (32)
	Colombia	BIO-MZn01 (34.5), SGBIOH2 (33), SGBIOH6 (32)
Zn Sorghum	India	Parbhani Shakti (32)
Fe Drybeans	Zimbabwe	Jasmine (86)
	Zambia	Mbereshi (92)
	Uganda	Roba (64), NARO bean 4C(78), NARO Bean 2 (76), NARO Bean 1 (71), NARO Bean 3 (70), NARO Bean 5C (67)
	Tanzania	Selian 14(85), Selian 15 (81), TARI BEAN 2 (87.2), TARI BEANS 4 (78.3), TARI BEAN 5 (90.5), TARI 19 (71)
	Rwanda	MAC (44), RWV 1129(77), CAB 2 (8), RWR 2245 (76), RWR 2154 (71), RWV 3316 (87), RWV 3317 (74), RWC 3006 (78), MAC 42 (91), RWV 2887 (85), RW-CB 184 (113), RWCB 181 (119), Rw-BB 183, (89), RW-CB 186 (91.2), RW- BB181 (94), Rw-BB 186 (96), RW-B3184 (98)
	DR Congo	RW-CB189 (95-113), RWR 2245 (66), RWR 2154 (71), COD MLB 001 (66), HM 21-7 (62), COD MLV 059 (84), VCB 81013 (69), Pigeon Vert (80), M 211 (77), PVA 1438 (79), Nain de Kyondo (76), COD MLB 032 (76), Cuarentino 0817 (100), Namulenga (76), Ecapan 021 (72), NUA 100 (78), MNC 488-2 (79), NUV 119-4 (72), Nyiramuhondo (72), NUA 99 (78.6)
	Burundi	MAC 44 (69), RWV 1129 (92), RWR 2245 (76), RWR 2154 (71)

(Contd.)

Table 1: *(Contd.)*

Vitamin/Micronutrient	Country	Biofortified Varieties Released (ppm)
	Panama	IDIAP NUA 24 (78), IDIAP NUA 27 (71), IDIAP P-09-11 (87), IDIAP P-13-38 (78)
	Nicaragua	INTA Nutritivo (70), INTA Ferroso (79), INTA BIOF100 (80), INTA BIODOR (76), INTA Rojo Bio-Apante (91)
	Honduras	Honduras Nutritivo (71), Rojo Fortificado (78)
	Guatemala	ICTA Peten-ACM (70), ICTA Superchiva-ACM (74), ICTA Chorti-ACM (95)
	El Salvador	CENTA Ferromas (77)
	Colombia	Corpoica Rojo 39 (86), Corpoica Rojo 43 (95), BIO-101 (83)
	Brazil	BRS Agreste (78.8), BRS 9435 Cometa (80), BRS Pontal (77), BIO-107 (82), BIO-102 (86)
	Bolivia	Fortaleza (87)
Fe Cowpea	Zimbabwe	CBC6 (90)
	Brazil	BRS Araca (62), BRS Tumucumaque (61), BRS Xiquexique (77)
	India	Pant Lobia-1 (82), Pant Lobia-2 (95), Pant Lobia-3 (97), Pant Lobia-4 (52), Pant Lobia-5 (66), Pant Lobia-7 (43)
Fe Lentil	Syria	Idlib-2, Idlib-3
	India	L 4717 (70), IPL 220 (75)
	Bangladesh	Barimasur-4 (51), Barimasur-5 (59), Barimasur-6 (63), Barimasur-7 (61), Barimasur-8 (60)
	Nepal	ILL 7723 (62), Rasuwa black (TBD)

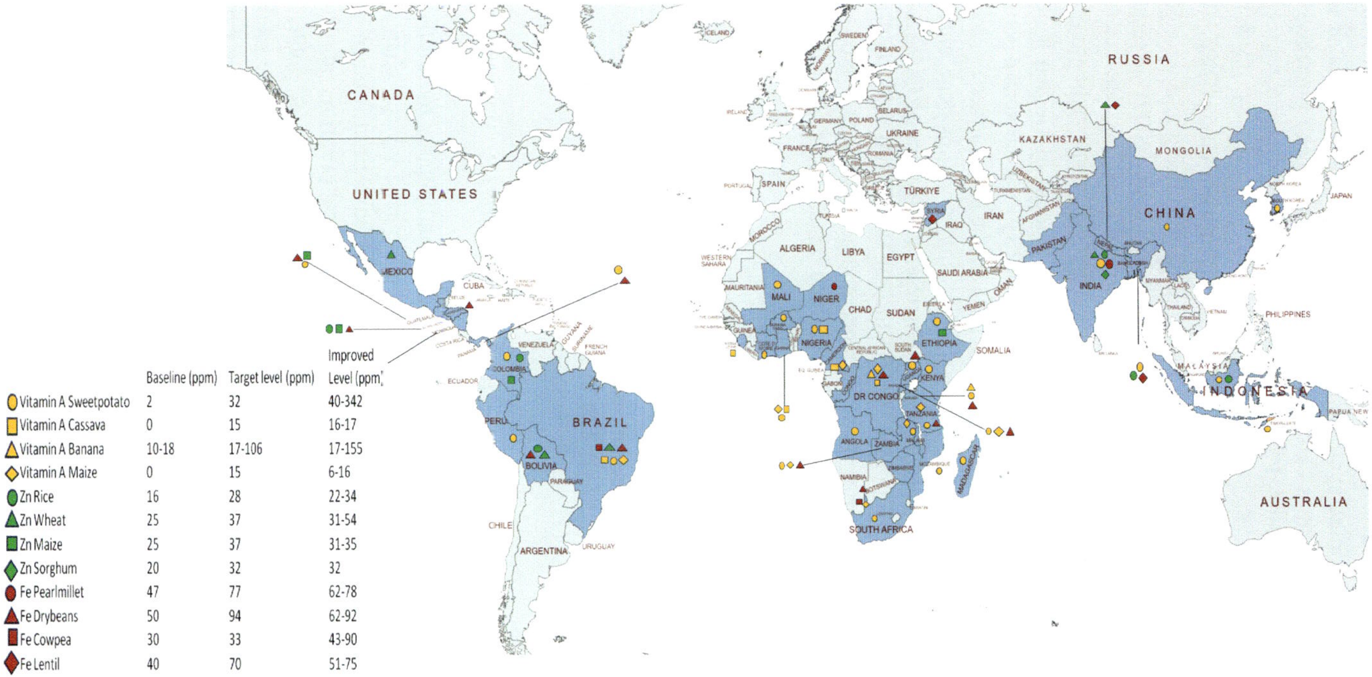

Figure 1: Global progress in biofortification programs: Released cultivars with enhanced micronutrient content.

genes or selected pathways. The primary target for biofortification for minerals is ferritin and nicotinamine synthase (NAS); for essential fatty acids, it is Δ6 desaturase; for vitamins, it is carotene desaturase and lycopene β-cyclase; and for essential amino acids, it is PSY (Garg et al., 2018).

In recent years, significant advancements have been made in the field of biofortification to address micronutrient deficiencies in staple crops. Trijatmiko et al. (2016) demonstrated an increase in Fe concentration in rice grains through the targeted manipulation of NAS genes. Meanwhile, Lee et al. (2009) successfully enhanced Zn content in genetically modified rice by overexpressing OsIRT1, marking a crucial stride in addressing nutritional deficiencies. The pioneering work of Beyer et al. (2002) focused on the development of Golden Rice, by expressing genes responsible for phytoene synthase (PSY) and carotene desaturase. Through targeted genetic modifications, the level of the beta-carotene precursor, phytoene, was notably increased, up to 23-fold by targeting gene encoding carotene desaturase. The approval for commercial cultivation of Golden Rice in Bangladesh and the Philippines stands out as a significant milestone. This development not only signifies the acceptance and endorsement of biofortified crops but also highlights the potential far-reaching impact these innovations can have on global nutrition.

Wheat with enhanced Fe content has been developed by targeting TaVIT2 (Connorton et al., 2017) and NAS2 (Beasley et al., 2019). Naqvi et al. (2009) reported multivitamin corn by engineering three distinct metabolic pathways. Lipkie et al. (2013) assessed the bioaccessibility of carotenoids from transgenic provitamin A biofortified Sorghum and reported improvement of pro-vitamin A by expression of Homo188-A. The overexpression of IbOr-Ins increased the carotenoid content of sweet potato by up to 7-fold (Park et al., 2015).

5. Genome Editing for Biofortification

CRISPR-based genome-editing is the most effective and popular biofortification method because it allows quick, precise, and multiplexed genetic change. To combat vitamin A deficiency, CRISPR-Cas systems have been used to create genome-edited golden crops like golden flesh melons (Tzuri et al., 2015), tomatoes (Li et al., 2018), golden bananas (Kaur et al., 2020) and golden rice (Dong et al., 2020). For instance, the 5.2-kb carotenogenesis cassette made up of the CrtI and maize PSY genes was used to generate the Golden rice cultivar Kitaake. The endosperm of Kitaake has 7.9 μg/g dry weight of beta-carotene (Dong et al., 2020). The CRISPR-Cas9 technique has also been used to produce crops that are biofortified with Zn and Fe. In rice, the OsVIT2 gene was knocked down through genome editing for increased Fe availability (Ludwig and Slamet-Loedin, 2019). This approach helps to improve Fe and Zn accumulation in wheat grain by targeting the genes Inositol Pentakisphosphate 2-Kinase 1 (TaIPK1) (Ibrahim et al., 2022).

6. Bioavailability of Micronutrients

The bioavailability of food defined in relation to human nutrition is the amount of nutrients in a diet of an individual that can be utilized and absorbed for different metabolic processes (Welch and Graham, 2004). Bioavailability of micronutrients in crop plants is low and consequently, people who rely on vegetarian diets are deficient in essential micronutrients. Therefore, assessing the bioavailability of micronutrients in newly developed varieties is a key aspect of biofortification programs.

6.1. Nutritional Efficacy of Biofortified Varieties

About 8.5 million African, Asian, and Latin American farms are growing biofortified crops. In malnourished people, efficacy trials have demonstrated an improvement in the status of Zn, Fe and provitamin A, cognitive function, work performance, and morbidity. Conventional plant breeding produces biofortified crop varieties that 20 million people worldwide consume, to reach one billion by 2030 (Bouis, 2018). The impact of biofortified crops on nutritional status is measured by efficacy trials under controlled conditions to ensure the bioavailability of the targeted traits by utilizing models and direct study in human beings.

Vitamin A: Nutritional efficacy trials depicted an increase in the uptake of provitamin A from biofortified varieties resulting in moderate upliftment of nutritional status by more circulation of beta carotene in the blood. Talsma et al. (2015) studied the efficacy of biofortified yellow cassava in Kenyan children for improving vitamin A. School children between 5–13 revealed significant enhancement of serum concentration of retinol and beta carotene. There are reports of a positive effect on vitamin A stores by consumption of sweet potatoes (Van Jaarsveld et al., 2005). Palmer et al. (2018) studied the impact of the consumption of biofortified maize on serum carotenoid concentration and reported improved serum β carotene concentration in maternal and child nutrition. Afolami et al. (2020) reported improved serum retinol and hemoglobin concentrations in the preschool children served with biofortified cassava. The interaction of pro-vitamin A-rich sweet potato on the Fe bioavailability of biofortified cowpeas was studied by Corrêa et al. (2020) using *in vitro* Caco-2 cells and *in vivo* depletion repletion rat model. The *in vitro* study presented higher bioavailability of Fe with or without sweet potato whereas the *in vivo* study exhibited that combination of rice and beans with carotenoid-rich sweet potatoes enhanced the expression of genes related to Fe metabolism and improved its bioavailability.

Fe: Research on nutritional efficacy studies of Fe demonstrated the upliftment of the nutritional status of the targeted population. The consumption of biofortified bean significantly has a positive effect on hemoglobin, serum ferritin, and body Fe in Rwandan women (Luna et al., 2020; Wenger et al., 2019). A positive effect on cognitive performance in 18 to 27-year-old female students by consumption of biofortified beans has also been observed (Murray-Kolb et al., 2017). In India, a randomized efficacy trial of biofortified pearl millet in 140 boys and girls aged 12–16 years significantly improved the daily intake of Fe (Scott et al., 2018). Fe absorption from Fe biofortified sweet potato is 1.9-fold higher in Malawian women (Jongstra et al., 2020).

Zn: Efficacy studies on rice biofortified with Zn demonstrated an increase in the bioavailability and absorption of Zn in adults (Brnićet et al., 2016). Zn absorption study in wheat in Mexico suggested substantially higher absorption of Zn in biofortified wheat as compared to non-biofortified wheat. Lowe et al. (2020) studied the impact of consuming Zn biofortified flour (Zincol-2016) (40 mg/kg) for 12 months on 482 households in Peshawar (presence of both an adolescent girl, aged 10–16 years and a child aged 1–5 years) and reported substantial improvement in nutritional status of the studied children.

7. Dissemination Challenges of Biofortified Crops

7.1. Increasing the Awareness of Health Benefits

A critical challenge is to make stakeholders, including policymakers, acknowledge the role of biofortification in enhancing public health. Educating both rural and urban populations about the health advantages of biofortified crops is crucial for increasing their adoption. Greater market demand, driven by attractive prices, will encourage farmers to cultivate biofortified varieties. Additionally, value-added products from these crops offer opportunities for rural entrepreneurs to establish small-scale industries.

7.2. Increasing Acceptance of Nutritious Food with Altered Appearance

Biofortified crops may have altered appearances that consumers may initially reject. For example, yellow/orange maize in Africa faced resistance initially. In Sub-Saharan Africa, white sweet potatoes were preferred over biofortified varieties. However, through extension efforts, policy support, and nutritional efficacy trials, these biofortified options gained popularity. Steur et al. (2017) found that farmers will pay 20% more for biofortified crop varieties in a meta-analysis of 23 studies with 9507 respondents however more consumer acceptance trials are still needed.

7.3. Policy Support

Government policy support is essential for the acceptance and adoption of biofortified crops. Several countries in Latin America, Africa, and Asia have developed strategies to promote biofortified crops. Asian nations like India, Bangladesh, China, Indonesia, Vietnam, and the Philippines have regulations for genetically modified crops and cultivate biofortified varieties. In contrast,

Latin America lacks such regulations, with only biofortified beans being planted and consumed. Initiatives like India's National Nutrition Strategy 2017 supports biofortification. Policies such as strengthening the seed chain, subsidizing seeds and inputs, ensuring minimum support prices, and providing loans and subsidies to rural entrepreneurs accelerate biofortified crop adoption.

7.4. Research Support

For addressing micronutrient deficiencies, National and International Agricultural Research Institutes must mainstream biofortification and recognize the minimum levels of micronutrients and vitamins (depending on baseline for each crop in different regions) for the release of crop varieties. Biofortification needs adequate monetary support to end malnutrition. CG centres in collaboration with National programs should take up on large-scale multilocation assessment of cultivated and wild germplasm of different crops for micronutrient concentration. Breeding programs must also address issues related to micronutrient bioavailability. Upscaling biofortified crops requires a better partnership between the public and private sectors. Multilateral institutions, the private sector, and NGOs must join hands to increase the area under biofortified varieties.

8. Conclusion

The current global scenario of biofortification is very encouraging and the achievement lifts the hopes for nutritional security for all. Sufficient genetic variation for micronutrient content is available in the germplasm with remarkable success in breeding micronutrient rich varieties in major crops. However, micronutrient concentration is a complex quantitative trait and is highly influenced by genotype, environment, and their interactions. Therefore, there is a need for molecular dissection of Fe and Zn concentrations to identify QTLs associated with uptake and transport. In conjunction with micronutrient content, natural variation relating to inhibitors and promoters that control the bioavailability of important micronutrients should concurrently be assessed. As biofortification of food crops is a sustainable approach to fight hidden hunger, there is an urgent need to mainstream biofortification of One CGAR breeding and national research programs to ensure increase mineral and vitamin content a mandatory breeding objective. Integrating biofortification traits into both public and private breeding programs and raising awareness among donors about the importance of funding and policy backing are crucial steps in this direction. CG centers hold global collections of crop germplasm which, in collaboration with national programs, should be analyzed for micronutrient concentration in different crops. Increasing bioavailability by reduction of antinutrients and increasing the concentration of promoters need strengthening. Modern gene-editing technologies need to be utilized extensively to accelerate biofortification programs. Future research requires stacking of multiple minerals using metabolic engineering, and their bioavailability. A clear understanding of genetic, physiological and molecular processes determining mineral nutrient absorption by plants and its subsequent transportation in plants will improve the nutritional performance of crops. Crop biofortification through gene stacking, introgression breeding, and a combination of conventional breeding and metabolic engineering hold promise for delivering biofortified crop varieties.

References

Afolami, I., M.N. Mwangi, F. Samuel, E. Boy, P. Ilona, E.F. Talsma, and A. Melse-Boonstra. 2021. Daily consumption of pro-vitamin A biofortified (yellow) cassava improves serum retinol concentrations in preschool children in Nigeria: A randomized controlled trial. *Am. J. Clin. Nutr.*, 113: 221–231._

Beasley, J.T., J.P. Bonneau, J.T. Sánchez-Palacios, L.T. Moreno-Moyano, D.L. Callahan, E. Tako, and A.A. Johnson. 2019. Metabolic engineering of bread wheat improves grain iron concentration and bioavailability. *Plant Biotechnol. J.*, 178: 1514–1526.

Bollinedi, H., A.K. Yadav, K.K. Vinod, S. Gopala Krishnan, P.K Bhowmick, M. Nagarajan, and A.K. Singh. 2020. Genome-wide association study reveals novel marker-trait associations (MTAs) governing the localization of Fe and Zn in the rice grain. *Front. Genet.*, 11: 213.

Bouis, H.E. 2018. Biofortification: An agricultural tool to address mineral and vitamin deficiencies. *Food Fortification in a Globalized World*, 69–81. Academic Press.

Brnić, M., R. Wegmüller, A. Melse-Boonstra, T. Stomph, C. Zeder, F.M. Tay, and R.F. Hurrell. 2016. Zinc absorption by adults is similar from intrinsically labeled zinc-biofortified rice and from rice fortified with labeled zinc sulfate. *J. Nutr.*, 146: 76–80.

Conaway, H.H., P. Henning, and U.H. Lerner. 2013. Vitamin a metabolism, action, and role in skeletal homeostasis. *Endocr. Rev.*, 34: 766–797.

Connorton, J.M., E.R. Jones, I. Rodríguez-Ramiro, S. Fairweather-Tait, C. Uauy, and J. Balk. 2017. Wheat vacuolar iron transporter TaVIT2 transports Fe and Mn and is effective for biofortification. *Plant Physiol.*, 174: 2434–2444.

Corrêa, S.R., P. Brigide, M.D.G. Vaz-Tostes, and N.M.B. Costa. 2020. Cultivars of biofortified cowpea and sweet potato: Bioavailability of iron and interaction with vitamin A *in vivo* and *in vitro*. *J. Food Sci.*, 85: 816–823.

De Steur, H., J. Wesana, D. Blancquaert, D. Van Der Straeten, and X. Gellynck. 2017. Methods matter: A meta-regression on the determinants of willingness-to-pay studies on biofortified foods. *Ann. N.Y. Acad. Sci.*, *1390*: 34–46.

Dong, O.X., S. Yu, R. Jain, N. Zhang, P.Q. Duong, C. Butler, and P.C. Ronald. 2020. Marker-free carotenoid-enriched rice generated through targeted gene insertion using CRISPR-Cas9. *Nat. Commun.*, 11: 1178.

Garg, M., N. Sharma, S. Sharma, P. Kapoor, A. Kumar, V. Chunduri, and P. Arora. 2018. Biofortified crops generated by breeding, agronomy, and transgenic approaches are improving lives of millions of people around the world. *Front. Nutr.*, 12.

Jongstra, R., M.N. Mwangi, G. Burgos, C. Zeder, J.W. Low, G. Mzembe, and R. Wegmüller. 2020. Iron absorption from iron-biofortified sweet potato is higher than regular sweet potato in Malawian women while iron absorption from regular and iron-biofortified potatoes is high in Peruvian women. *J. Nutr.*, 150: 3094–3102.

Kaur, N., A. Alok, P. Kumar, N. Kaur, P. Awasthi, S. Chaturvedi, and S. Tiwari. 2020. CRISPR/Cas9 directed editing of lycopene epsilon-cyclase modulates metabolic flux for β-carotene biosynthesis in banana fruit. *Metab. Eng.*, 59: 76–86.

Khokhar, J.S., J. King, I.P. King, S.D. Young, M.J. Foulkes, J. De Silva, and M.R. Broadley. 2020. Novel sources of variation in grain Zinc (Zn) concentration in bread wheat germplasm derived from Watkins landraces. *PloS One*, 15: 0229107.

Khokhar, J.S., S. Sareen, B.S. Tyagi, G. Singh, L. Wilson, I.P. King, and M.R. Broadley. 2018. Variation in grain Zn concentration, and the grain iron in field-grown Indian wheat. *PLoS One*, 13: 0192026.

Kumar, S., H.K. Dikshit, G.P. Mishra, A. Singh, M. Aski, and P.S. Virk. 2022. Biofortification of staple crops: Present status and future strategies. *Biofortification of Staple Crops*, 1–30. Singapore: Springer Singapore.

Lee, S., and G. An. (2009). Over-expression of OsIRT1 leads to increased iron and zinc accumulations in rice. *Plant Cell Environ.*, 32: 408–416.

Li, X., Y. Wang, S. Chen, H. Tian, D. Fu, B. Zhu, and H. Zhu. 2018. Lycopene is enriched in tomato fruit by CRISPR/Cas9-mediated multiplex genome editing. *Front. Plant Sci.*, 9: 559.

Lowe, N.M., M. Zaman, M.J. Khan, A.K. Brazier, B. Shahzad, U. Ullah, and S. Tishkovskaya. 2022. Biofortified wheat increases dietary zinc intake: A randomized controlled efficacy study of zincol-2016 in rural Pakistan. *Front. Nutr.*, 8: 809783.

Ludwig, Y., and I.H. Slamet-Loedin. 2019. Genetic biofortification to enrich rice and wheat grain iron: From genes to product. *Front. Plant Sci.*, 10: 833.

Luna, S.V., L.M. Pompano, M. Lung'aho, J.B. Gahutu, and J.D Haas. 2020. Increased iron status during a feeding trial of iron-biofortified beans increases physical work efficiency in Rwandan women. *J. Nutr.*, 150: 1093–1099.

Marshall, N.E., B. Abrams, L.A. Barbour, P. Catalano, P. Christian, J.E. Friedman, and K.L. Thornburg. 2022. The importance of nutrition in pregnancy and lactation: Lifelong consequences. *AJOG*, 226: 607–632.

Mishra, G.P., H.K. Dikshit, Priti, B. Kukreja, M. Aski, D.K. Yadava, and S. Kumar. 2022. Historical overview of biofortification in crop plants and its implications. *Biofortification of Staple Crops*, 31–61. Singapore: Springer Singapore.

Murray-Kolb, L.E., M.J. Wenger, S.P. Scott, S.E. Rhoten, M.G. Lung'aho, and J.D. Haas. 2017. Consumption of iron-biofortified beans positively affects cognitive performance in 18- to 27-year-old Rwandan female college students in an 18-week randomized controlled efficacy trial. *J. Nutr.*, 147: 2109–2117.

Naqvi, S., C. Zhu, G. Farre, K. Ramessar, L. Bassie, J. Breitenbach, and P. Christou. 2009. Transgenic multivitamin corn through biofortification of endosperm with three vitamins representing three distinct metabolic pathways. *PNAS*, 106: 7762–7767.

Palmer, A.C., N.E. Craft, K.J. Schulze, M. Barffour, J. Chileshe, W. Siamusantu, and K.P. West Jr. 2018. Impact of biofortified maize consumption on serum carotenoid concentrations in Zambian children. *Eur. J. Clin. Nutr.*, 72: 301–303.

Park, S.C., S.H. Kim, S. Park, H.U. Lee, J.S. Lee, W.S. Park, and S.S. Kwak. 2015. Enhanced accumulation of carotenoids in sweet potato plants overexpressing IbOr-Ins gene in purple-fleshed sweet potato cultivar. *Plant Physiol. Biochem.*, 86: 82–90.

Pucher, A., H. Høgh-Jensen, J. Gondah, C.T. Hash, and B.I. Haussmann. 2014. Micronutrient density and stability in West African pearl millet: Potential for biofortification. *Crop Sci.*, 54: 1709–1720.

Scott, S.P., L.E. Murray-Kolb, M.J. Wenger, S.A. Udipi, P.S. Ghugre, E. Boy, and J.D. Haas. 2018. Cognitive performance in Indian school-going adolescents is positively affected by consumption of iron-biofortified pearl millet: A 6-month randomized controlled efficacy trial. *J. Nutr.*, 148: 1462–1471.

Singh, A., V. Sharma, H.K. Dikshit, M. Aski, H. Kumar, N. Thirunavukkarasu, and A. Sarker. 2017. Association mapping unveils favorable alleles for grain iron and zinc concentrations in lentil (Lens culinaris subsp. culinaris). *PLoS One*, 12: 0188296.

Skalny, A.V., L. Rink, O.P. Ajsuvakova, M. Aschner, V.A. Gritsenko, S.I. Alekseenko, and A.A. Tinkov. 2020. Zinc and respiratory tract infections: Perspectives for COVID-19. *Int. J. Mol. Med.*, 46: 17–26.

Talsma, E.F., I.D. Brouwer, H. Verhoef, G.N. Mbera, A.M. Mwangi, A.Y. Demir, and A. Melse-Boonstra. 2016. Biofortified yellow cassava and vitamin A status of Kenyan children: A randomized controlled trial. *Am. J. Clin. Nutr.*, 103: 258–267.

Tzuri, G., X. Zhou, N. Chayut, H. Yuan, V. Portnoy, A. Meir, and Y. Tadmor. 2015. A 'golden' SNP in CmOr governs the fruit flesh color of melon (*Cucumis melo*). *Plant J.*, 82: 267–279.

Van Jaarsveld, P.J., M. Faber, S.A. Tanumihardjo, P. Nestel, C.J. Lombard, and A.J.S. Benadé. 2005. β-carotene–rich orange-fleshed sweet potato improves the vitamin A status of primary school children assessed with the modified-relative-dose-response test. *Am. J. Clin. Nutr.*, 81: 1080–1087.

Vignesh, M., F. Hossain, T. Nepolean, S. Saha, P.K. Agrawal, S.K. Guleria, and H.S. Gupta. 2012. Genetic variability for kernel b-carotene and utilization of crtRB1 3'TE gene for biofortification in maize (*Zea mays* L.). *Indian J. Genet. Plant Breed.*, 72: 189–194.

Welch, R.M., and R.D. Graham. 2004. Breeding for micronutrients in staple food crops from a human nutrition perspective. *J. Exp. Bot.*, 55: 353–364.

Wenger, M.J., S.E. Rhoten, L.E. Murray-Kolb, S.P. Scott, E. Boy, J.B. Gahutu, and J.D. Haas. 2019. Changes in iron status are related to changes in brain activity and behavior in Rwandan female university students: Results from a randomized controlled efficacy trial involving iron-biofortified beans. *J. Nutr.*, 149: 687–697.

CHAPTER

4

Grain Iron (Fe): Decades of Research and Results

Neeraja C.N.*[0000-0002-2418-986X], **Suman K.**[0000-0002-3711-7881], **Jaldhani V.**[0000-0002-5206-7309], **Mangrauthia S.K.**[0000-0001-6316-9848], **Kalyani M. Barbadikar**[0000-0002-7734-8916], **Papa Rao Vaikuntapu**[0000-0001-6527-1600] **and Sundaram R.M.**[0000-0002-9857-8251]

ICAR-Indian Institute of Rice Research, Rajendranagar, Hyderabad - 500030, India

1. Introduction

Globally, around two billion people are reported to be nutrient deficient, particularly in iron (Fe), zinc (Zn), iodine and vitamin A, especially in countries with low or middle-income groups (https://www.globalhungerindex.org/). Iron deficiency causes anemia, impairs cognitive and physical development, decreases immunity, and reduces work productivity (Stevens et al., 2013). Grain iron content is primarily found in the aleurone and embryo, making whole cereal grains an effective solution for acute iron deficiency. But through polishing as in rice, milling as in wheat, and other post-harvest processes, the Fe content is lost along with aleurone, embryo, and other tissues. While the Fe deficiency is being addressed by providing fortification of staple foods and supplementation and advocacy of dietary diversity, biofortification of major crops can be a sustainable and cost-effective approach. With the initiative of CGIAR, the HarvestPlus Challenge Program targets the biofortification of major staple cereals and vegetables (www.harvestplus.org).

2. Contribution to Human Health

Among the 22 essential trace elements for human health, Fe is a critical and an essential micronutrient for the normal functioning of several biological

*Corresponding author: cnneeraja@gmail.com

processes (https://www.who.int/). Depending on the age, body weight, and gender, the recommended dosage of Fe can be from 7–18 mg/day, whereas for pregnant women, 27 mg/day is advised (https://ods.od.nih.gov/factsheets/Iron-HealthProfessional/). With its redox function, Fe functions as a cofactor in DNA synthesis and repair, in addition to being a component of various proteins and enzymes associated with respiration, energy, collagen, and immune metabolisms.

3. Strategies Used for Fe Biofortification

Fe biofortification employs conventional breeding, genetic engineering, and agronomic strategies. Agronomic approaches introduce Fe *via* soil, foliar application, or seed treatment, boosting grain Fe content in rice and wheat. Conventional breeding harnesses genetic variability, combining high Fe and yield through crosses, selecting recombinants, and stabilizing them as biofortified varieties. Successes include pearl millet, wheat, beans, cassava, lentils, potato, and pomegranate varieties (Garg et al., 2018; Yadava et al., 2022; Burgos et al., 2023). Genetic engineering excels in low-genetic-variability crops, introducing genes to elevate Fe content and bioavailability (Malik and Maqbool, 2020). Transgenics mitigate anti-nutrients, modulate transcription factors, and enhance metal-chelating molecules for Fe mobilization to endosperm, bolstering bioavailability. Emerging genome editing methods like TALENs, ZFNs, and CRISPR-Cas are explored for biofortification. With regulatory backing, this transgene-free technology holds global promise.

4. Target Crops

The baseline for Fe content of target tissues such as grains, tubers, and roots in various food crops has been studied. Genetic variability exists for grain Fe content *viz.,* only up to 6 ppm in polished rice, wheat has up to 32 ppm, maize has up to 30 ppm, pearl millet has up to 50 ppm and lentils have up to 50 ppm (Yadava et al., 2018). The Fe content is reported to be maximum in landraces and wild species (Stangoulis and Knez, 2022). Based on the baseline values, per capita consumption of the food crops, loss during cooking, and estimated bioavailability, the target Fe concentration for each food crop has been proposed (Table 1, Figure 1).

4.1. Rice

Rice is a vital staple for over half of the global population, but polished rice, the preferred form for consumption, is a poor source of Fe. Common rice varieties have Zn content in polished rice below 6 ppm. Brown rice, on the other hand, exhibits wide genetic variability in grain Fe content, reaching up to 60 ppm. Unfortunately, ~80% of these nutrients are lost during polishing when the embryo and aleurone layers are removed (Suman et al., 2021). In the early

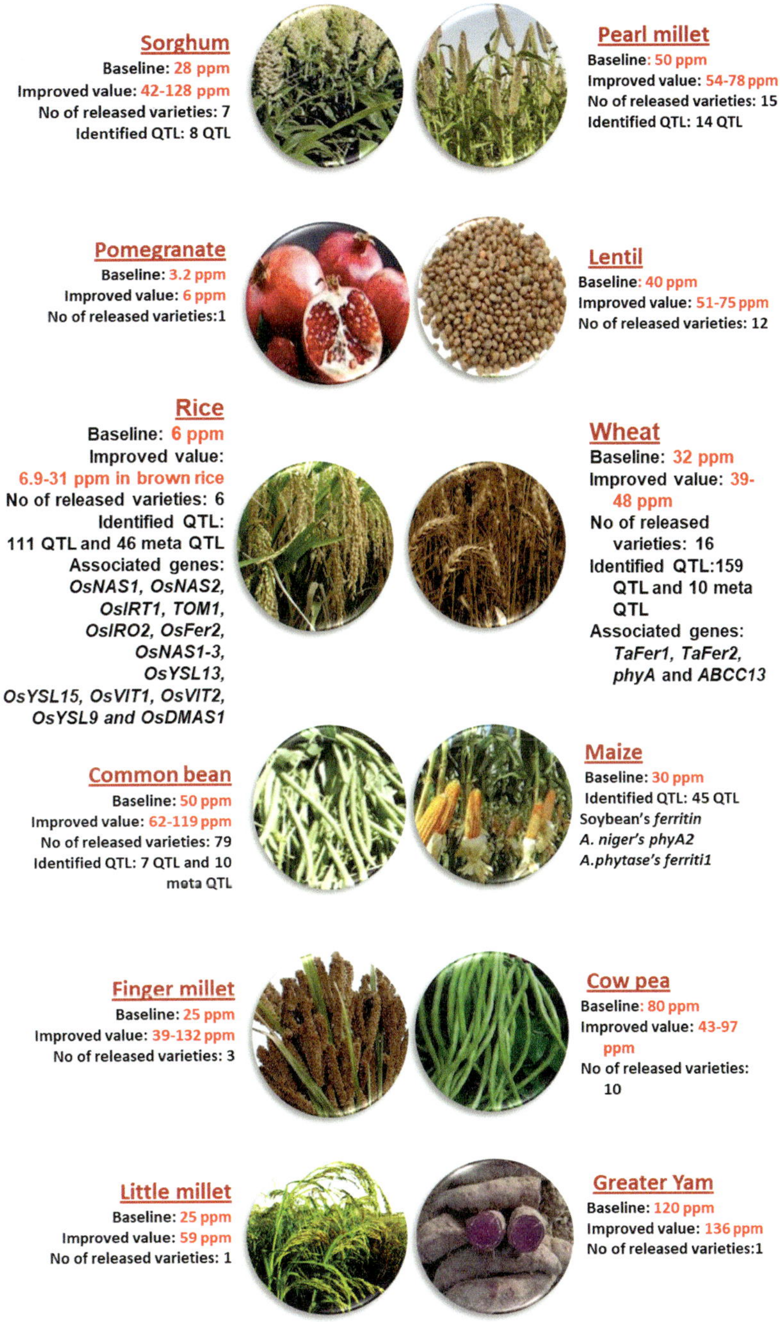

Figure 1: Tracking the progress of iron (Fe) biofortification.

Table 1: Iron (Fe) baseline, genetic variability, and target levels achieved in various crops

Crop	Tissue	Baseline (ppm)	Genetic Variability (ppm)	Levels Achieved (ppm)
Rice	Grain	6	60	>20 in brown rice
Wheat	Grain	28-32		>38
Maize	Kernel		12.5-30.8	
Sorghum	Grain	25-28	12-68	
Pearl Millet	Grain	45-50	34-90	>70
Finger Millet	Grain	25	39-132	>38
Small Millet	Grain	25		>55
Lentil	Grain	40	31.55-136.91	>70
Greater Yam	Tuber	70-120		>135
Common Bean	Grain	50		>62
Pomegranate	Fruit	2.7-3.2[#]		>5[#]

[#]mg/100 g
Adopted from Yadava et al., 2022; Bouis and Welch, 2010b; Islam et al., 2002.

2000s, IRRI developed an improved line, IR68144-3B-2-2-3, with roughly 21 ppm of Fe in brown rice. This variety was released as high-Fe rice, NSIC Rc172 (MS13), but faced challenges in adoption for other reasons.

From 23 published articles, 111 stable QTLs for grain Fe content were identified, spanning various populations, seasons, locations, and environments. Within the QTL regions, numerous candidate genes exist, with unclear roles in Fe metabolism yet to be fully understood. In another study encompassing 22 QTLs for grain Fe and 22 for grain Zn, 48 meta-QTLs were identified across all 12 chromosomes, including two meta-QTLs for Fe and Zn individually, and 44 meta-QTLs for both Fe and Zn (Raza et al., 2019).

To enhance grain Fe, rice single/multiple gene/s *viz., OsNAS1* and *OsNAS2, OsIRT1, TOM1, OsIRO2, OsFer2, OsNAS1* to *OsNAS3, OsYSL13*, and *OsYSL15* were overexpressed in low Fe cultivars and/or *OsVIT1* and *OsVIT2*, *OsYSL9* and *OsDMAS1* genes were knocked down (Ludwig and Slamet-Loedin, 2019). For Fe biofortification, the combination of promising genes from multiple crops *viz.,* soybean *ferritin*, barley *NAS, HAA*, and *NAAT*, Arabidosis *NAS, IRT, NRAMP* and *FRD*, maize *PSY*, and pea *ferritin* were incorporated into rice genome under the control of specific promoters to get the high Fe levels in rice grains (Table 2).

4.2. Wheat

Wheat, a key staple crop for global food security, typically contains 28–32 ppm of Fe in adapted varieties. For biofortification, the target Fe level was set at 45 ppm. For Fe, wide genetic variability exists for wheat as wheat cultivars and

Table 2: List of reported genes for iron concentration exploited through transgenic approaches

Gene	Fe (ppm)		Cultivar
	Wild	Transgenic	
Rice			
Over expression of rice gene in contrasting rice			
OsIRT1	10	12	Dongjin[$]
OsYSL15	12	14	Dongjin[$]
OsNAS3	4	12	EYI105[$$]
OsNAS1	12	19	Xiushui110[$]
OsYSL2	1.8	7.5	Tsukinohikari[$]
OsNAS1, OsNAS2, OSNAS3	4.5	19	Nipponbare[$]
TOM1	15	18	Tsukinohikar[$]
OsIRO2	6	15.5	Tsukinohikari[$]
OsFer2			Pusa-SugandhII[$$]
OsNAS2	4	10	Dongjin[$]
OsNAS1	20	40	EYI105[$$]
OsYSL13	11	15	Zhonghua11[$]
Over expression of different species gene in rice			
SoyferH1	14.3	38	Kitaake[$]
SoyferH1	18	18	M12[$$]
SoyferH1	10	37	IR68144[$$]
SoyferH1	17	25	Kitaake[$]
SoyferH1	3.8	9.2	BR29[$$]
HvNAS1	4	8.5	Tsukinohikari[$]
SoyferH1	6.7	16	Swarna[$$]
SoyferH1	3.3	7.6	IR64[$$]
HvYS1	4	9	EYI105[$$]
AtIRT1	2.2	4.8	Taipei309[$]
Rice gene silencing/knockdown			
OsVIT1	20	26	Zhonghua 11[$]
OsVIT2	20	28	Dongjin[$]
OsVIT2	5	8	Dongjin[$]
OsYSL9	1	2.5	Tsukinohikari[$]

Table 2: (*Contd.*)

Gene	Fe (ppm)		Cultivar
	Wild	Transgenic	
OsDMAS1	5	5	Dongjin$
Over expression of multicrop genes in rice			
PyFerritin+rgMT+phyA	10	22	Taipei309$
HvNAS1, HvNAS1+ HvNAAT, IDS3	5.8	7.3	Tsukinohikari$
AtNAS1+ Pvferritin+ Afphytase	1	7	Taipei309$
OsYSL2+ SoyFerH2+ HvNAS1	1	4	Tsukinohikari$
HvNAS1+ OsYSL2+ SoyFerH2	3.2	6.3	Paw San Yin$$$
GmFERRITIN+OsNAS2	2.5	15	IR64$$
AtIRT1+PvFERRITIN+AtNAS1	2.7	10.4	Nipponbare$
AtNAS1+PvFER+ZmPSY+PaCRT1	1.8	6	Nipponbare$
OsNAS1+HvHAATb	4	18	EYI105$$
OsNAS1+HvNAATb	20	55	EYI105$$
AtNAS1+AtFRD3+PvFer	2.05	11.08	Nipponbare$
AtNAS1+PvFer+AtNRAMP3	2.7	13.65	IR64$$
Wheat			
Over expression of wheat gene in contrasting wheat			
Ta FERRITIN	72	130	Bobwhite##
Ta VIT			
OsNAS2	40	80	Bobwhite##
Over expression of multi crop genes in wheat			
OsNAS2 +/or PvFERRITIN			Bobwhite##

$ *Japonica* cv.; $$ *Indica* cv.; $$$ *Tropical japonica* cv.; ## *T. aestivum* cv.

landraces, primary gene pool as progenitors of hexaploid wheat, and secondary gene pool comprising the wild and domesticated tetraploid wheat. Several biofortified wheat varieties with enhanced grain iron content have been released in India. Biofortified wheat varieties with high Zn developed by CIMMYT also have shown high Fe (Thapa et al., 2022).

Many QTLs have been identified using bi-parental populations derived from bread wheat, synthetic wheat, or *T. spelta* (Wani et al., 2022). Among 20 QTLs identified for grain Fe in wheat, a major was QTL *QGFe.cimmyt-4A_P2* on chromosome 4A. Pleiotropic or tightly linked QTLs were also found on chromosome 3B. Out of 159 QTLs reported for grain Fe and Zn in 12 studies, 10 identified meta-QTLs were distributed on five different chromosomes (2D, 5A, 5B, 6A, and 7A), including 3 MQTLs for Fe only and 7 MQTLs for both

Zn and Fe. These genomic regions are promising in MAS for the improvement of grain Fe. The MTAs have been identified in six studies in wheat and *Ae. tauschii* using GWAS. Genetic engineering of ferritin gene from soybean and TaFer1-A has enhanced grain Fe (Borg et al., 2012). The bioavailability of grain Fe has been increased by overexpression of *PHYA* and silencing of *ABCC13* (Bhati et al., 2016).

4.3. Maize

The grain Fe content in maize has limited natural genetic variation of 12.5–30.8 ppm (Hossain et al., 2021), thus breeding efforts are minimal for biofortification.

In maize, a total six published articles identified 45 QTL for Fe using bi-parental populations with PV% from 6.8% to 43.7%. SNPs were identified for Fe through GWAS, which explained high PV% with genomic prediction ability of 0.324–0.734 across different environments (Velu et al., 2016). To enhance grain Fe in maize, soybean *FERRITIN* and *A. niger PHYA2* were over-expressed and *ABCC* knocked out.

4.4. Millets

Pearl millet is a highly nutritious cereal with higher Fe and is one of the ideal staple food crops for biofortification. Since pearl millet is a highly cross-pollinated crop, open-pollinated varieties, and hybrids can be developed with high-grain nutrients. Genetic variability for grain Fe and zinc was found for pearl millet (Govindaraj et al., 2016; Satyavathi et al., 2021). ICRISAT, NARS partners, and HarvestPlus have developed ten varieties enriched with Fe and Zn. Recently, the benchmark levels for optimum Fe (>42 ppm) and Zn (>32 ppm) contents were fixed for varietal release in pearl millet in India (Satyavathi et al., 2021). Similarly, high grain Fe finger millet (ranging from 39–132 ppm over popular varieties with 25 ppm) and little millet varieties (with 59 ppm over popular varieties with 25 ppm) have been released in India (Yadava et al., 2022). Fourteen QTLs for Fe have been identified in pearl millet using SSRs in the RIL population evaluated in three environments (Singhal et al., 2021).

4.5. Sorghum

Sorghum, naturally enriched with more nutrients than major staples like rice and wheat, is being promoted for its enhanced consumption as a nutri-cereal. In sorghum germplasm, a wide genetic variability ranging from 12 to 68 ppm was observed for grain Fe. Several varieties and hybrids with high iron content have been released in India and Nigeria. Eight QTLs for grain Fe, with PV% ranging from 5.9% to 6.8%, were identified in sorghum (Kotla et al., 2019).

4.6. Legumes

Globally, legumes or pulses are known as protein sources and are rich in various micronutrients (Thavarajah et al., 2022). Common beans, which are consumed in several countries with low purchasing power, have been a focal point for developing Fe biofortified varieties, with the potential to increase grain Fe content by approximately 60–80%. This effort is led by HarvestPlus and other international agencies (Glahn and Noh, 2021). More than 60 Fe biofortified varieties have been developed in Latin American and African countries (Table 3). In 7 QTL studies, researchers identified 10 meta-QTLs for Fe content in beans (Izquierdo et al., 2018). The efficacy of biofortified beans has been demonstrated in iron-depleted women in Rwanda following their consumption (Haas et al., 2016).

Lentils, adapted to several dryland regions across the world, have been bred for their high Fe content, with collaborative efforts from ICARDA, HarvestPlus, and NARS, resulting in the release of varieties in Bangladesh, Nepal, India, Ethiopia, and Syria. QTLs for Fe content in lentils were identified through biparental and association mapping (Kumar et al., 2019). Cowpea, known for its high protein content, has also been developed with high Fe content by NARS in India and Brazil (Table 3).

Table 3: List of Fe biofortified varieties/hybrids released across the world

Crop/Variety	Region	Year	Fe (ppm)
Rice			
NSIC Rc172	Philippines	2013	13-21 (Brown rice)
BU Aromatic Hybrid Dhan-1	Bangladesh	2016	10
BU Aromatic Dhan-2			10
Binadhan20		2017	20-31
BRRI Dhan84			10.1
CENTA A-Nutremas	El Salvador	2019	6.9
Wheat			
HI1605	Asia	2017	43
HI8759			42.1
WB02			40
HPBW01			40
HI8777		2018	48.7
MACS4028			46.1
DBW173			40.7
HI1633		2020	41.6
DBW187			43.1
DDW47			40.1
HI8805			40.4

HD3249			42.5
MACS4058			39.5
HD3298			43.1
HI1633			41.6
MP(JW)1358		2021	40.6
Pearl Millet			
Dhanashakti	Asia	2014	72
Shakti1201		2015	74
HHB 299		2017	73
AHB1200-Fe			77
ABV04		2018	70
HHB311			62
RHB234			62
AHB1269Fe			73
RHB233			64
Phule Mahashakti			78
MotiShakti		2019	74
HHB 67 Improved-2		2021	54.8
CO10		2023	59
Chakti	Africa	2018	65
LCIC-MV5		2023	74.2
Finger Millet			
VR929	Asia	2020	131.8
CFMV1			58
CFMV2			39
Little Millet			
CLMV1	Asia	2020	59
Greater Yam			
Da340	Asia	2020	136.2
Pomegranate			
Solapur Lal	Asia	2017	5.6-6.1[#]
Beans			
BRSPontal	LAC	2004	77
BRS9435Cometa		2007	80
BRSAgreste		2009	78.8

(*Contd.*)

Table 3: (*Contd.*)

Crop/Variety	Region	Year	Fe (ppm)
Fortaleza			87
ICTA-Peten-ACM		2010	70
ICTA-Superchiva-ACM		2011	74
CENTA-Ferromas			77
IDIAPNUA24		2012	78
IDIAPNUA27			71
INTANutritivo			70
INTAFerroso		2014	79
Corpoica-Rojo39		2015	86
Corpoica-Rojo43			95
Honduras Nutritivo		2016	71
Bio-101			83
BIO-107			82
INTABIOF100		2017	80
INTABIODOR			76
RojoFortificado			78
ICTAChorti-ACM			95
IDIAPP-09-11		2018	87
IDIAPP-13-38			78
INTARojoBio-Apante		2019	91
Bio-102			86
MAC44	Africa	2010	78
RWV1129			77
CAB2			81
RWR2245			76
RWR2154			71
RWR2245		2011	66
RWR2154			71
Mbereshi		2012	92
RWV3316			87
RWV3006			78
RWV3317			74
MAC42			91

RWV2887		85
CODMLB001		66
HM21-7		62
CODMLV059		84
VCB81013		69
PigeonVert		80
M211		77
Roba1	2013	64
PVA1438		79
Nainde-Kyondo		76
CODMLB032		76
Cuarentino0817		100
Namulenga		76
Ecapan021		72
MAC44	2015	69
RWV1129		92
RWR2245		76
RWR2154		71
NARO-Bean4C	2016	78
NARO-Bean2		76
NARO-Bean1		71
NARO-Bean3		70
NARO-Bean5C		67
NUA100		78
MNC488-2		79
NUV119-4		72
Nyiramuhondo		72
Selian14	2018	85
Selian15		81
NUA99		78.6
Jasmine	2019	86
TARIBEAN-2	2020	87.2
TARIBEAN-4		78.3
TARIBEAN-5		90.5
TARI-19		71

(*Contd.*)

Table 3: (*Contd.*)

Crop/Variety	Region	Year	Fe (ppm)
RW-CB184		2021	113
RW-CB181			119
Rw-BB183			89
RW-CB186			91.2
RW-BB181			94
Rw-BB186			96
RW-BB184			98
RW-CB189			95-113
Cowpea			
BRS Xiquexique	LAC	2008	77
BRS Araca		2009	62
BRS Tumucumaque			61
Pant Lobia-1	Asia	2008	82
Pant Lobia-2		2010	95
Pant Lobia-3		2013	97
Pant Lobia-4		2014	52
Pant Lobia-5		2015	66
Pant Lobia-7		2019	43
CBC6(C/87/6/13)	Africa	2021	90
Lentil			
Idlib-2	Asia	2010	TBD
Idlib-3			TBD
Barimasur-4			51
Barimasur-5			59
Barimasur-6			63
Barimasur-7		2012	61
ILL7723		2013	62
L4704			74
Barimasur-8		2015	60
L4717		2017	65
IPL220		2018	75
Rasuwablack		2020	TBD

[#]mg/100 mg

Adopted from Yadava et al., 2022; Senguttuvel et al., 2023, https://bcr.harvestplus.org/

4.7. Vegetables

The International Potato Center (CIP) and HarvestPlus have harnessed genetic variation in potatoes to develop high-Fe breeding lines. The National Institute for Agrarian Innovation (INIA) has developed the high-Fe INIA 321 Kawsay variety in Peru. In a randomized feeding trial involving women from the Peruvian Highlands, it was found that the total iron absorbed from high-Fe potatoes was higher than that from non-biofortified potatoes (Burgos et al., 2023).

Cassava, known as a major staple food worldwide and with tolerance to multiple stresses, has had its Fe content increased through a genetic engineering approach (Narayanan et al., 2019). A Greater Yam variety with 136 ppm of Fe has been developed, surpassing popular varieties with 70–120 ppm of Fe (Yadava et al., 2022).

4.8. Fruits

Pomegranate high-Fe (5.6–6.1 mg/100g) variety has been developed in India while, popular varieties have lower Fe 2.7–3.2 mg/100g (Yadava et al., 2022).

5. Contribution of Environmental Factors

Plants acquire iron (Fe) from the rhizosphere. While Fe is highly abundant in the Earth's crust, its availability to plants is reported to be influenced by factors such as clay content, cation exchange capacity, soil organic matter content, redox potential, and soil pH (Nguyen and Marschner, 2016). The inoculation of rhizosphere microbes and arbuscular mycorrhizal fungi has been shown to enhance Fe uptake in durum wheat (Ercoli et al., 2017; Morrissey and Guerinot, 2009).

6. Bioavailability

Bioavailability refers to the proportion of the ingested nutrient that is absorbed and available for normal physiological functions. It is usually lower than 15% for plant-based foods. Measuring bioavailability is not straightforward because it depends on factors such as the food matrix, dietary factors, chemical form, health, age, excretory losses, nutrient-nutrient interactions, and life stages like pregnancy (Diego Quintaes et al., 2017). Developing biofortified crops with high bioavailability of nutrients in the human gut is a major concern (Neeraja et al., 2017). Antinutrients such as phytic acid inhibit mineral absorption and hinder Fe bioavailability. Engineering the synthesis of plant metal chelators to increase Fe bioavailability is a promising approach. For instance, in wheat, overexpression of the *NAS* gene and enhanced Fe bioavailability in rice (JT Beasley, 2019). The sourdough process in whole wheat bread increased Fe bioavailability in a Caco-2 cell model, but only when combined with other Fe sources (Rodriguez-Ramiro et al., 2017). Cognitive performance, especially search efficiency and memory task retrieval speed, improved in women aged 18–25 who consumed Fe biofortified beans (86.1 ppm) than those consuming normal beans (Murray-Kolb et al., 2017). Four independent studies have clearly demonstrated the positive effects of consuming biofortified Fe pearl millet in target populations in India (Mehta et al., 2017; Scott et al., 2018).

7. Economics of Fe Biofortification

The Global Burden of Disease Report 2017 and Human Development Report 2017, that looked at health problems caused by not getting enough iron, found that over time, the overall burden of iron deficiency has been going down worldwide. But at the same time, the differences in how it affects different people have been getting bigger. It turns out that iron deficiency is still more common in women compared to men. However, there have been some positive changes in East Asia and some parts of Asia since 1990, meaning things are improving there (Wang et al., 2022).

8. Success Stories

8.1. Fe Beans

The release of high-Fe bean varieties in South America and Africa is one of the significant success stories of Fe biofortification. An increase of haemoglobin and total body Fe was observed in Fe exhausted university women in Rwanda after consumption of Fe beans for 4.5 months in Rwanda (Haas et al., 2016). As per Ex-ante analyses, with high-Fe beans, around 20–66$ per day can be averted in the pessimistic scenario and 439$ in the optimistic scenario (Meenakshi et al., 2010).

8.2. Fe Pearl Millet

The release and adoption of high-Fe pearl millet represent a significant success story. Among its clinical trials, a study involving 20 women from Benin showed that consuming iron-biofortified millet nearly doubled the amount of iron absorbed (Cercamondi et al., 2013). Subsequent studies have also indicated a positive impact of consuming biofortified Fe pearl millet among children and adolescents in India (Mehta et al., 2017; Scott et al., 2018). Therefore, high-Fe millet is likely to be highly effective in addressing iron deficiency in millet-consuming populations.

9. Challenges

Fe biofortification in crops faces multidimensional challenges in the development of varieties, their efficacy, and their adoption.

9.1. Developers

Based on the target edible tissue, availability of genetic variability for Fe content is a major issue for the breeders for identification of donor material, e.g., Fe in polished rice which is mainly present in the seed coat and is removed during milling (Swamy et al., 2021). Fe content is influenced by environment, especially of soil (micronutrients) and its management is a major concern for consistency of Fe content in the target tissues of the crops. For genetic engineering approach, the major problem of developing fortified crops is the cost of research and of regulatory compliance, due to the extreme precautionary regulation of transgenic crops. Though several genes and genomic regions have been identified for Fe content across the crops, their utilization in the breeding programs is still slow because of the complex nature of genetics and the involvement of minor QTL.

9.2. Farmers

As there is no special price for the Fe biofortified crops, so there is no incentive for farmers to take up their cultivation. As most of the time, the trait of high Fe is invisible in target tissue, farmers are not aware of the high nutrients in the crop even for their own consumption

9.3. Target Consumers

Lack of availability of Fe biofortified foods and awareness about the Fe biofortified foods is the major issue for the target consumers with poor purchase power.

9.4 Bioavailability

In case of Fe, mere intake is not adequate, the role of anti-nutritional factors and promoters of Fe absorption is also critical for its efficacy.

9.5. Climate Change

Recent reports of climate change pose serious challenges for existing nutritive levels. According to a recent report, elevated CO_2 could lead to deficiencies of Zn for 175 million people, protein for 122 million, and Fe for 1.4 billion women of child-bearing age and young children (Smith and Myers, 2019).

10. Conclusion

With the prevalence of Fe deficiency across the world, more so in vulnerable groups like children and women, all possible kinds of interventions are needed to address the issue. Fe biofortified crops offer potential solution at least to alleviate the acute Fe deficiency and need to be integrated in the existing food systems. Systematic studies on nutritional impact of Fe biofortified varieties in the form of clinical trials as a proof of concept have been conducted in a few crops, but more studies are the need of the hour for driving the importance of Fe biofortified crops to the policymakers. Not only the agricultural community, but public health nutrition community should also be convinced that biofortification can be a cost-effective nutrition intervention (Bouis and Saltzman, 2017). Convergence of crop breeding, nutrition, policy, and popularization could only realize the last mile of biofortification reaching the target consumers.

References

Beasley, J.T., J.P. Bonneau, J.T. Sanchez-Palacios, L.T. Moreno-Moyano, D.L. Callahan, E. Tako, R.P. Glahn, E. Lombi and A.A. Johnson. 2019. Metabolic engineering of bread wheat improves grain iron concentration and bioavailability. *Plant Biotechnol. J.*, 17(8): 1514–1526.

Bhati, K.K., A. Alok, A. Kumar, J. Kaur, S. Tiwari, and A.K Pandey. 2016. Silencing of ABCC13 transporter in wheat reveals its involvement in grain development, phytic acid accumulation, and lateral root formation. *J. Exp. Bot.*, 67(14): 4379–4389.

Borg, S., H. Brinch-Pedersen, B. Tauris, L.H. Madsen, B. Darbani, S. Noeparvar, and P.B. Holm. 2012. Wheat ferritins: Improving the iron content of the wheat grain. *J. Cereal Sci.*, 56(2): 204–213.

Bouis, H.E., and A. Saltzman. 2017. Improving nutrition through biofortification: A review of evidence from HarvestPlus, 2003 through 2016. *Glob. Food Sec.*, 12: 49–58.

Bouis, H.E., and R.M Welch. 2010b. Biofortification: A sustainable agricultural strategy for reducing micronutrient malnutrition in the global south. *Crop Sci.*, 50: 20.

Burgos, G., R. Liria, C. Zeder, P.A. Kroon, G. Hareau, M. Penny, and S. Fairweather-Tait. 2023. Total iron absorbed from iron-biofortified potatoes is higher than from non-biofortified potatoes: A randomized trial using stable iron isotopes in women from the Peruvian highlands. *J. Nutr.*, 153(6): 1710–1717.

Cercamondi, C.I., I.M. Egli, E. Mitchikpe, F. Tossou, C. Zeder, J.D. Hounhouigan, and R.F. Hurrell. 2013. Total iron absorption by young women from iron-biofortified pearl millet composite meals is double that from regular millet meals but less than that from post-harvest iron-fortified millet meals. *J. Nutr.*, 143(9): 1376–1382.

Diego Quintaes, K., R. Barberá, and A. Cilla. 2017. Iron bioavailability in iron-fortified cereal foods: The contribution of *in vitro* studies. *Crit. Rev. Food Sci. Nutr.*, 57(10): 2028–2041.

Ercoli, L., A. Schüßler, I. Arduini, and E. Pellegrino. 2017. Strong increase of durum wheat iron and zinc content by field-inoculation with arbuscular mycorrhizal fungi at different soil nitrogen availabilities. *Plant and Soil*, 419(1–2): 153–167.

Garg, M., N. Sharma, S. Sharma, P. Kapoor, A. Kumar, V. Chunduri, and P. Arora. 2018. Biofortified crops generated by breeding, agronomy, and transgenic approaches are improving lives of millions of people around the world. *Front. Nutr.*, 5.

Glahn, R.P., and H. Noh. 2021. Redefining bean iron biofortification: A review of the evidence for moving to a high Fe bioavailability approach. *Front. Sustain.*, 1–9.

Govindaraj, M., K.N. Rai, and P. Shanmugasundaram. 2016. Intra-population genetic variance for grain iron and zinc contents and agronomic traits in pearl millet. *Crop J.*, *4*(1): 48–54.

Haas, J.D., S.V. Luna, M.G. Lung'aho, M.J. Wenger, L.E. Murray-Kolb, S. Beebe, and I.M. Egli. 2016. Consuming iron biofortified beans increases iron status in Rwandan women after 128 days in a randomized controlled feeding trial. *J. Nutr.*, 146(8): 1586–1592.

Hossain, F., S. Rakshit, B. Kumar, J.J. Amalraj, V. Muthusamy, B. Prakash, and M.C. Kamboj. 2021. Molecular breeding for increasing nutrition quality in maize: Recent progress. *Molecular Breeding in Wheat, Maize and Sorghum: Strategies for Improving Abiotic Stress Tolerance and Yield*, 360–379.

Islam, F.M.A., K.E. Basford, C. Jara, R.J. Redden, and S. Beebe. 2002. Seed compositional and disease resistance differences among gene pools in cultivated common bean. *Genet. Resour. Crop Evol.*, *49*(3): 285–293.

Izquierdo, P., C. Astudillo, M.W. Blair, A.M. Iqbal, B. Raatz, and K.A. Cichy. 2018. Meta-QTL analysis of seed iron and zinc concentration and content in common bean (*Phaseolus vulgaris* L.). *TAG. Theor. Appl. Genet.*, 131(8): 1645–1658.

Kotla, A., R. Phuke, K. Hariprasanna, S.P. Mehtre, A. Rathore, S. Gorthy, and A.A. Kumar. 2019. Identification of QTLs and candidate genes for high grain Fe and Zn concentration in sorghum (*Sorghum bicolor* (L.) *Moench*). *J. Cereal Sci.*, 90: 102850.

Kumar, H., A. Singh, H.K. Dikshit, G.P. Mishra, M. Aski, M.C. Meena, and S. Kumar. 2019. Genetic dissection of grain iron and zinc concentrations in lentil (*Lens culinaris* Medik.). *J. Genet.*, 98.

Ludwig, Y., and I.H. Slamet-Loedin. 2019. Genetic biofortification to enrich rice and wheat grain iron: From genes to product. *Front. Plant Sci.*, 1–10.

Malik, K.A., and A. Maqbool. 2020. Transgenic crops for biofortification. *Front. Sustain. Food Syst.*, 1–15.

Meenakshi, J.V., N.L. Johnson, V.M. Manyong, H. De Groote, J. Javelosa, D.R. Yanggen, and E. Meng. 2010. How cost-effective is biofortification in combating micronutrient malnutrition? An ex ante sssessment. *World Dev.*, 38(1): 64–75.

Mehta, S., J.L. Finkelstein, S. Venkatramanan, S.L. Huey, S.A. Udipi, P. Ghugre, and J.D. Haas. 2017. Effect of iron and zinc-biofortified pearl millet consumption on growth and immune competence in children aged 12–18 months in India: Study protocol for a randomized controlled trial. *BMJ*, 7(11).

Morrissey, J., and M. Lou Guerinot. 2009. Iron uptake and transport in plants: The good, the bad, and the ionome. *Chem. Rev.*, 109(10): 4553–4567.

Murray-Kolb, L.E., M.J. Wenger, S.P. Scott, S.E. Rhoten, M.G. Lung'aho, and J.D. Haas. 2017. Consumption of iron-biofortified beans positively affects cognitive performance in 18- to 27-year-old Rwandan female college students in an 18-week randomized controlled efficacy trial. *J. Nutr.*, 147(11), 2109–2117.

Narayanan, N., G. Beyene, R.D. Chauhan, E. Gaitán-Solís, J. Gehan, P. Butts, and N.J. Taylor. 2019. Biofortification of field-grown cassava by engineering expression of an iron transporter and ferritin. *Nat. Biotechnol.*, 37(2): 144–151.

Neeraja, C.N., V. Ravindra Babu, Sewa Ram, Firoz Hossain, K. Hariprasanna, B.S. Rajpurohit, Prabhakar, T. Longvah, K.S. Prasad, J.S. Sandhu, and Swapan K. Datta. 2017. Biofortification in cereals: Progress and prospects. *Curr. Sci.*, 113(6), 1050–1057.

Nguyen, T.T., and P. Marschner. 2016. Soil respiration, microbial biomass and nutrient availability in soil after repeated addition of low and high C/N plant residues. *Biol. Fertil. Soils*, 52(2): 165–176.

Raza, Q., A. Riaz, M. Sabar, R.M. Atif, and K. Bashir. 2019. Meta-analysis of grain iron and zinc associated QTLs identified hotspot chromosomal regions and positional candidate genes for breeding biofortified rice. *Plant Sci.*, 288.

Rodriguez-Ramiro, I., C.A. Brearley, S.F.A. Bruggraber, A. Perfecto, P. Shewry, and S. Fairweather-Tait. 2017. Assessment of iron bioavailability from different bread-making processes using an *in vitro* intestinal cell model. *Food Chem.*, 228: 91–98.

Satyavathi, C.T., S. Ambawat, V. Khandelwal, and R.K. Srivastava. 2021. Pearl millet: A climate-resilient nutricereal for mitigating hidden hunger and provide nutritional security. *Front. Plant Sci.*, 12: 659938.

Scott, S.P., L.E. Murray-Kolb, M.J. Wenger, S.A. Udipi, P.S. Ghugre, E. Boy, and J.D. Haas. 2018. Cognitive performance in Indian school-going adolescents is positively affected by consumption of iron-biofortified pearl millet: A 6-month randomized controlled efficacy trial. *J. Nutr.*, 148(9): 1462–1471.

Senguttuvel, P., G. Padmavathi, C. Jasmine, D. Sanjeeva Rao, C.N. Neeraja, V. Jaldhani, and M. Govindaraj. 2023. Rice biofortification: Breeding and genomic approaches for genetic enhancement of grain zinc and iron contents. *Front. Plant Sci.*, 14.

Singhal, T., C.T. Satyavathi, S.P. Singh, A. Kumar, S.M. Sankar, C. Bhardwaj, and N. Singh. 2021. Multi-environment quantitative trait loci mapping for grain iron and zinc content using bi-parental recombinant inbred line mapping population in pearl millet. *Front. Plant Sci.*, 12.

Smith, M.R., and S.S. Myers. 2019. Global health implications of nutrient changes in rice under high atmospheric carbon dioxide. *GeoHealth*, 7: 190–200.

Stangoulis, J.C.R., and M. Knez. 2022. Biofortification of major crop plants with iron and zinc: Achievements and future directions. *Plant Soil*, 474(1–2): 57–76.

Stevens, G.A., M.M. Finucane, L.M. De-Regil, C.J. Paciorek, S.R. Flaxman, F. Branca, and M. Ezzati. 2013. Global, regional, and national trends in haemoglobin concentration and prevalence of total and severe anaemia in children and pregnant and non-pregnant women for 1995–2011: A systematic analysis of population-representative data. *Lancet Glob. Health*, 1(1).

Suman, K., C.N. Neeraja, P. Madhubabu, S. Rathod, S. Bej, K.P. Jadhav, and S.R. Voleti. 2021. Identification of promising RILs for high grain zinc through genotype × environment analysis and stable grain zinc QTL using SSRs and SNPs in rice (*Oryza sativa* L.). *Front. Plant Sci.*, 12: 587482.

Swamy, B.P.M., B. Marathi, A.I.F. Ribeiro-Barros, M.I.C. Calayugan, and F.K. Ricachenevsky. 2021. Iron biofortification in rice: An update on quantitative trait loci and candidate genes. *Front. Plant Sci.*, 12: 1–11.

Thapa, D.B., M. Subedi, R.P. Yadav, B.P. Joshi, B.N. Adhikari, K.P. Shrestha, and V. Govindan. 2022. Variation in grain zinc and iron concentrations, grain yield, and associated traits of biofortified bread wheat genotypes in Nepal. *Front. Plant Sci.*, 13: 1–13.

Thavarajah, D., S. Powers, G. Vandermark, C.R. Johnson, E. Shipe, and P. Thavarajah. 2022. Pulse crop biofortification toward human health, targeting prebiotic carbohydrates, protein, and minerals. *Biofortification of Staple Crops*, 205–224. Singapore: Springer Singapore.

Velu, G., J. Crossa, R.P. Singh, Y. Hao, S. Dreisigacker, P. Perez-Rodriguez, and G.S. Mavi. 2016. Genomic prediction for grain zinc and iron concentrations in spring wheat. *Theor. Appl. Genet.*, 129(8): 1595–1605.

Wang, M., H. Gao, J. Wang, C. Cao, X. Ying, Y. Wei, and M. Yang. 2022. Global burden and inequality of iron deficiency: Findings from the Global Burden of Disease datasets 1990–2017. *Nutr. J.*, 21(1): 16.

Wani, S.H., K. Gaikwad, A. Razzaq, K. Samantara, M. Kumar, and V. Govindan. 2022. Improving zinc and iron biofortification in wheat through genomics approaches. *Mol. Biol. Rep.*, 49(8): 8007–8023.

Yadava, D.K., P.R. Choudhury, F. Hossain, and D. Kumar. 2022. Biofortified Varieties: Sustainable Way to Alleviate Malnutrition. ICAR, New Delhi.

Yadava, D.K., F. Hossain, P.R. Choudhury, and D.S.A. Kumar. 2018. Biofortification of crops: A sustainable tool for nutritional security in India. *Indian J. Med. Res.*, 148(5): 626–631.

CHAPTER
5

Zinc (Zn): A Prime and High-Yield Resource

Dinesh Pandey[a*[0009-0009-8284-260X]], **Apoorv Tiwari**[b[0000-0003-2294-0688]], **H. Punetha**[c] **and Anil Kumar**[d[0000-0002-3174-5203]]

[a] Department of Molecular Biology & Genetic Engineering, College of Basic Sciences & Humanities, G.B. Pant University of Ag. & Tech. Pantnagar
[b] National Agricultural Biotechnology Institute, Mohali
[c] Department of Biochemistry, College of Basic Sciences & Humanities, G.B. Pant University of Ag. & Tech. Pantnagar
[d] Rani Laxmi Bai Central Agriculture University, Jhansi

1. Introduction

Zn deficiency is a critical global health concern affecting approximately two billion individuals worldwide, with an estimated 17.3% of the global population suffering from it due to the consumption of nutritionally poor cereal crops. Zn is an essential trace mineral crucial for various physiological processes and enzymatic reactions within the body (Roohani et al., 2013). It plays a vital role in immune system support, wound healing, antioxidant protection, hormone regulation, reproductive health, cognitive function, taste and smell perception, metabolism, skin health, and bone health. Zn deficiency can result in a wide range of symptoms, including decreased immunity, delayed wound healing, altered taste and smell leading to appetite loss, skin and hair problems, growth issues, cognitive impairment, reproductive health problems, digestive disturbances, and vision changes (Kiani et al., 2022).

The effective treatment of Zn micronutrient deficiency depends on its root causes, often related to poor diets, particularly in economically poor Asian and African populations who depend on locally available staple crops. Historically, breeding aimed at enhancing crop productivity leading to a decrease in

*Corresponding author: dineshpandeymbge@gmail.com

grain Zn concentration, known as the "dilution effect". To address this issue, strategies such as biofortification, promoting Zn-rich foods, and providing supplements have been employed. In addition, the use of Zn-fertilized soils and environment friendly microbial inoculants is being explored to combat fertilizer pollution and promote the biofortification of cereals like rice, wheat, and maize (Maqbool et al., 2019). However, associated challenges related to fertilizer pollution and the social acceptability of genetically modified crops needs to be addressed.

2. Biofortification Strategies to Develop Zn Biofortified Crops

Understanding the physiological mechanism of Zn accumulation or distribution in food crop grains is necessary for Zn biofortification. The micronutrient content in the grain of food crops depends on soil and plant factors. If specialized fertilizers are not used, extensive farming of cereals depletes soil micronutrients like Zn. In such soils, crops don't grow properly because Zn is limited and their grains are Zn-deficient. Consumption of such cereals by people living in areas where the soil has already low Zn content tends to suffer more from its deficiency. Biofortification aims to improve and enhance the relationship between soil, crops, and people so as to ensure that people obtain enough Zn from food produce and maintain good health (Garg et al., 2018).

However, plant factors include root architecture, soil micronutrient uptake, translocation to vegetative tissues, and retro-translocation from vegetative to reproductive tissues. Further efficiency of the plant to distribute it into grains decides the final content of the micronutrient. Understanding the physiological and molecular mechanisms of micronutrient uptake, translocation, and accumulation in edible plant parts will help change genotypes to accumulate more Zn (Etienne et al., 2018).

2.1. Agronomic Strategies of Zn Biofortification

Agronomic biofortification of Zn, which involves enhancing the Zn content in crops through agricultural practices, offers a promising solution. Applying fertilizers strategically during crop growth boosts mineral content in grains and fruits. Soil application, foliar application, and seed priming are three main methods as described below (Table 1).

2.1.1. Soil Application

The effectiveness of agronomic biofortification through Zn application is closely linked to soil conditions:

Soil pH: Zn availability is higher in acidic soils, while alkaline soils have low Zn availability. Liming acidic soils can reduce Zn availability. In calcareous soils,

Table 1: Agronomic biofortification of Zn

Crop	Treatment	Reference
Maize	Soil application of $ZnSO_4.7H_2O$ @ 2.5, 5, and 7.5 kg/ha	Faujdar et al., 2014
Aromatic rice	Soil and foliar application (at flowering) of Zn as $ZnSO_4.7H_2O$	Shivay Y.S., 2015
Basmati rice	Different Zn sources were applied at the rate 5 kg Zn/ha	Singh and Shivay, 2019
Rice	Soil and foliar sprays (two) at maximum tillering and panicle initiation stage with different sources of Zn at different rates	Ghasal et al., 2018
	Soil and foliar application of $ZnSO_4.7H_2O$, and Zn-coated urea	Prasad and Shivay, 2018
Wheat	Soil and foliar sprays (two) at maximum tillering and booting stage with different sources of Zn at different rates	Ghasal et al., 2017
	Soil and foliar application of Zn as $ZnSO_4.H_2O$	Firdous et al., 2018
	Dextran sulfate (DEX (SO_4))-coated ZnO nanoparticles	Elhaj and Unrine et al., 2018
	Different rates and source of Zn at 1/3-sowing, 1/3–3 weeks after sowing (WAS), and 1/3–6 WAS	Prasad and Shivay, 2018
Bread wheat, durum wheat and triticale	Foliar application of 0.5% $ZnSO_4.7H_2O$ at maximum tillering, flower initiation, milk, and dough stages The maximum percent increase was 145.9% in wheat, 178.1% in durum wheat, and 157.4% in triticale	Dhaliwal et al., 2019
Pulses		
Chickpea	Soil application (5 kg/ha) through $ZnSO_4.7H_2O$ Increase in concentration of Zn by 5.4 $mg\,kg^{-1}$	Tabassum et al., 2014
Pigeon pea	Soil application (20 kg/ha) $ZnSO_4.7H_2O$ Increase in concentration of Zn by 10.6 $mg\,kg^{-1}$	Behera et al., 2020

Lentil	Foliar spray of 0.5% $ZnSO_4.7H_2O$ at the pre-flowering stage Increase in concentration by 10.5 mg kg^{-1} in grains	Dhaliwal et al., 2021
Oilseeds		
Safflower	Soil application (5 kg/ha) through $ZnSO_4.7H_2O$	Roy and Ghosh, 2020
Linseed	Foliar application of $ZnSO_4.7H_2O$ and Zn-EDTA (0, 0.25, 0.50, and 0.75%) at flowering and capsule formation stages	Dhaliwal et al., 2023
Vegetables		
Brassicaceae microgreens (Arugula, red cabbage, and red mustard)	Soilless media: $ZnSO_4$ (5–10 mg/L) 75–281% increase in Zn content in crop	Gioia et al., 2019
Arugula (Broad leaf)	Foliar application (Zn @ 1.5 kg/ha at 25 DAE) 279% increase in Zn content	Rugeles-Reyes et al., 2019
Lettuce	Different Zn doses (0, 5, 10, 20, 30 mg kg^{-1})	Almeida et al., 2020
Broccoli	0.25% $ZnSO_4.7H_2O$ @ 15 mL per pot	Rivera-Martin et al., 2020
Fruits		
Pear	Foliar spray of 1.5% ZnEDTA	Liu et al., 2023
Banana	Injection of $ZnSO_4.7H_2O$ in pseudostem @ 1, 2, and 4% Triple Zn content in fruit than control	Guevara et al., 2021

Adopted from Avnee et al., 2023.

Zn deficiency can result from factors like elevated pH due to calcium carbonate ($CaCO_3$) and the formation of insoluble calcium zincate (Chahal et al., 2023).

Soil moisture content: Soil moisture plays a vital role in Zn transport from soil to roots through diffusion. Low soil moisture content limits this process, making Zn deficiency more pronounced in rainfed conditions (Scavo et al., 2019).

Interactions with other nutrients: Zn interacts positively with nitrogen and negatively with phosphorus. This interaction affects nutrient uptake and utilization in plants, impacting crop yield and quality. Negative interactions with elements like iron (Fe), manganese (Mn), and copper (Cu) emphasize the need for considering nutrient relationships in biofortification efforts (Maares and Haase, 2020).

Soil microbes: Plant growth-promoting rhizobacteria and mycorrhiza can enhance Zn mobilization and absorption by plants, influencing its availability (Wang et al., 2020).

Genetic characteristics: The effectiveness of Zn application depends on the genetic traits of the crop, as well as the source and dosage of Zn in the soil. Optimizing Zn dosage is essential to prevent toxicity in plants (Wang et al., 2020).

2.1.2. Foliar Application

Foliar application can be a highly effective method, especially considering its advantages:

Low fertilizer use: Foliar application requires lower fertilizer quantities compared to soil application.

Independence from soil factors: Foliar application is not influenced by soil conditions or antagonist nutrients, ensuring consistent results.

The timing of foliar application: Applying Zn during specific growth phases, such as the heading and early milk stages, can result in higher Zn enrichment in grains. The method has shown significant success in increasing grain Zn concentration across multiple locations (Awal and Rahman, 2022). It is particularly valuable in regions prone to drought conditions, as it can enhance crop yield and provide stress resistance through improved defense mechanisms against oxidative cell damage. Wang et al. (2023) reported that foliar spraying of Zn oxide nanoparticles effectively boosted Zn content in rice grains.

2.1.3. Seed Priming

The seed priming approach is attractive since it is simple to implement, harmless to the surrounding ecosystem, cheap to implement, and effective against a wide range of environmental stresses. Seed priming with Zn-infused solutions enhances seedling vigor and crop yields making high-Zn seeds especially beneficial in rainfed environments (Rashid et al., 2019). Choukri et al. (2022) reported that Zn-priming, in comparison to soil treatment of the required amount, is an effective choice for agronomic biofortification of maize grain. Priming maize grain with Zn in 0.5% Zn sulfate for 24 hours raised the Zn content by 15%.

2.2. Breeding-based Biofortification

Successful breeding based biofortification requires high genetic variation, identifying desirable genetic traits related to Zn acquisition, translocation, remobilization, and bioavailability, extensive crossing and backcrossing, ensuring that the desired traits remain stable across various environmental conditions, adaptability to regional agricultural practices and consumer preferences (Velu et al., 2018). Breeding programs take advantage of grain Zn concentration variation in modern wheat genotypes and their wild relatives.

HarvestPlus has been a pioneer in the field of plant breeding-based biofortification strategies to enhance the nutritional content of crops. One of their key approaches involves systematically developing target concentrations for essential micronutrients, such as Fe and Zn, across a diverse range of staple crops. Their strategy typically begins with an assessment of whether stable genetic variability for these micronutrients exists within a particular crop. This assessment involves screening different types of germplasm, varieties, and elite breeding lines to identify and evaluate genetic variations and heritability of desired micronutrient in the edible part of plant.

This analysis is crucial in understanding how environmental factors influenced micronutrient levels and whether the genetic aspect was robust enough to support a successful breeding program. Notably, in specific cases like wheat, the valuable high Zn trait was discovered in wild relatives, necessitating a more intricate breeding strategy. To introduce the high Zn alleles from these wild relatives, innovative approaches such as creating synthetics were employed. These synthetics were then integrated into conventional breeding programs. By 2016, HarvestPlus partners had already introduced over 140 biofortified varieties of 10 essential crops in 26 countries, highlighting the substantial reach and impact of their initiative. A few successful examples of Plant breeding based Zn biofortification are given in Table 2.

Challenges in breeding for Zn biofortification include maintaining productivity. The increasing Zn levels should not compromise crop productivity. Trait stability across diverse environments is another major challenge. Genotype should efficiently absorb and transport Zn to the grain when adequate plant available Zn is present in the soil. Consumer acceptance and ensuring that biofortified varieties meet consumer preferences (Wani et al., 2022) is also critical. Assessing the bioavailability of micronutrients for traditional consumption is critical. Balancing bioavailability and anti-nutrients is another major challenge. Biofortification efforts must strike a balance between reducing anti-nutritional factors that limit Zn bioavailability, such as phytic acid, and maintaining compounds that offer health benefits. Evaluating Zn bioavailability post-harvest, during processing, and cooking is essential to assess the true impact of biofortification (Wang et al., 2021).

2.3. Transgenic Approaches for Zn Biofortification

Transgenic methods offer advanced strategies for developing crop varieties with elevated Zn content. Identifying the genes involved in Zn uptake, translocation, and accumulation in edible plant parts is important before modifying them in target crop plants for Zn biofortification (Figure 1). Plants absorb Zn from rhizospheric soil and use it in intracellular metabolic processes. Zn is transported through both apoplastic and symplastic pathways across the root to the xylem,

Table 2: Examples of plant breeding based Zn biofortification of crop plants

Type of Cereal	Type of Biofortification	Status	Variety/Country	Reference
Rice	Zn, Fe	Released	Bangladesh: BRRIdhan 62, BRRIdhan 72, BRRIdhan 64	CIAT, HarvestPlus
		Traditional variety/Research	Jalmagna	Gregorio et al., 2000.
Wheat	Zn	Released	India: BHU 1, BHU 3, BHU 5, BHU 6, BHU 17, BHU 18 Pakistan: NR 419, 42, 421, Znol	CIAT, CIMMYT, HarvestPlus
	Zn, Fe	Released	India: WB2	Indian Institute of Wheat and Barley Research, India
	Zn	Released	India: PBW1Zn	
	Zn, Fe	Research		Cakmak et al., 2010
Sorghum	Zn	Research		Reddy et al., 2005
Millets	Fe, Zn (Pearl Millet)	Released	India: Dhanashakti Hybrid ICMH 1201 (Shakti-1201)	ICRISAT, HarvestPlus
	Fe, Zn	Research		Rai et al., 2012
Lentils	Fe, Zn	Released	Bangladesh: Barimasur-4, Barimasur-5, Barimasur-6, Barimasur-7, Barimasur-8 Nepal: ILL 7723- Khajurah-1, Khajurah-2, Shital, Sisir, Shekhar and Simal India: L4704 and Pusa Vaibhav Ethiopia: Alemaya Syria: Idlib-2 and Idlib-3	ICARDA, HarvestPlus
Beans	Zn	Research	-	Blair et al., 2009
Potato	Zn, Fe	Research	-	Brown et al., 2010

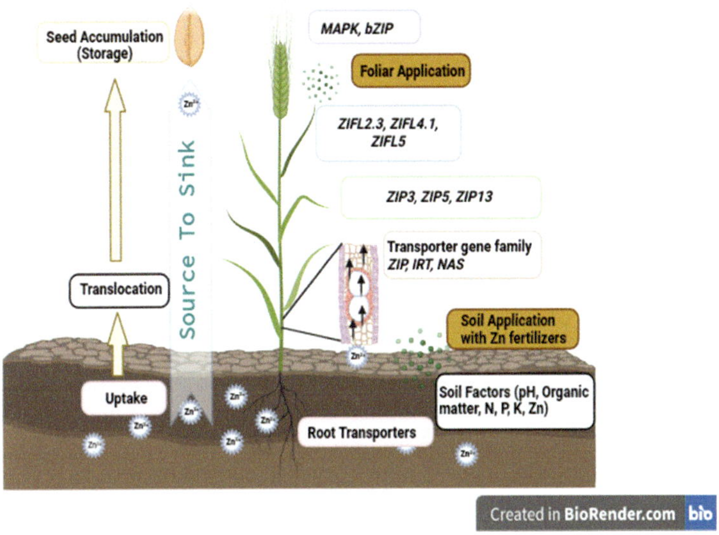

Figure 1: Schematic diagram showing the role of molecular players involved in Zn homeostasis in wheat.

although a substantial fraction may traverse the root and reach the xylem via the apoplast. The plasma membrane of root cells can absorb Zn as Zn2+ or Zn-phytosiderophores (PS). It was previously thought that mugineic acid (MA) and nicotianamine only affect Fe homeostasis, but PS also affects Zn translocation. DMA synthase (DMAS) secretes MAs in graminaceous species (rice, wheat, and barley) under Zn deficiency, while NAS synthesizes NA in all higher plants under Fe deficiency (Nozoye et al., 2019). The expression of HvNAS and HvNAAT-B in barley shoots increased under Zn deficient conditions. Moreover, the genes HvNAS1, HvNAAT-A, and HvNAAT-B also showed increased relative expression in Fe- and Zn-deficient roots but the relative expression of these genes in Fe-deficient shoots was not observed very significant. This suggested that DMA synthesis in shoots is induced under Zn deficiency whereas MA synthesis in barley roots is only induced when the plant is grown under both Fe and Zn deficient conditions. Henceforth, DMA, NA, and other transporters are suggested to play a key role in Zn mobilization in plants. Several Zn-regulated transporters (ZIP) gene families have been identified and predicted to be involved in Fe and Zn transport in plants. For example, in rice, several ZIP family genes have been characterized, including OsIRT1, OsIRT2, and OsZIP1-8 (Suzuki et al., 2006).

Use of ZIP family Fe and Zn transporter proteins can enhance grain micronutrient concentrations by facilitating the uptake and transport of cationic micronutrients like Zn. For instance, the introduction of a Zn transporter gene from Arabidopsis thaliana into barley roots resulted in increased Zn concentrations in the grain (Kumar et al., 2019). TaZIP transporters transport Zn from wheat roots to shoots, helping maintain ion balance and increasing Zn tolerance. The study found that TaZIP5, TaZIP6, and TaZIP13 help plants to absorb Zn, while several other TaZIP genes help Zn translocate. When the wheat plants experienced a deficiency in Zn, it led to the activation of specific transcription factors from group FbZIP and 14 TaZIP transporters (Li et al., 2022). Functional analysis of Group FbZIPs revealed cysteine-histidine-rich motifs that may bind Zn2+ when the supply of Zn is limited. ZIPs with Zn-deficiency-response elements (ZDREs) could translocate Zn, including TaZIP3, TaZIP5, TaZIP7, and TaZIP13. The conservation of ZDREs varied among ZIPs, suggesting that each ZIP transporter may play a different role in Zn homeostasis (Evens et al., 2017).

Besides, YSL proteins may also load Zn into the phloem, where Zn is transported as a Zn-NA complex, or as a complex with small proteins, to target sink tissues). Although there is not yet any specific information available about the Zn uptake and translocation in higher plants, however, till now it was believed that plants follow the same strategies for Zn-uptake as it was involved in Fe homeostasis. Understanding the molecular pattern of expression of Zn transporters in different tissues could offer significant information so that we can understand the genetic and molecular basis of Zn absorption and translocation in plants (Hamzah et al., 2022).

The research aimed to dissect the genetic pathways responsible for regulating Zn concentration in wheat grains, with a focus on identifying genes associated with increased Zn accumulation. By genotyping European wheat cultivars, previously unknown genomic regions linked to Zn uptake and transport were uncovered. Additionally, the study identified candidate genes related to Zn concentration in wheat grain, specifically those associated with mitogen-activated protein kinase (MAPK) and bZIP (Alomari et al., 2018). Ishimaru et al. (2005) isolated OsIRT1, OsZIP1, OsZIP2, OsZIP3, OsZIP4, OsZIP5, OsZIP6, and OsZIP7 genes from Fe and Zn-deficient rice roots' cDNA library. These genes were highly homologous to Arabidopsis and could support YH003 and ZHY3 yeast mutant growth on Fe and Zn-depleted media. OsIRT1 was up-regulated in Fe-deficient roots and shoots, while OsZIP1, OsZIP2, OsZIP3, OsZIP4, OsZIP5, OsZIP6, and OsZIP7 were up-regulated in Zn-deficient roots and shoots.

Furthermore, the investigation revealed that an abundant nitrogen supply had a positive effect on Zn concentration, enhancing the rate at which Zn was transported from the roots to the shoots in winter wheat grains. Nitrogen supply also led to the upregulation of the TaZIP3 and TaZIP7 genes in various plant tissues. In a separate study in nutrient-deficient conditions, overexpressed genes TaZIP3 and TaZIP7 were found to significantly increase the accumulation and translocation of both Zn and Fe in the roots and shoots of wheat plants.

Biofortified transgenic crops usually contain Fe and Zn. To accumulate Fe and Zn in rice endosperm, Banakar et al. (2017) created transgenic rice plants that expressed nicotianamine and 2′-deoxymugenic acid (DMA). Transgenic rice plants could obtain 4-fold more Fe and 2-fold more Zn in the endosperm. Kavitha et al. (2015) created transgenic Arabidopsis lines that overexpressed ZmZIP7 and found higher Fe and Zn concentrations in roots, shoots, and grains. They also examined maize ZmZIP7 expression profiles and found that root, stem, leaf, and grain sequestered more Fe and Zn than young ear and tassel.

Trijatmiko et al. (2016) investigated rice Fe and Zn enrichment molecularly using transgenics. Transgenic rice was created by cloning soybean ferritin (SferH-1) and rice nicotianamine synthase (OsNAS2) under endosperm-specific promoters. Endosperm Fe and Zn concentrations increased significantly.

3. Genome-Wide Association Study (GWAS)

GWAS is a powerful approach for identifying genes associated with high Zn levels in cereals. By genotyping numerous samples and analyzing Zn data, researchers can pinpoint genetic markers consistently linked to elevated Zn content. These markers may correspond to specific genes or regulatory elements influencing Zn uptake, transport, accumulation, or regulation within the organism. GWAS results provide valuable insights into the genetic foundation of high Zn concentration, opening doors for further research into Zn metabolism mechanisms and their applications in crop improvement and nutritional enhancement (Pujar et al., 2020). The results of this GWAS offer valuable insights into the genetic basis of high Zn concentration, potentially enabling further research into the mechanisms and pathways involved in Zn metabolism and providing opportunities for crop improvement, nutritional enhancement, or other applications related to Zn content in plants or animals. There have been conducted numerous studies for the identification of loci linked with high Zn content in cereals. Finger millet, a high-Fe and Zn crop, is a good model system for studying the complex genetic and molecular mechanisms of grain Fe and Zn accumulation. Genotyping by Sequencing was used to analyze 202 finger millet genotypes' genetic diversity and population structure. GBS SNPs were used for association mapping to find grain Fe and Zn markers. Five SNP markers, three for grain Fe (TP413281 associated with ferritin (Fer1), TP566120 with Fe-regulated transporter-like protein (IRT2), and TP1431395 with yellow stripe-like 2 proteins (YSL2), and two for grain Zn (TP1316808 associated with ZIP1 protein and TP784188 associated with ZTP29-like protein), showed homology to candidate genes of *Oryza sativa* and *Setaria italica*, which might play an important role in grain Fe and Zn homeostasis process in finger millet (Chandra et al., 2020).

4. Conclusion

The biofortification approach holds remarkable potential in addressing the issue of Zn malnutrition. This approach doesn't necessitate any dietary alterations and can be swiftly embraced by communities. Agronomic biofortification not only enhances the Zn concentration in grains, delivering health benefits, but it can also play a role in reducing Zn deficiency, especially in regions characterized by intensive cropping practices where micronutrient supplementation is often overlooked. Moreover, with the advancements in genetic engineering and molecular tools, Zn biofortification can be accelerated. This paves the way for the development of superior Zn efficient and Zn rich varieties, making it comparatively easier to achieve this goal.

References

Almeida, H.J., V.M. Vergara Carmona, I.M. Ferreira, A.E. Furtini, A.B. Cecílio, and M. Mauad. 2020. Soil type and zinc doses in agronomic biofortification of lettuce genotypes. *Agronomy*, 10: 124.

Alomari, D.Z., K. Eggert, N. von Wirén, A.M. Alqudah, A. Polley, A. Plieske, M.W. Ganal, K. Pillen, and M.S. Röder. 2018. Identifying candidate genes for enhancing grain Zn concentration in wheat. *Front Plant Sci.*, 9: 1313.

Awal, M.A., T. Islam, and M.H. Rahma. 2022. Zn biofortification of wheat grain through agronomic approaches in Bangladesh. *Discovery*, 58(324): 1415–1428.

Blair, M.W., C. Astudillo, M.A. Grusak, R. Graham, and S.E. Beebe. 2009. Inheritance of seed iron and zinc concentrations in common bean (*Phaseolus vulgaris* L.). *Mol. Breed.*, 23(2): 197–207.

Brown, C.R., K.G. Haynes, M. Moore, M.J. Pavek, D.C. Hane, S.L. Love, R.G. Novy, and J.C. Miller. 2010. Stability and broad-sense heritability of mineral content in potato: Iron. *Am. J. Potato Res.*, 87(4): 390–396.

Cakmak, I. 2010. Biofortification of cereals with zinc and iron through fertilization strategy. *19th World Congress of Soil Science, Soil Solutions for a Changing World*, 5: 1–6.

Chahal, S.K., M.G. Hettiarachchi, O.N. Nathan, and J.G. Mary. 2023. Fate and plant uptake of different Zn fertilizer sources upon their application to an alkaline calcareous soil. *ACS Agric. Sci. Technol.*, 3(9): 725–737.

Chandra, A.K., D. Pandey, A. Tiwari, D. Sharma, A. Agarwal, S. Sood, and A. Kumar. 2020. An omics study of iron and Zn homeostasis in finger millet: Biofortified foods for micronutrient deficiency in an era of climate change. *Omics: J. Integr. Biol.*, 24(12): 688–705.

Choukri, M., A. Abouabdillah, R. Bouabid, O.H. Abd-Elkader, O. Pacioglu, F. Boufahja, and M. Bourioug. 2022. Zn application through seed priming improves productivity and grain nutritional quality of silage corn. *Saudi J. Biol. Sci.*, 29(12): 103456.

Dhaliwal, S.S., V. Sharma, V. Verma, H. Singh, P. Singh, and K. Kaur. 2023. Biofortification of linseed (*Linum usitatissimum* L.) through mineral and chelated forms of Zn on yield, Zn accumulation, quality parameters, efficiency indices, and economic under low Zn soils of northwestern India. *J. Plant Nutr.*, 46: 356–359.

Dhaliwal, S.S., V. Sharma, A.K. Shukla, J. Kaur, V. Verma, P. Singh, H. Singh, S.H. Abdel-Hafez, S. Sayed, A. Gaber, R. Ali, and A. Hossain. 2021. Enrichment of zinc and iron micronutrients in lentil (*Lens culinaris* Medik.) through biofortification. *Molecules*, 26: 7671.

Dhaliwal, S.S., H. Ram, A.K. Shukla, and G.S. Mavi. 2019. Zinc biofortification of bread wheat, triticale, and durum wheat cultivars1 by foliar zinc fertilization. *J. Plant Nutr.*, 42: 813–822.

Elhaj, B.Z., and J.M. Unrine, 2018. Functionalized-ZnO-nanoparticle seed treatments to enhance growth and Zn content of wheat (*Triticum aestivum*) seedlings. *J. Agric. Food Chem.*, 66: 12166–12178.

Etienne, P., S. Diquelou, M. Prudent, C. Salon, A. Maillard, and A. Ourry. 2018. Macro and micronutrient storage in plants and their remobilization when facing scarcity: The case of drought. *Agriculture*, 8(1): 14.

Evens, N.P., P. Buchner, L.E. Williams, and M.J. Hawkesford. 2017. The role of ZIP transporters and group FbZIP transcription factors in the Zn-deficiency response of wheat (*Triticum aestivum*). *Plant J.*, 92(2): 291–304.

FAO, IFAD, UNICEF, WFP, and WHO. 2022. Repurposing food and agricultural policies to make healthy diets more affordable. *The State of Food Security and Nutrition in the World 2022*. Rome, FAO.

Faujdar, R.S., M. Sharma, R.L. Solanki, and R.C. Dangi. 2014. Effect of FYM, biofertilizers and zinc on yield and micronutrients uptake in maize. *Asian J. Soil Sc.*, 9: 121–125.

Firdous, S., B.K. Agarwal, and V. Chhabra. 2018. Zinc-fertilization effects on wheat yield and yield components. *J. Pharmacog Phytochem.*, 7: 3497–3499.

Garg, M., N. Sharma, S. Sharma, P. Kapoor, A. Kumar, V. Chunduri, and P. Arora. 2018. Biofortified crops generated by breeding, agronomy, and transgenic approaches are improving lives of millions of people around the world. *Front Nutr.*, 5: 12.

Ghasal, P.C., Y.S. Shivay, V. Pooniya, M. Choudhary, and R.K. Verma. 2018. Zinc partitioning in basmati rice varieties as influenced by Zn fertilization. *Crop J.*, 6: 136–137.

Ghasal, P.C., Y.S. Shivay, V. Pooniya, M. Choudhary, and R.K. Verma. 2017. Response of wheat genotypes to zinc fertilization for improving productivity and quality. *Arch. Agron. Soil Sci.*, 63: 1597–1612.

Gioia, F.D., S.A. Petropoulos, M.O. Hampton, K. Morgan, and E.N. Rosskopf. 2019. Zinc and iron agronomic biofortification of brassicaceae microgreens. *Agronomy*, 9(11): 677–678.

Gregorio, G.B., D. Senadhira, H. Htut, and R.D. Graham. 2021. Breeding for trace mineral density in rice. *Food Nutr. Bull.*, 21: 382–386.

Guevara, J.E.M., A.G. Sanches, V.M.D. Pedrosa, M.C.P. Cruz, J.A. Silva, and G.H. Teixeira. 2021. The injection of zinc sulfate into banana tree pseudostem can triple the zinc content and it is an effective method for fruit biofortification. *J. Food Compos. Anal.*, 102: 104020.

Hamzah, Mohd. S.M., K. Usman, M. Rizwan, H.A. Jabri, and M. Alsafran. 2022. Functions and strategies for enhancing Zn availability in plants for sustainable agriculture. *Front. Plant Sci.*, 13: 1033092.

Ishimaru, Y., M. Suzuki, T. Kobayashi, M. Takahashi, H. Nakanishi, S. Mori, and N.K. Nishizawa. 2005. OsZIP4, a novel zinc-regulated zinc transporter in rice. *J. Exp. Bot.*, 56(422): 3207–3214.

Kavitha, P.G., S. Kuruvilla, and M.K. Mathew. 2015. Functional characterization of a transition metal ion transporter, OsZIP6 from rice (*Oryza sativa* L.). *Plant Physiol. Biochem.*, 97: 165–174.

Kiani, A.K., K. Dhuli, K. Donato, B. Aquilanti, V. Velluti, G. Matera, A. Iaconelli, S.T. Connelly, F. Bellinato, P. Gisondi, and M. Bertelli. 2022. Main nutritional deficiencies. *J. Prev. Med. Hyg.*, 63(2): E93–E101.

Kumar, S., A. Palve, C. Joshi, R.K. Srivastava, and Rukhsar. 2019. Crop biofortification for iron (Fe), Zn, and vitamin A with transgenic approaches. *Heliyon*, 5(6): e01914.

Maares, M., and H. Haase. 2020. A guide to human Zn absorption: General overview and recent advances of *in vitro* intestinal models. *Nutrients*, 12(3): 762.

Maqbool, M.A., and A. Beshir. 2019. Zn biofortification of maize (*Zea mays* L.): Status and challenges. *Plant Breed.*, 138: 1–28.

Prasad, R., and Y.S. Shivay. 2018. Relative grain zinc loading ability of cereals (rice, wheat, maize) and a grain legume (chickpea). *Indian J. Agric Sci.*, 88: 307–313.

Pujar, M., S. Gangaprasad, M. Govindaraj, S.S. Gangurde, A. Kanatti, and H. Kudapa. (2020). Genome-wide association study uncovers genomic regions associated with grain iron, Zn, and protein content in pearl millet. *Scientific Reports*, 10(1): 19473.

Rai, K.N., M. Govindraj, and A.S. Rao. 2012. Genetic enhancement of grain iron and zinc content in pearl millet. *Crops Food*, 4(3): 119–125.

Rashid, A., H. Ram, C.Q. Zou, B. Rerkasem, A.P. Duarte, S. Simunji, A. Yazici, S. Guo, M. Rizwan, R.S. Bal, Z. Wang, S.S. Malik, N. Phattarakul, R.S. de Freitas, O. Lungu, V.L.N.P. Barros, and I. Cakmak. 2019. Effect of Zn-biofortified seeds on grain yield of wheat, rice, and common bean grown in six countries. *J. Plant Nutr. Soil SC.*, 182: 791–804.

Reddy, B.V.S., S. Ramesh, and T. Longvah. 2005. Prospects of breeding for micronutrients and β-carotene-dense sorghums. *Int. Sorghum Millets Newsl.*, 46: 10–14.

Rivera-Martin, A., M.R. Broadley, and M.J. Poblaciones. 2020. Soil and foliar zinc application to biofortify broccoli (*Brassica oleracea* var. *italica* L.): Effects on the zinc concentration and bioavailability. *Plant Soil Environ.*, 66: 113–118.

Roohani, N., R. Hurrell, R. Kelishadi, and R. Schulin. 2013. Zn and its importance for human health: An integrative review. *J. Res. Med. Sci.: The Official Journal of Isfahan University of Medical Sciences*, 18(2): 144–157.

Roy, N., and K. Ghosh. 2020. Enhancing nutrient availability, yield, and quality of safflower (*Carthamus tinctorius* L.) through zinc and sulphur in Alfisol. *J. Pharmacog Phytochem.*, 9: 693–697.

Rugeles-Reyes, S.M., M.A. Lopez Aguilar, and P.H. Silva. 2019. Foliar application of zinc in the agronomic biofortification of arugula. *Food Sci. Technol.*, 39: 1011–1017.

Scavo, A., C. Abbate, and G. Mauromicale. 2019. Plant allelochemicals: Agronomic, nutritional, and ecological relevance in the soil system. *Plant Soil*, 442: 23–48.

Shivay, Y.S. 2015. Role of agronomic biofortification in alleviating malnutrition. *Int. J. Econ. Plants*, 2: 153–158.

Singh, A., and Y.S. Shivay. 2019. Effects of green manures and zinc fertilizer sources on DTPA-extractable zinc in soil and zinc content in basmati rice plants at different growth stages. *Pedosphere*, 29: 504–505.

Suzuki, M., M. Takahashi, T. Tsukamoto, S. Watanabe, and N.K. Nishizawa. 2006. Biosynthesis and secretion of mugineic acid family phytosiderophores in Zn-deficient barley. *Plant J.*, 48: 85–97.

Tabassum, S., S. Jeet, R. Kumar, C.M. Dev, P. Kumar, and R. Rehana. 2014. Effect of organic manure and zinc fertilization on zinc transformation and biofortification of crops in Vertisols of Central India. *J. Agric. Sci.*, 6: 221–231.

Trijatmiko, K.R., C. Dueñas, N. Tsakirpaloglou, L. Torrizo, F.M. Arines, A. Johnson, … and I.H. Slamet-Loedin. 2016. Biofortified indica rice attains iron and Zn nutrition dietary targets in the field. *Sci. Rep.*, 6: 19792.

Velu, G., R.P. Singh, L. Crespo-Herrera, P. Juliana, S. Dreisigacker, R. Valluru, J. Stangoulis, V.S. Sohu, G.S. Mavi, V.K. Mishra, A. Balasubramaniam, R. Chatrath, V.

Gupta, G.P. Singh, and A.K. Joshi. 2018. Genetic dissection of grain Zn concentration in spring wheat for mainstreaming biofortification in CIMMYT wheat breeding. *Sci. Rep.*, 8: 13526.

Wang, S., F. Ruotong, Y. Xijun, C. Jie, M. Kailiang, W. Rui, Z. Haipeng, and Z. Hongcheng. 2023. Foliar spraying of ZnO nanoparticles enhanced the yield, quality, and zinc enrichment of rice grains. *Foods*, 19: 3677.

Wang, S., X. Tian, and Q. Liu. 2020. The effectiveness of foliar applications of Zn and biostimulants to increase Zn concentration and bioavailability of wheat grain. *Agronomy*, 10: 178.

Wang, Y., Y. Meng, Y. Ma, L. Liu, D. Wu, X. Shu, L. Pan, and Q. Lai. 2021. Combination of high Zn density and low phytic acid for improving Zn bioavailability in rice (*Oryza stavia* L.) grain. *Rice* (New York, N.Y.), 14(1): 23.

Wani, S.H., K. Gaikwad, A. Razzaq, K. Samantara, M. Kumar, and V. Govindan. 2022. Improving Zn and iron biofortification in wheat through genomics approaches. *Mol. Biol. Rep.*, 49: 8007–8023.

Yu, H., L. Liu, Z. Ma, J. Zhou, S. Tang, L. Wu, G. Fu, Y. Li, and M. Xu. 2023. Spraying ZnEDTA at high concentrations: An ignored potential for producing zinc-fortified pear (*Pyrus* spp.) fruits without causing leaf and fruitlet burns. *Sci. Hortic.*, 322: 112380.

CHAPTER

6

Phytochemicals: An Emerging Application

Rohit Kumar[a*[0000-0002-5811-8816]], Era Chaudhary[b,c], Satveer Kaur[b], Amneet Kaur Badwal[b] and Lakshmi Giri[b]

[a] Clemson University – Plant and Environmental Sciences, South Carolina
[b] National Agri Food Biotechnology Institute, Mohali, Punjab, India
[c] Regional Centre for Biotechnology, Faridabad, Haryana, India

1. Introduction

Phytochemicals are components of a plant-based diet with defensive or disease-preventing properties. There is evidence linking their consumption to a decreased likelihood of developing various chronic ailments, such as cancer, diabetes, atherosclerosis, inflammation, and obesity (Xiao et al., 2011; Tiwari et al., 2023). Despite the effectiveness of the body's natural antioxidant defense system, exogenous antioxidants are still necessary to avoid harmful free radicals. Anthocyanins in dietary plants protect against oxidative stress. Flavonoids, the largest, most diversified, and most investigated phytochemical category, involved in many developmental and biochemical processes as well as environmental responses (Shen et al., 2022; Tiwari and Husain, 2017). As a result, flavonoids play a significant role in the defense against various illnesses that affect the human body. Based on the heterocyclic C ring oxygenation pattern, they have been further subcategorized into a total of six primary subclasses. These subclasses are as follows: flavones, flavonols, flavanones, flavanols, isoflavones, and anthocyanidins (Liu et al., 2021). The diverse pharmacological actions of flavonoids can be attributed to their polyphenolic structure (Panche et al., 2016).

The antioxidant action of flavonoids' functional hydroxyl groups eliminates free radicals in the human body. This aids in preventing radical production,

*Corresponding author: mohank@clemson.edu

which damages biomolecules and contributes to oxidative stress and other diseases (Kumar et al., 2013). There is a high demand for crops with increased phytochemical content due to the urgent need to improve human health. Yet, phytochemicals' contribution to human food is not solely determined by their content in specific plant tissue. These phytochemicals must be circulated from the component foods matrix while digestion, absorbed by the bloodstream, and transported to their target tissues. Only a small percentage of phytochemicals released from consumed plant tissue eventually becomes available for absorption and is bioaccessible. Increasing the plant phytochemical bioaccessibility is a viable and important target for agronomic techniques to increase the nutritional quality of various staple food crops (Shahidi and Pan, 2022; Palafox-Carlos et al., 2011; Velderrain-Rodríguez et al., 2014).

The introduction of biofortification techniques in various crops holds effectiveness in mitigating micronutrient deficiencies and enhancing the nutritional and health conditions of global populations. Various agronomic and genetic approaches have been used to obtain phytochemical biofortified crops (Joshi-Saha et al., 2022). Several studies found that consuming colored maize reduces the risk of non-communicable diseases, which would be a good way to manage them (Toufektsian et al., 2008). It has already been shown that phytochemicals-based maize (blue, red, and purple) also reduces colorectal cancer in male rats (Hagiwara et al., 2001). Much work has been done to improve wheat's antioxidant properties and anthocyanins content (Garg et al., 2016; Sharma et al., 2018; Gamel et al., 2023; Sharma et al., 2023). Many breeding programs worldwide have used the colored wheat trait (black, blue, and purple), which is linked to strong antioxidant qualities, to boost the nutritional content of wheat (Garg et al., 2016; Žofajová et al., 2017). In addition, a traditional breeding strategy was used to generate an anthocyanin biofortified tomato called "Sun Black". This variety has dark purple fruit coloration due to a higher anthocyanin accumulation in the fruit's peel (Mazzucato et al., 2008). This book chapter explores the enhancement of phytochemical content in various crops to improve nutritional quality of food sources. It covers historical developments and recent advancements and highlighting strategies to increase phytochemical levels.

2. Biosynthesis of Flavonoids

Flavonoids constitute a class of polyphenolic chemicals synthesized via the phenylpropanoid pathway. The flavonoid biosynthesis pathway includes a number of enzymes and intermediates that vary by plant species (Figure 1). The initial step in its biosynthesis is catalyzed by the enzyme phenylalanine ammonia lyase (PAL), which converts phenylalanine to cinnamic acid. The enzyme 4-coumaroyl, CoA ligase (4CL), converts cinnamic acid to p-coumaroyl-CoA, and chalcone synthase (CHS) mediates the conversion of 4-Coumaroyl-CoA to naringenin chalcone. After that, naringenin chalcone is converted to different flavonoids such as flavones, flavonols, flavanones, anthocyanins, and proanthocyanidins by specific enzymes such as flavanone 3-hydroxylase (F3H), flavone synthase (FNS), flavonol synthase (FLS), anthocyanidin synthase (ANS), and dihydroflavonol 4-reductase (DFR) (Liu et al., 2021). The final step involves glycosylation, adding sugar molecules to the flavonoid aglycone to form glycosides. The whole pathway is controlled by regulatory complex formed by transcription factors, basic helix-loop-helix (bHLH) and myeloblastosis (MYB) that include WDR (WD40 repeats) proteins. Several factors, including light, temperature, hormones, and environmental stresses, regulate the flavonoid biosynthesis pathway. Flavonoids fulfill significant functions within the plant kingdoms, encompassing pigmentation, UV shielding, pollinator attraction, and defense mechanisms against pathogens and herbivores (Zhang et al., 2023; Kaur et al., 2022).

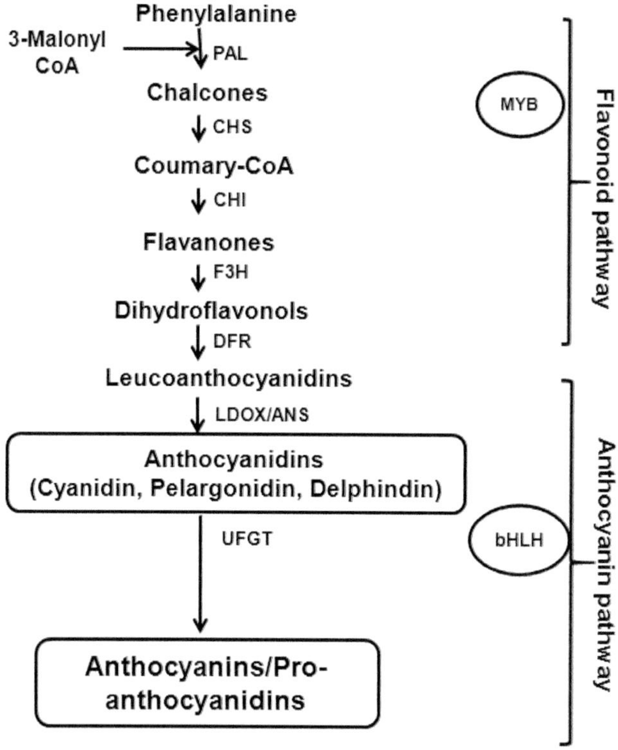

Figure 1: The biosynthetic pathway of flavonoids highlighting the key genes utilized in biofortification.

3. Phytochemical Biofortification in Cereals

Cereals are the primary energy source for the global population, serving as a staple food. The bran layers of cereals are rich in proteins, fibers, minerals, oil, several vitamins, and phytochemicals. Among cereals, rice, wheat, and maize are three widely used crops with various phytochemicals like flavonoids, phenols, and anthocyanins in their layers as shown in Figure 2.

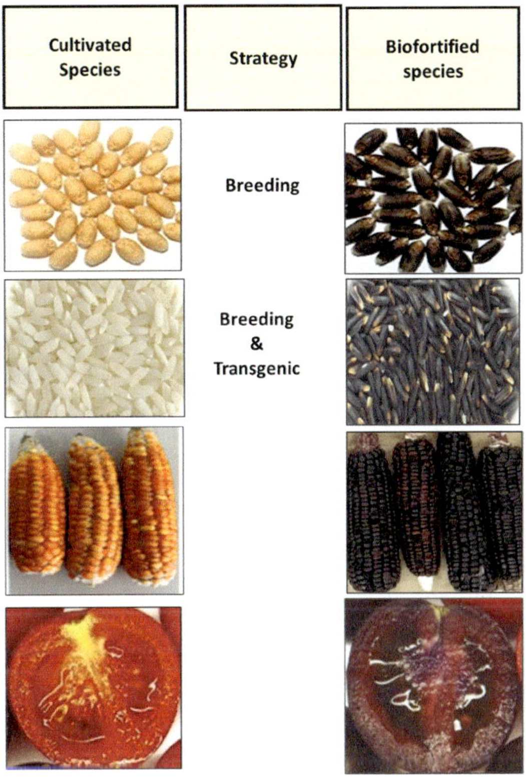

Figure 2: Increasing anthocyanin content leads to purple to black crops with enhanced antioxidant properties.

3.1. Rice

Red and black rice varieties are known for their richness in anthocyanins, a group of water-soluble purple to reddish flavonoids extensively reported in grains (Abdel et al., 2006). The color of the pericarp varies from white, brown, and red to black. Red rice (*Oryza punctate*), also known as red-kernelled rice, contains tannin pigments that give the rice a red or brownish-red appearance. Black rice (*Oryza sativa* L. *indica*) also goes by alternative names like emperor's rice, purple rice, king's rice, Forbidden rice, and fortune rice (Ito et al., 2019). Anthocyanins are abundant in black rice kernels, while pro-anthocyanidins are present in red rice kernels. Most of these pigments either disappear completely or remain in trace amounts during milling operations (Goffman and Bergman, 2004; Lamberts et al., 2007). The analysis of a diverse assemblage of rice germplasm, encompassing a variety of colors such as white, red, and black rice, reveals a notable association between the color attributes of rice grains and key characteristics such as flavonoid contents, total phenolics, and antioxidant capacity (Shen et al., 2009). Black rice contains 18 different anthocyanin species, with cyanidin-3-O-glucoside (C3G) and peonidin-3-O-glucoside (P3G) being the two most prevalent anthocyanins (Shao et al., 2018).

The allelic loci Pp (Prp-a), P, and Pb (Prp-b) are linked to three chromosomal loci, namely Kala1, Kala3, and Kala4. These chromosomal loci are responsible for encoding DFR, MYB, and bHLH, respectively (Maeda et al., 2014) and control the development of black pericarp. Homologous recombination and insertional genetic rearrangement in the Kala4 gene promoter are required to develop the black pericarp in rice grains. The emergence and incorporation of the black grain (or purple grain) trait into different rice varieties (*indica, japonica, aus, aromatic*) lacks conclusive genetic evidence. Black rice is intensely black and typically turns deep purple when cooked, possibly due to anthocyanins being converted into new colored phenolic compounds. The utilization of breeding techniques can facilitate the development of novel black rice cultivars that exhibit enhanced yields and possess the desired phytochemical composition (Oikawa et al., 2015). Few reports exist regarding the reproductive processes and inheritance patterns associated with the black grain characteristic in descendant populations. The integration of genome-wide sequencing and transcriptomic sequencing techniques in studying black rice plants led to identifying various genes associated with anthocyanin biosynthesis, such as F3H, DFR, UGT, and ANS1 (Oh et al., 2018). A study involved the development of the transgenic rice cultivar capable of synthesizing different types of flavonoids in its seeds. This was achieved by introducing genes PAL and CHS that led to the development of rice with an accumulation of naringenin (a class of flavonoids). The target flavonoids genistein, apigenin, and kaempferol were also produced by adding FLS, F3H, IFS, and FNS genes, along with PAL and CHS (Ogo et al., 2013). Black rice contains anthocyanins in its seed layers, which are typically removed during the milling process. Consequently, introducing endosperm-specific expression of anthocyanins would offer greater advantages compared to conventional rice varieties. The engineering of pathways of metabolism for biofortification poses challenges due to the intricate genetic regulatory network regulating its biosynthetic pathway. A highly efficient vector-based system was devised with the intent of transgene stacking, which was subsequently employed to manipulate anthocyanin biosynthesis in the endosperm of rice. A construct that consisted of eight regulatory and structural genes associated with anthocyanin biosynthesis was created, sourced from maize and coleus. The expression of these genes was controlled by endosperm-specific promoters. The integration of this particular construct into rice led to the creation of a novel biofortified germplasm referred to as “Purple Endosperm Rice” or Chinese ‘Zijingmi’. This variety exhibits high levels of anthocyanins and antioxidant activity within the endosperm (Zhu et al., 2017).

3.2. Wheat

Wheat inherently lacks potent antioxidants like anthocyanins. Developing biofortified colored wheat is a relatively recent research area, focusing on enhancing nutritional and functional properties by incorporating pigments such as anthocyanins (Garg et al., 2022; Dhua et al., 2021). Wheat cultivars with colorful grains purple and blue have been cultivated for decades. These wheat cultivars loaded with bioactive anthocyanin compounds are believed to be useful germplasm in breeding programs due to their scavenging ability for free radicals (Zhang et al., 2022; Garg et al., 2022). A consequence of breeding, the black wheat has been developed by combining purple pericarp trait with the blue aleurone trait (Garg et al., 2022). Many research efforts have been conducted to ascertain the specific genes responsible for the synthesis of anthocyanin in wheat seeds, both in terms of their structural composition and regulatory function. Additionally, these studies aim to elucidate the consequential effects of this biosynthetic pathway on human health (Sharma et al., 2020; Sharma et al., 2023; Kapoor et al., 2023).

3.3. Maize

Maize is regarded as one of nature's most diverse grain crops. In Mesoamerica, various types of maize, including colored maize, have been used as a nutritious dietary staple for thousands of years. Due to their vibrant appearance, multicolored, red, purple, blue, and black maize kernels are now manufactured in limited quantities, often for specialized meals or as ornaments. Phytochemical biofortification in maize can potentially improve this staple crop's nutritional quality. Anthocyanins are found in the pericarp of purple maize kernels, similar to wheat and barley, and in the aleurone layer of blue varieties. Both aleurone and pericarp layers of dark red and black kernels contain anthocyanins (Francavilla and Joye, 2020). The anthocyanin biosynthesis pathway's enzymes are encoded by a group of structural genes, *R locus* (chromosome 10), *B locus* (chromosome 2), and regulatory genes, *C1,* and *Pl.* These genes encode two sets of functionally identical proteins, RIB and ClIPI (Dooner at al., 1991). For instance, R-S and CI are needed to produce anthocyanin in the endosperm's aleurone layer. Lago et al. (2014a) conducted a cross between the 'Scagliolo' polenta maize variety and a donor containing dominant B1 and Pl1 anthocyanin regulatory genes. These genes are responsible for the coloration of the seed pericarp (Lago et al., 2014a, b). The successful integration of 11 genes into transgenic maize resulted in the production of germplasm with a high concentration of anthocyanin-rich purple embryos. In HiII embryos, this genetic modification successfully reconstructed the anthocyanin biosynthesis pathway. This study's findings confirm that the regulatory genes ZmC1 and ZmR2 play an important role in reconstructing the anthocyanin production pathway in HiII embryos (Liu X et al., 2018).

4. Phytochemical Biofortification in Vegetables and Fruits

Fruits and vegetables are widely recognized as primary sources of phytochemicals. Consuming colored fruits and vegetables could easily ensure a regular anthocyanin supply without altering eating habits, providing extra benefits naturally from all their phytochemical components, none of which are obtainable from supplementary sources. Colored fruits like berries, plums, grapes etc are all well known for their health promoting effects. The antioxidants in staple vegetables like potatoes benefit human health, improving immunity against various diseases. These benefits are known to be enhanced by the purple outer layer and yellow flesh color (Luthra et al., 2020). There are various varieties of vegetables and fruits developed with enhanced agronomic and nutritional traits such as purple cauliflower, purple-fleshed sweet potato, red potato tuber, purple pepper etc. (Matto et al., 2022).

4.1. Potato

The implementation of biofortification techniques in potatoes has the potential to effectively mitigate micronutrient deficiencies. This is particularly significant as potatoes rank among the most prominent sources of polyphenolic compounds, following apples and oranges (Ezekiel et al., 2013). Several South American potato varieties have been suggested as intriguing sources of flavonoids in breeding programs to increase anthocyanin content. For example, the purple-skinned speciality potato cultivar Kufri Neelkanth with yellow flesh has been developed in India. On a molecular basis, the ternary complex MBW, comprising of bHLH, MYB, and WD40 proteins, regulates anthocyanin accumulation. The locus D (Developer) containing the StAN1 gene produces an R2R3 MYB transcription factor that directly triggers the anthocyanin biosynthesis pathway (Payyavula et al., 2013). Single or combined overexpression of the anthocyanin biosynthesis genes *CHI*, *CHS*, and *DFR* led to a substantial increase in flavonoids and anthocyanins and decreased starch and glucose levels. Transgenic potato plants overexpressing UDP-glucose: flavonoid-3-O-glucosyltransferase (3GT) contained significantly higher quantities of anthocyanins, namely, sinapic acid, chlorogenic acid, kaempferol, and proanthocyanins in the potato skin than wild (Wei et al., 2012; Yoshihara et al., 2005). It was found that purple-skinned and fleshed potatoes had 48% fewer anthocyanins when exposed to high temperatures, indicating that genetic and environment interaction also affects the biosynthesis of genes associated with anthocyanin production (Liu et al., 2019).

4.2. Tomato

Tomato fruits produce various secondary metabolites, including polyphenols, carotenoids, volatiles, and alkaloids. Genetic engineering involves two methods using structural or regulatory DNA, sometimes in combination as shown in

Figure 2. Certain wild tomato species, specifically *Solanum chilense* and *S. chesmaniae*, can produce anthocyanins in their fruit's subepidermal tissue. The genotypes of these alleles have been successfully incorporated into a novel tomato cultivar known as Sun Black (SB). The richly pigmented skin of this particular tomato variety is ascribed to its elevated levels of anthocyanins within the peel (Blando et al., 2019). The tomato exhibits a typical red pulp and possesses a flavor profile consistent with that of a conventional tomato. The tomato's recessive gene ATV (atroviolacea) was developed through crossbreeding with *S. cheesmaniae*, resulting in the introgression of the dominant gene AFT (anthocyanin fruit) into a cultivated variety through the process of crossing. The plants that emerged from this study exhibited a double mutant phenotype (Aft/ Aft atv/atv), characterized by the accumulation of anthocyanins in the fruit peel during the ripening process. This accumulation was notably triggered by both biotic and abiotic stress factors (Povero et al., 2011). The flavanol concentration of the peel increased due to the overexpression of the Petunia CHI gene. By silencing the DET1 regulatory gene ((De-etiolated) using an RNAi technique with fruit-specific promoters, flavonoid levels were also elevated (Bovy et al., 2007). Transgenic lines, CHI × PAP1 lines, were generated through the fusion of tomato lines that exhibited overexpression of the CHI gene from onion and Arabidopsis PAP1 lines that induced the upregulation of early biosynthetic genes. Compared to wild-type tomatoes, the developed tomato line collected more anthocyanins in the fruit skin (Lim et al., 2017). The accumulation of flavonoids was primarily limited to the pericarp of the fruit. Various attempts for anthocyanin rich pulp tomatoes, "purple tomatoes" was made by ectopic expression of *Delila* and *Rosea 1* from snapdragon. The upregulation of these genes led to an increased accumulation of anthocyanins in both the peel and pulp tissues (Maligeppagol et al., 2013; Butelli et al., 2008).

4.3. Carrot

Carrots are a nutrient-dense food known for their pharmacological effects and health benefits. They are rich in natural bioactive compounds, including phenolics, carotenoids, polyacetylenes, and ascorbic acid. Some varieties have high anthocyanin concentrations, such as purple or black carrots (*Daucus carota* ssp. *sativus* var. atrorubens Alef), with dark purple to black roots. Black carrot anthocyanins have recently garnered attention as potential sources of natural dietary fibre. The characterization of genes responsible for regulating anthocyanin biosynthesis has been conducted in this study (Xu et al., 2019, 2020).

4.4. Other Fruits and Vegetables

Many released vegetable varieties with enhanced phytochemical content have been developed mainly through traditional breeding methods. For instance, the purple-fleshed radish variety Pusa-Jamuni, released in India, has high anthocyanin content. Bhu Krishna is a sweet potato with high anthocyanin content (90 mg per

100 g). Pusa Asita is a black-colored carrot with high anthocyanin content (Sil et al., 2023). Similarly, the litchi genetic stock NRCL-29 has fruit pericarp has abundant anthocyanins (94.35 mg per 100 g) (Lal et al., 2021). A red-colored radish Kasi Lohith, is a good source of antioxidants (80–100% higher than white radish), vitamin C, and phytochemicals like anthocyanin and phenolics (Singh et al., 2021). Transgenic "Mexican" lime plants have been successfully developed, overexpressing the Ruby MYB transcription factor or VvmybA1 within the fruit juice sac to produce anthocyanins in a citrus variety (Thilmony et al., 2022).

5. Conclusion

In summary, the implementation of biofortification techniques to enhance the phytochemical content in various crops is a crucial approach to augment the nutritional value of our food resources. Phytochemicals, including flavonoids and anthocyanins, are naturally occurring compounds present in plants that have demonstrated potential health advantages for human beings. Biofortification involves breeding crops to increase their levels of these beneficial phytochemicals. This can be achieved through traditional breeding methods, as well as through genetic engineering and biotechnology. The development of biofortified crops has the potential to address micronutrient deficiencies, which affects millions of people worldwide. It may also reduce the risk of chronic diseases such as cancer and cardiovascular disease. Several major crops, including wheat, rice, maize, and sweet potato, have been biofortified with phytochemicals with great success, and research is continuing for developing biofortified varieties of other crops. Overall, biofortification of phytochemicals in different crops is an important strategy for improving global health and nutrition, and it is a promising area of research for the future.

References

Abdel-Aal, E.S.M., J.C. Young, and I. Rabalski. 2006. Anthocyanin composition in black, blue, pink, purple, and red cereal grains. *J. Agric. Food Chem.*, 54: 4696–4704.

Blando, F., H. Berland, G. Maiorano, M. Durante, A. Mazzucato, M.E. Picarella, ... and Ø.M. Andersen. 2019. Nutraceutical characterization of anthocyanin-rich fruits produced by "Sun Black" tomato line. *Front Nutr.*, 6: 133.

Bovy, A., E. Schijlen, and R.D. Hall. 2007. Metabolic engineering of flavonoids in tomato (*Solanum lycopersicum*): The potential for metabolomics. *Metabolomic*, 3: 399–412.

Butelli, E., L. Titta, M. Giorgio, H.P. Mock, A. Matros, S. Peterek, ... and C. Martin. 2008. Enrichment of tomato fruit with health-promoting anthocyanins by expression of select transcription factors. *Nat. Biotechnol.*, 26: 1301–1308.

Dhua, S., K. Kumar, Y. Kumar, L. Singh, and V.S. Sharanagat. 2021. Composition, characteristics and health promising prospects of black wheat: A review. *Trends Food Sci.*, 112: 780–794.

Dooner, H.K., T.P. Robbins, and R.A. Jorgensen. 1991. Genetic and developmental control of anthocyanin biosynthesis. *Annu. Rev. Genet.*, 25: 173–199.

Ezekiel, R., N. Singh, S. Sharma, and A. Kaur. 2013. Beneficial phytochemicals in potato: A review. *Food Res. Int.*, 50: 487–496.

Francavilla, A., and I.J. Joye. 2020. Anthocyanins in whole grain cereals and their potential effect on health. *Nutrients*, 12: 2922.

Gamel, T.H., S.M.G. Saeed, R. Ali, and E.S.M Abdel-Aal. 2023. Purple wheat: Food development, anthocyanin stability, and potential health benefits. *Foods*, 12: 1358.

Garg, M., S. Kaur, A. Sharma, A. Kumari, V. Tiwari, S. Sharma, ... and M. Krishania. 2022. Rising demand for healthy foods - anthocyanin biofortified colored wheat is a new research trend. *Front Nutr.*, 9: 878221.

Garg, M., M. Chawla, V. Chunduri, R. Kumar, S. Sharma, N.K. Sharma, and S.P. Singh. 2016. Transfer of grain colors to elite wheat cultivars and their characterization. *J. Cereal Sci.*, 71: 138-144.

Goffman, F.D., and C.J. Bergman. 2004. Rice kernel phenolic content and its relationship with antiradical efficiency. *J. Sci. Food Agric.*, 84: 1235–1240.

Hagiwara, A., K. Miyashita, T. Nakanishi, M. Sano, S. Tamano, T. Kadota. ... and T. Shirai. 2001. Pronounced inhibition by a natural anthocyanin, purple corn color, of 2-amino-1-methyl-6-phenylimidazo [4, 5-b] pyridine (PhIP)-associated colorectal carcinogenesis in male F344 rats pretreated with 1,2-dimethylhydrazine. *Cancer Lett.*, 171: 17–25.

Ito, V.C., and L.G. Lacerda. 2019. Black rice (*Oryza sativa* L.): A review of its historical aspects, chemical composition, nutritional and functional properties, and applications and processing technologies. *Food Chem.*, 301: 125304.

Joshi-Saha, A., S.K. Sethy, G. Misra, G.P. Dixit, A.K. Srivastava, and A. Sarker. 2022. Biofortified legumes: Present scenario, possibilities, and challenges. *Field Crops Res.*, 279: 108467.

Kapoor, P., S. Sharma, A. Tiwari, S. Kaur, A. Kumari, H. Sonah, ... and M. Garg. 2023. Genome–transcriptome transition approaches to characterize anthocyanin biosynthesis pathway genes in blue, black, and purple wheat. *Genes*, 14: 809.

Kaur, S., V. Tiwari, A. Kumari, E. Chaudhary, A. Sharma, U. Ali, and M. Garg. 2022. Protective and defensive role of anthocyanins under plant abiotic and biotic stresses: An emerging application in sustainable agriculture. *J. Biotechnol.*, 361: 12–29.

Kumar, S., A. Mishra, and A.K Pandey. 2013. Antioxidant mediated protective effect of Parthenium hysterophorus against oxidative damage using *in vitro* models. *BMC Complement Altern. Med.*, 13: 1–9.

Lamberts, L., E. De Bie, G.E. Vandeputte, W.S. Veraverbeke, V. Derycke, W. De Man, and J.A. Delcour. 2007. Effect of milling on colour and nutritional properties of rice. *Food Chem.*, 100: 1496–1503.

Lago, C., E. Cassani, C. Zanzi, M. Landoni, R. Trovato, and R. Pilu. 2014a. Development and study of a maize cultivar rich in anthocyanins: Coloured polenta, a new functional food. *Plant Breed.*, 133: 210–217.

Lago, C., M. Landoni, E. Cassani, S. Atanassiu, E. Cantaluppi, and R. Pilu. 2014b. Development and characterization of a colored sweet corn line as a new functional food. *Maydica*, 59: 191–200.

Lal, N., and V. Nath. 2020. Potential and tolerant genotypes: Sources for improving productivity in litchi. *Editorial Board*. 1.

Lim, W., and J. Li. 2017. Synergetic effect of the Onion CHI gene on the PAP1 regulatory gene for enhancing the flavonoid profile of tomato skin. *Sci. Rep.*, 7: 12377.

Liu, X., W. Yang, B. Mu, S. Li, Y. Li, X. Zhou, ... and R. Chen. 2018. Engineering of "Purple Embryo Maize" with a multigene expression system derived from a bidirectional promoter and self-cleaving 2A peptides. *Plant Biotechnol. J.*, 16: 1107.

Liu, Y., K. Lin-Wang, R.V. Espley, L. Wang, Y. Li, Z. Liu, ... and A.C. Allan. 2019. StMYB44 negatively regulates anthocyanin biosynthesis at high temperatures in tuber flesh of potato. *Plant Biotechnol. J.*, 70: 3809–3824.

Liu, Weixin, Yi Feng, Suhang Yu, Zhengqi Fan, Xinlei Li, Jiyuan Li, and Hengfu Yin. 2021.The flavonoid biosynthesis network in plants. *Int. J. Mol. Sci.*, 22: 12824.

Luthra, S.K., V.K. Gupta, J.K. Tiwari, V. Kumar, V. Bhardwaj, S. Sood, … and S.K. Chakrabarti. 2020. Potato Breeding in India. *CPRI, Shimla Technical Bulletin*, 74: 214.

Maeda, H., T. Yamaguchi, M. Omoteno, T. Takarada, K. Fujita, K. Murata, ... and T. Ebitani. 2014. Genetic dissection of black grain rice by the development of a near isogenic line. *Breed. Sci.*, 64: 134–141.

Maligeppagol, M., G.S. Chandra, P.M. Navale, H. Deepa, P.R. Rajeev, R. Asokan, ... and N.K. Kumar. 2013. Anthocyanin enrichment of tomato (*Solanum lycopersicum* L.) fruit by metabolic engineering. *Curr. Sci.*, 72–80.

Mattoo, A.K., S.L. Dutt, S. Dwivedi, B. Singh, M. Garg, and R. Ortiz. 2022. Anthocyanin-rich vegetables for human consumption—Focus on potato, sweet potato, and tomato. *Int. J. Mol. Sci.*, 23: 2634.

Mazzucato, A., R. Papa, E. Bitocchi, P. Mosconi, L. Nanni, V. Negri, ... and F. Veronesi. 2008. Genetic diversity, structure, and marker-trait associations in a collection of Italian tomato (*Solanum lcopersicum* L.) landraces. *Theor. Appl. Genet.*, 116: 657–669.

Ogo, Y., K. Ozawa, T. Ishimaru, T. Murayama, and F. Takaiwa. 2013. Transgenic rice seed synthesizing diverse flavonoids at high levels: A new platform for flavonoid production with associated health benefits. *Plant Biotechnol. J.*, 11: 734–746.

Oh, J.H., Y.J. Lee, E.J. Byeon, B.C. Kang, D.S. Kyeoung, and C.K. Kim. 2018. Whole-genome resequencing and transcriptomic analysis of genes regulating anthocyanin biosynthesis in black rice plants. *3 Biotech*, 8: 1–13.

Oikawa, T., H. Maeda, T. Oguchi, T. Yamaguchi, N. Tanabe, K. Ebana, ... and T. Izawa. 2015. The birth of a black rice gene and its local spread by introgression. *The Plant Cell*, 27: 2401–2414.

Palafox-Carlos, H., J.F. Ayala-Zavala, and G.A. González-Aguilar. 2011. The role of dietary fiber in the bioaccessibility and bioavailability of fruit and vegetable antioxidants. *J. Food Sci.*, 76: R6–R15.

Panche, A.N., A.D. Diwan, and S.R. Chandra. 2016. Flavonoids: An overview. *J. Nutr. Sci.*, 5: e47.

Payyavula, R.S., R.K. Singh, and D.A. Navarre. 2013. Transcription factors, sucrose, and sucrose metabolic genes interact to regulate potato phenylpropanoid metabolism. *J. Exp. Bot.*, 64: 5115–5131.

Povero, G., S. Gonzali, L. Bassolino, A. Mazzucato, and P. Perata. 2011. Transcriptional analysis in high-anthocyanin tomatoes reveals synergistic effect of Aft and Atv genes. *J. Plant Physiol.*, 168: 270–279.

Shahidi, F., and Y. Pan. 2022. Influence of food matrix and food processing on the chemical interaction and bioaccessibility of dietary phytochemicals: A review. *Crit. Rev. Food Sci. Nutr.*, 62: 6421–6445.

Shao, Y., Z. Hu, Y. Yu, R. Mou, Z. Zhu, and T. Beta. 2018. Phenolic acids, anthocyanins, proanthocyanidins, antioxidant activity, minerals, and their correlations in non-pigmented, red, and black rice. *Food Chem.*, 239: 733–741.

Sharma, A., M. Yadav, A. Tiwari, U. Ali, M. Krishania, M. Bala, ... and M. Garg. 2023. A comparative study of colored wheat lines across laboratories for validation of their phytochemicals and antioxidant activity. *J. Cereal Sci.*, 112: 103719.

Sharma, S., P. Khare, A. Kumar, V. Chunduri, A. Kumar, P. Kapoor, ... and M. Garg. 2020. Anthocyanin-biofortified colored wheat prevents high fat diet-induced alterations in mice: Nutrigenomics studies. *Mol. Nutr. Food Res.*, 64: 1900999.

Sharma, S., V. Chunduri, A. Kumar, R. Kumar, P. Khare, K.K. Kondepudi, and M. Garg. 2018. Anthocyanin biofortified colored wheat: Nutritional and functional characterization. *PloS One*, 13: e0194367.

Shen, Y., L. Jin, P. Xiao, Y. Lu, and J.S. Bao. 2009. Total phenolics, flavonoids, antioxidant capacity in rice grain and their relations to grain color and grain size. *J. Cereal Sci.*, 49: 106–111.

Sil, D., T. Bhattacharjee, and S. Saha. 2023. Biofortification of vegetable crops. *Vigyan Varta*, 4: 137–140.

Singh, B.K. 2021. Radish (*Raphanus sativus* L.): Breeding for higher yield, better quality, and wider adaptability. *Advances in Plant Breeding Strategies: Vegetable Crops: Bulbs, Roots and Tubers*, 8: 275–304.

Thilmony, R., K. Dasgupta, M. Shao, D. Harris, J. Hartman, L.A. Harden, ... and J.G. Thomson. 2022. Tissue-specific expression of Ruby in Mexican lime (*C. aurantifolia*) confers anthocyanin accumulation in fruit. *Front. Plant Sci.*, 13: 945738.

Tiwari, S.C., and N. Husain. 2017. Biological activities and role of flavonoids in human health–A. *Indian J. Sci. Res.*, 12: 193–196.

Tiwari, V., S. Sharma, A. Tiwari, B. Sheoran, S. Kaur, A. Sharma, ... and M. Garg. 2023. Effect of dietary anthocyanins on biomarkers of type 2 diabetes and related obesity: A systematic review and meta-analysis. *Crit. Rev. Food Sci. Nutr.*, 1–18.

Toufektsian, M.C., M. de Lorgeril, N. Nagy, P. Salen, M.B. Donati, L. Giordano, ... and C. Martin. 2008. Chronic dietary intake of plant-derived anthocyanins protects the rat heart against ischemia-reperfusion injury. *J. Nutr.*, 138: 747–752.

Velderrain-Rodríguez, G.R., H. Palafox-Carlos, A. Wall-Medrano, J.F. Ayala-Zavala, C.O. Chen, M. Robles-Sánchez, ... and G.A. González-Aguilar. 2014. Phenolic compounds: Their journey after intake. *Food Funct.*, 5: 189–197.

Wei, Q., Q.Y. Wang, Z.H. Feng, B. Wang, Y.F. Zhang, and Q. Yang. 2012. Increased accumulation of anthocyanins in transgenic potato tubers by overexpressing the 3GT gene. *Plant Biotechnol. Rep.*, 6: 69–75.

Xiao, Z.P., Z.Y. Peng, M.J. Peng, W.B. Yan, Y.Z. Ouyang, and H.L. Zhu. 2011. Flavonoids health benefits and their molecular mechanism. *Mini-Rev. Med. Chem.*, 11: 169–177.

Xu, Z.S., Q.Q. Yang, K. Feng, X. Yu, and A.S. Xiong. 2020. DcMYB113, a root-specific R2R3-MYB, conditions anthocyanin biosynthesis and modification in carrot. *Plant Biotechnol. J.*, 18: 1585–1597.

Xu, Z.S., Q.Q. Yang, K. Feng, and A.S. Xiong. 2019. Changing carrot color: Insertions in DcMYB7 alter the regulation of anthocyanin biosynthesis and modification. *Plant Physiol.*, 181: 195–207.

Yoshihara, N., T. Imayama, M. Fukuchi-Mizutani, H. Okuhara, Y. Tanaka, I. Ino, and T. Yabuya. 2005. cDNA cloning and characterization of UDP-glucose: Anthocyanidin 3-O-glucosyltransferase in Iris hollandica. *Plant Sci.*, 169: 496–501.

Zhang, Q., S.S. Gangurde, X. Yang, and C. Zhao. 2023. Roles of flavonoids in crop quality improvement and response to stresses. *Front. Plant Sci.*, 14: 1210666.

Zhang, S., F. Sun, C. Zhang, M. Zhang, W. Wang, C. Zhang, and Y. Xi. 2022. Anthocyanin biosynthesis and a regulatory network of different-colored wheat grains revealed by multiomics analysis. *J. Agric. Food Chem.*, 70: 887–900.

Zhu, Q., S. Yu, D. Zeng, H. Liu, H. Wang, Z. Yang, X. Xie, R. Shen, J. Tan, H. Li, X. Zhao, Q. Zhang, Y. Chen, J. Guo, L. Chen, and Y.G. Liu. 2017. Development of "Purple Endosperm Rice" by engineering anthocyanin biosynthesis in the endosperm with a high-efficiency transgene stacking system. *Mol. Plant.*, 10: 918–929.

Žofajová, A., M. Havrlentová, M. Ondrejovič, M. Juraška, B. Michalíková, and Ľ. Deáková. 2017. Variability of quantitative and qualitative traits of colored winter wheat. *Agriculture*, 63: 102.

CHAPTER
7

Vitamin A: A Major HarvestPlus Target for Outstanding Results

M.T. Labuschagne[0000-0003-0593-2678]
Department of Plant Sciences, University of the Free State, Bloemfontein, South Africa

1. Introduction

Vitamin A deficiency (VAD) remains a nutrient deficiency that occurs widely in resource-poor populations in developing countries, where it is a major cause of preventable blindness and impaired immune systems (Palmer et al., 2017). It is therefore crucial to increase the provitamin A intake of vulnerable people. The major form of vitamin A in plants is provitamin A carotenoids (Harrison, 2022). Carotenoids such as β-carotene, β-cryptoxanthin, and α-carotene have provitamin A activity. The β-carotenes produce the most provitamin A, and therefore most provitamin A biofortification projects have been focused on increased β-carotene accumulation in foods (Eggersdorfer and Weiss, 2018). Biofortification of some crops such as maize, cassava, and sweet potato with provitamin A has been highly successful. The aim of this paper was to review (1) the current knowledge on carotenoids, including their biosynthesis, accumulation and degradation; (2) the effects of vitamin A deficiency on humans; (3) biofortification through conventional and transgenic approaches; (4) bioaccessibility and bioavailability of carotenoids; (5) post-harvest stability of carotenoids; (6) gene-editing as a strategy for carotenoid biofortification; and (7) maize, cassava and sweet potato as specific case studies for provitamin A carotenoid biofortification.

2. Carotenoids and Their Synthesis

Carotenoids occur naturally in plants and are a group of pigmented lipophilic isoprenoids that are structurally diverse (Martínez-Cámara et al., 2021).

*Corresponding author: labuscm@ufs.ac.za

Biosynthesis of carotenoids are localized in plastids, where they are a part of plant photosynthetic systems. In photosynthetic tissue, the carotenoids are mainly located in the chloroplasts, and in non-photosynthetic tissue in the chromoplasts. They play an essential role in light harvesting and photoprotection from oxidation (Rodriguez-Concepcion et al., 2018).

The general carotenoid biosynthesis pathway is indicated in Figure 1 (Badejo, 2018). Regulatory nodes within the carotenoid biosynthetic pathway influence the control of the movement of metabolites into and through the pathway. Biosynthetic steps within the pathway can limit flux, and in many crop species these steps can be targeted for the genetic engineering of carotenoid pathways (Giuliano, 2017).

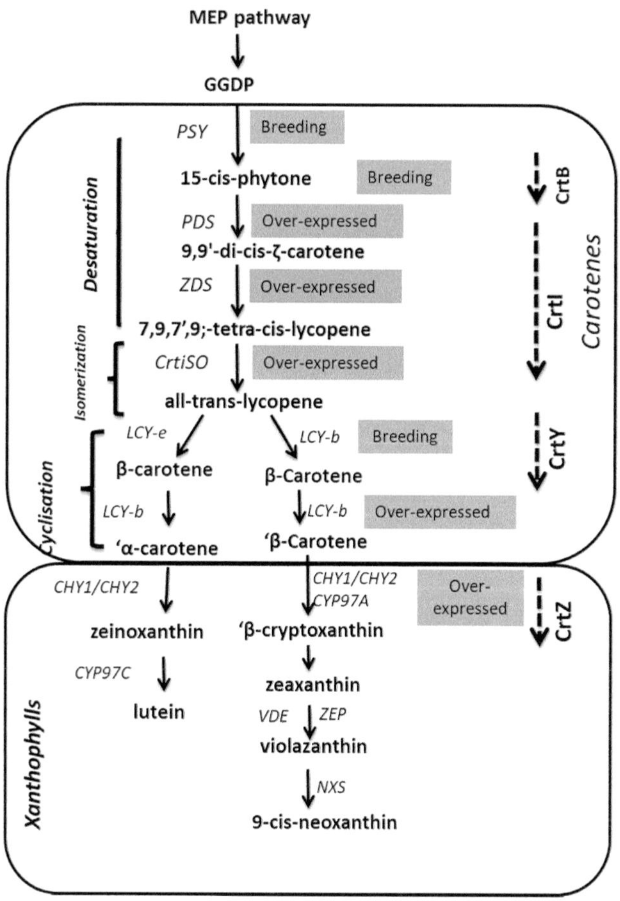

Figure 1: The carotenoid biosynthesis pathway found in most plants. GGDP – Geranylgeranyl diphosphate is derived from the plastidial MEP (Methylerythritol 4 phosphate) pathway; PSY – Phytoene synthase; PDS – Phytoene desaturase; ZDS – z-carotene desaturase; CRITSO – Carotenoid isomerase; LCY-b – (lycopene epsilon cyclase-b), Lycopene b-cyclase (Adapted from Badejo, 2018, with permission).

Genes that are targeted for manipulation of the biosynthesis pathway include *CrtB* and *LCY* (in maize for example) (Liu et al., 2015).

Classification of provitamin A carotenoids is done based on chemical structure, where they are divided into two main classes: xanthophylls and carotenes. The carotenes comprise of β-carotene, α-carotene, and lycopene, which are non-oxygenated hydrocarbons; and also the xanthophylls which contain oxygen functional groups of carotenes. They include β-cryptoxanthin, lutein, and zeaxanthin (Rodriguez-Amaya, 2016; Harrison and Curley, 2016). Typically, carotenes are orange pigments while xanthophylls are yellow. Several types of xanthophylls exist, such as α- and β-cryptoxanthin, zeaxanthin, lutein, flavoxanthin, neoxanthin, and violaxanthin (Berman et al., 2017). The main carotenes are α-, β-, and γ-carotene. The β-carotene has about 50% higher vitamin A activity than α-carotene and β-cryptoxanthin (Eggersdorfer and Wyss, 2018; Drapal et al., 2019). Vitamin A is fat-soluble, consisting of a group of three unsaturated hydrocarbon compounds, these being retinol, retinal, and retinoic acid. These are the active forms of vitamin A. Humans and animals cannot produce these three compounds and have to obtain them from a dietary source (Figure 1) (Morelli et al., 2022).

3. Vitamin A Deficiency

Provitamin A carotenoids provide up to 35% (in western societies) and 80% (in developing countries) of vitamin A intake, showing its importance as source of vitamin A (Rodriguez-Amaya, 2016). Worldwide VAD is amongst the most prominent micronutrient deficiencies. It is a significant problem in the global public health in low- and middle-income countries. Early childhood, pregnancy, and breastfeeding are phases in the human life with high nutritional demand, and the effects of VAD are the most prominent at these stages (Wiseman et al., 2017). Countries characterized by food insecurity, poverty, high levels of infectious diseases, and poor infrastructure are usually more prone to VAD (Tanumihardjo et al., 2016).

VAD is the main cause of diseases linked to poor immune systems, development disorders, as well as preventable blindness (Palmer et al., 2017). VAD is one of the most important causes of night blindness and xerophthalmia, which impair vision. It also reduces responses of the immune system, leading to high infection rates of diseases such as measles and diarrhea (Watkins and Pogson, 2020). VAD in the diet contributes significantly to infectious diseases in children, leading to high morbidity and mortality (Milani et al., 2017).

4. Accumulation of Carotenoids in Crops

Carotenoid accumulation in plastids is determined by the balance between their biosynthesis, sequestration, and degradation. Suitable structures that are available for carotenoid storage and appropriation contribute to their accumulation

(Sadali et al., 2019; Torres-Montilla and Rodriguez-Concepcion, 2021; Morelli et al., 2022).

The largest amount of carotenoids is accumulated in two of the plastid types, chloroplasts and chromoplasts. In the chloroplast, the carotenoids are mostly stored in protein and photosystem complexes, which are associated with the harvesting of light. In the chromoplasts, which are the non-photosynthetic plastids, carotenoids are often stored as lipid droplets in specialized sub-organullar compartments. Chloroplasts and chromoplasts have the common feature that they are specialized for the production and storage of carotenoids, but chromoplasts have a diverse ultrastructure and carotenoid composition, in contrast with chloroplasts. Chromoplasts occur as macro components in vegetables, fruits, and storage organs in non-photosynthetic tissues, where they cause the characteristic colors of fruit and vegetables, while also contributing to their flavor and aroma (Sadali et al., 2019).

Carotenoids also accumulate in other plastid types, such as xanthophylls in cereal grain and starch-storage plastids and amyloplasts in tubers. In orange sweet potato and carrots, carotenoids are deposited and stored in solid-crystalline form, which leads to highly stable β-carotene compared with various fruits, green leafy vegetables and cereal grains. In contrast to this, β-carotene in Golden Rice was found to have high bioavailability due to it being present in lipid droplets. This was, however, associated by high rates of carotenoid degradation (Schaub et al., 2017).

5. Carotenoid Degradation

Carotenoids are highly prone to oxidation because they are unsaturated compounds. This causes a loss of vitamin A activity (Nkhata et al., 2020). Light and heat contribute to degradation. Carotenoids are oxidatively degraded by reactive oxygen species (ROS) due to the interaction of lipoxygenases with polyunsaturated fatty acids. On the other hand, carotenoid oxidative degradation is prevented when ROS are scavenged or quenched by tocopherols. Carotenoids esterified with fatty acids lead to an increase in the lipophilicity of the molecules. They can also be incorporated into plastoglobules and fibrils, which are rich in lipids. The presence of fibrillin protein, which is a structural component of carotenoid-sequestering fibrils, is associated with the accumulation of carotenoids (Watkins and Pogson, 2020).

6. Provitamin A Biofortification

6.1. The Application of Conventional Breeding for Biofortification

For the biofortification by conventional breeding, sufficient genetic variation in carotenoids is necessary. Accessions with high carotenoid concentrations can be

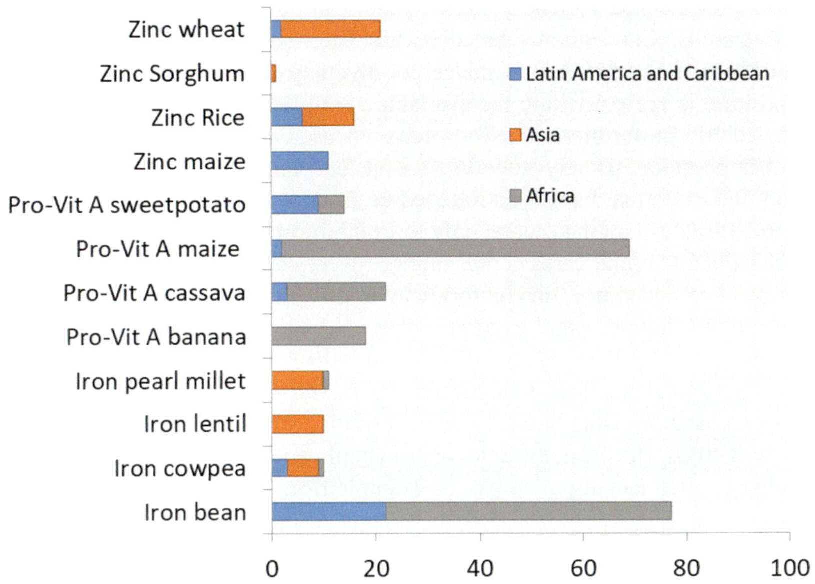

Figure 2: Vitamin A biofortified crop varieties released in the world by 2021 by region/continent (harvestplus.org/biofortification-hub).

Figure 3: Yellow cassava: Rich in β-carotene, a vitamin A precursor.

crossed with adapted varieties, and selection can be done to obtain high levels of carotenoids. Conventional breeding has been successfully applied for the biofortification of cassava and maize, as diversity for carotenoid content and composition is present within the available germplasm (Giuliano, 2017; Siwela et al., 2020). Biofortification efforts have focused on three micronutrients of global importance, namely vitamin A, iron, and zinc. Figure 2 shows all the biofortified crops that have been released in the world by 2021, with provitamin A maize varieties ranking second only to iron biofortified beans, and provitamin A banana/plantain and cassava and orange sweet potato featuring prominently. In terms of provitamin A, this biofortification drive has led to the official release in several developing countries of provitamin A rich orange maize, orange-flesh sweet potato and yellow cassava (Bouis and Saltzman, 2017) (Figure 3).

Conventional breeding strategies for provitamin A biofortification so far have largely focused on using natural genetic variation to increase the provitamin A content of crops. However, the amount of natural variation required for successful breeding is often lacking in crops, for example rice.

6.2. Biofortification through Biotechnology

Genetic engineering usually represents metabolic engineering of the biochemical pathway, but it requires knowledge of the metabolism and promotors of the specific vitamin for biofortification (Wurtzel, 2019; Jaramillo et al., 2022). The knowledge currently available and modern molecular tools have enabled a large number of individual, or combined uses of these tools for biofortification of various plant-derived foods, and the metabolic engineering of carotenoids, as reviewed in several papers (Giuliano, 2017; Rodriguez-Concepcion et al., 2018; Feder et al., 2019; Watkins and Pogson, 2020; Zheng et al., 2020). This is very useful in crops such as rice and wheat, where there is no natural variation for provitamin A content (Guiliano, 2017).

Golden Rice is the best example of a transgenic provitamin A content biofortified crop. Initially a rice line was developed which expressed a daffodil phytoene synthase (*PSY*) gene. This caused phytoene accumulation in the rice endosperm (Paine et al., 2005). Golden Rice was the first transgenic provitamin A enriched crop, which lead to the transgenic biofortification of other staple crops such as maize, wheat, and banana for provitamin A (Watkins and Pogson, 2020). Golden Rice is an example of successful carotenoid biotechnology, where this rice accumulates β-carotene in the seed endosperm (Paine et al., 2005). The efforts to biofortify more rice varieties for increased carotenoid content has since continued, by introgressing this trait into local rice varieties. However, so far no transgenic provitamin A crops have been approved for release to farmers (De Steur et al., 2017), due to public concerns regarding GM crops, as well as numerous regulatory issues. In the Philippines, Golden Rice was only recently approved for planting, a process which took two decades (Welsch and Li, 2022).

Related unintentional effects such as metabolic homeostasis and altered hormone levels due to genetic engineering of carotenoid content have been reported, which have created new challenges. This is caused by shared precursors of carotenoids with other important metabolites which are involved in regulation of plant development, photosynthesis, photoprotection, and interactions with the environment. Carotenoid-derived products such as abscisic acid and some apocarotenoids are also involved in environmental interactions and plant development regulation (Morelli et al., 2023).

Usually, in the design of biotechnological approaches to biofortify crops with carotenoids, two phases are considered. Firstly, there is the manipulation of carotenoid levels in plant tissues and secondly an evaluation of the shelf life (carotenoid degradation) and the bioaccessibility of the carotenoids in the product (Sun et al., 2021). It was reported that genotypes developed from β-carotene manipulation on whole plant level had several other positive characteristics such as improved tolerance to abiotic stress conditions, higher fruit yield and improved shelf life (Diretto et al., 2020; Mi et al., 2022). Unfortunately, biofortification is often not effective due to interfering with associated metabolic and hormonal networks, which often has unpredicted consequences with devastating effects. The regulation of the plant carotenoid pathway is often still poorly understood in the context of the total metabolism and the control networks of total development, and this problem needs to be overcome for effective carotenoid biofortification (Morelli et al., 2023).

7. Bioavailability and Bioaccessibility of Carotenoids

In the process of biofortification, it is not only a question of increasing the carotenoid content, as the bioavailability and bioaccessibility of carotenoids from foods are equally important and will determine the role they will play in human diets. The proportion of carotenoids consumed in the diet that can be absorbed, transported, and utilized or stored in the human body for normal physiological functions, is defined as the bioavailability (Rodriguez-Amaya, 2016). The carotenoids are lipophilic isoprenoids. To become bioavailable, the isoprenoids must be released from the food matrix and incorporated into lipid-containing intestinal micelles which can be interspersed in water. Bioaccessibility is a component of bioavailability. In the case of carotenoids, it will become bioaccessible if it is released from the food matrix during digestion. It becomes available for absorption and release into the blood stream after incorporation into micelles (Rodriguez-Concepcion et al., 2018; Meléndez-Martínez et al., 2022).

The matrix in which carotenoids occur in plant cells influence their bioaccessibility (Watkins and Pogson, 2020; Zheng et al., 2020). In general, carotenoids associated with lipids have a higher bioaccessibility because of easier micellization, while the crystalline or aggregated carotenoids are less bioaccessible. High stability is often equated with poor bioaccessibility because enhanced stability is caused by reduced effects of external factors that cause carotenoid degradation, but its release from the food matrix is also reduced (Andersen et al., 2021).

8. Post-harvest Stability

Free carotenoids are generally unstable compounds, and rapid decay of carotenoids is caused mainly by oxidative degradation. The cytosol within the cell reduces exposure and reactions of carotenoids with oxygen or oxygen radicals. If target carotenoids are produced and deposited in these structures, the carotenoid half-life can be increased (Andersen et al., 2021). Tocopherols are powerful antioxidants with vitamin E activity. They are metabolically related to carotenoids, and there is co-expression of genes for tocopherol and carotenoid biosynthetic enzymes. In seeds of sorghum and Arabidopsis, β-carotene accumulation and stability were increased with an increase in tocopherol expression (Sun et al., 2021). Plastoglobules are specialized storage compartments. Together with high tocopherol levels in artificial chromoplasts of phytoene synthase gene (*crtB*)-expressing leaves, it should increase the stability of β-carotene, in particular, and carotenoids in general (Morelli et al., 2022). There are also some other molecular tools available that promote formation of chromoplasts. The ORANGE (OR) chaperone is one of these tools, and it has been extensively studied in recent years (Feder et al., 2019; Torres-Montilla and Rodriguez-Concepcion, 2021). The OR protein prevents PSY degradation, thereby promoting its activity and accumulation. It also prevents the metabolism of β-carotene by unknown mechanisms, leading to increased β-carotene accumulation (Feder et al., 2019).

9. Gene-editing with CRISPR-Cas

The regulatory issues of GMO could be overcome by CRISPR–Cas-mediated genome editing. Genome editing products are being exempted from the existing restrictive GMO regulations in a number of countries (Sprink et al., 2022). CRISPR–Cas technique has already been successfully used for the improvement of carotenoid content by precise nucleotide changes (Zhu et al., 2020; Zheng et al., 2021).

A Golden Rice gene cassette was recently introduced into a rice chromosomal region with the use of CRISPR–Cas. There are no adverse effects on the host if transgenes are introduced into this chromosomal region (Dong et al., 2020). Bananas enriched with β-carotene were also developed using CRISPR-Cas. The tested ripe banana fruit tissue and Golden Rice had comparable β-carotene levels (Kaur et al. 2020).

10. Case Studies for Provitamin A Biofortification

10.1. Maize

In sub-Saharan Africa people consume large quantities of maize meal daily (up to 330 g/person/day). In poor communities, maize is often consumed at every meal. This creates the opportunity to substitute white maize meal with meal from provitamin A maize. This could contribute towards meeting the daily vitamin A requirement of these communities (Simpungwe et al., 2017). In yellow and orange maize varieties, provitamin A content ranges from less than two to 25 $\mu g\ g^{-1}$ (Nkhata et al., 2020), while commonly grown and consumed maize does not have the needed provitamin A carotenoids to manage VAD (Galani et al., 2020). Due to the continued consumption of maize with low levels of carotenoids in sub-Saharan countries, VAD remains a severe public health problem in this region (Zuma et al., 2018).

Newly developed biofortified maize cultivars should have at least 15 $\mu g\ g^{-1}$ of provitamin A, which is the target set by HarvestPlus. Biofortified maize should have substantial amounts of especially β-carotene, α-carotene, and β-cryptoxanthin. CIMMYT currently has hundreds of provitamin A genotypes available, with provitamin A levels varying between 2 $\mu g\ g^{-1}$ and >25 $\mu g\ g^{-1}$ (Prasanna et al., 2020).

In contrast to some other cereals, maize has the ability to synthesize provitamin A as well as non-provitamin A or xanthophylls. The presence of molecular markers for genes that confer increased provitamin A content allows the tracking of these favorable alleles. Grain provitamin A concentrations were found to be significantly correlated with *LcyE* and *CrtRB1* expression. Provitamin A and protein quality were increased in two QPM inbred lines through successful marker-assisted backcrossing of favorable alleles of *CrtRB1* from a provitamin A donor into these lines (Liu et al., 2015). Likewise, multinutrient maize hybrids with high levels of provitamin A, lysine and tryptophan content were developed through marker-assisted stacking of *CrtRB1*, *LcyE,* and *opaque-2* loci in a QPM genetic background (Zunjare et al., 2018).

Suwarno et al. (2014) assessed orange kernel color in maize, and the relative color intensity and found that phenotypic selection for kernel color should be relatively successful. They reported five common quantitative trait loci (QTLs) and six rare QTLs associated with orange colour intensity in maize. Half of these QTLs coincided with genes involved in carotenoid biosynthesis.

A number of genome-wide association studies (GWAS) of various carotenoids in maize grain have been reported. It was found that the MEP and carotenoid pathways in maize are encoded by 59 genes. Genome-wide associations with various carotenoid traits in maize grain was found with four of these genes, *LcyE*, *CrtRB1*, zeaxanthin epoxidase 1 (*zep1*), and e-ring hydroxylase (*lut1*). On a genome-wide level, variation in maize grain carotenoids have been associated with expression of four (*vp5, im1, dxs2,* and *dxs3*) of the five genes that occur upstream of lycopene cyclization (Azmach et al., 2018; Baseggio et al., 2020). Diepenbrock et al. (2021) reported that 75% of β-carotene variation attributed to QTL were explained by only four genes (*LcyE, CrtRB1, dxs2*, and *psy1*), while three of the four genes (*LcyE, dxs2,* and *psy1*) explained 52% of α-cryptoxanthin variation; and just two genes (*LcyE* and *dxs2*) explained 64% of α-carotene variation. Their study indicated that selection of the related alleles within these four major-effect genes can increase the β-carotene and β-cryptoxanthin levels simultaneously, and potentially also the α-carotene levels.

Provitamin A carotenoids in transgenic maize can be up to 169-fold more than that of normal maize, addressing 100% of daily vitamin A requirement, but commercial release of a transgenic line is expensive, tedious, and time-consuming (Pixley, 2010).

10.2. Cassava

Cassava (*Manihot esculenta* Crantz) is a major staple crop in tropical regions, where its starchy roots are consumed, but it lacks natural carotenoid content. In order to alleviate VAD, provitamin A biofortification of cassava is a continuous process (Jaramillo et al., 2022). Initial efforts to improve the carotenoid content in cassava was mainly done through conventional breeding where the focus was on identifying and characterizing useful genetic variation in the available germplasm. Nassar et al. (2007) reported high levels of ß-carotene (4 $\mu g\ g^{-1}$) and lycopene (5 $\mu g\ g^{-1}$) in some landraces in Brazil, which is the center of cassava diversity.

Conventional cassava breeding has only managed to achieve a moderate provitamin A carotenoid content, which has been limiting the contribution of cassava to providing adequate dietary vitamin A levels. This lead to the implementation of strategies to increase carotenoid content through marker assisted selection for genes from the carotenoid biosynthesis pathway. The ORANGE protein (OR) gene, which promotes the accumulation of carotenoids and their stability in several plants, has been emphasized in recent years. Jaramillo et al. (2022) reported four different OR variants in yellow cassava roots.

There has been limited progress in the development of provitamin A cassava varieties as breeding efforts in the past have focused mainly on developing high-yielding cassava with good agronomic performance. One of the constraints in developing provitamin A cassava is that dry matter content is reduced when carotenoid content in roots is increased. The genetics underlying this relationship is still unknown. Alternative breeding approaches, such as genetic transformation, have been suggested to overcome this problem, but have not been very successful (Telengech et al., 2014). Generally, farmers prefer high dry matter content cassava, which could reduce the farmers' preference for provitamin A cassava varieties. This causes a dilemma for breeders (Njukwe et al., 2013).

10.3. Sweet Potato

Sweet potato [*Ipomoea batatas* (L.) Lam.] ranks seventh on the list of most important food crops in the world (Rodriguez-Bonilla et al., 2014). Over the last 25 years there has been a concerted effort to breed and deliver and make people aware of orange fleshed sweet potato (OFSP), especially in sub-Saharan Africa (SSA). By consuming 120 g of biofortified OFSP per day, young children can meet their daily vitamin A requirement (Low et al., 2020). There is a sweet potato breeding network in SSA, and 14 of these countries have bred more than 100 new cultivars from 2009, of which 61% were OFSP (Okello et al., 2019). Sweet potato is a hexaploid and is also highly heterozygous. Due to the complexity of the genome, cross incompatibility and sterility, conventional breeding techniques have not always been successful. Molecular breeding and genetic engineering have been used by scientists in various countries as alternatives to improve quantity and quality of carotenoids in sweet potato (Zubair Khan et al., 2022). Genomic tools for sweet potato improvement were increased in recent years with the release of reference genomes of wild diploids *I. trifida* and *I. triloba* (Wu et al., 2018) and the release of the first draft genome of *I. batatas* (Yang et al., 2017).

Beta-carotene makes up 80–92% of the carotenoids in OFSP, but it also contains several epoxy carotenoids unique to sweet potato tubers (Zubair Khan et al., 2022). Kim et al. (2013) did some of the early work to enhance provitamin A by isolating and analyzing the orange (OR) gene. White fleshed sweet potato transformed by this gene had a 10-fold increase in the amount of β-carotene and total carotenoids. Recent genetic engineering of sweet potato with the *IbOr* gene showed that it can be effectively used for improvement of total carotenoids, and especially β-carotene. CRISPR-Cas9 was also suggested as an effective strategy for carotenoid biofortification (Zubair Khan et al., 2022). Low et al. (2020) suggested that an improvement of the value chain, starting from the planting materials through to processing and the consumer, and increased awareness of consumers across the board, integration of OFSP into local diets, and availability of OFSP-based products will contribute towards using this biofortified crop to its full potential.

11. Conclusion

Knowledge on all aspects of carotenoids such as their synthesis, storage, and roles in metabolic and hormonal networks of plants has significantly increased in recent years. Conventional breeding has allowed the successful release of provitamin A crops such as orange sweet potato and maize, and yellow bananas and cassava, amongst others. Transgenic approaches have led to the release of biofortified crops like Golden Rice, albeit after a difficult journey. A model engineering strategy could be tailored for vitamin A biofortification. There are still regulatory and consumer obstacles in terms of GM crops, and gene editing may be the most appropriate way to overcome this, as genome editing products are being exempted from existing GM regulations in many countries. It is important that post-harvest stability and bioaccessibility are monitored in the engineered products. Vitamin A biofortification has come a long way, and there are many new and exciting possibilities for biofortification strategies for a whole range of crops, that could contribute towards the alleviation of VAD, especially in the developing regions of the world.

References

Andersen, T.B., B. Llorente, L. Morelli, S. Torres-Montilla, G. Bordanaba-Florit, F.A. Espinosa, M.R. Rodriguez-Goberna, N. Campos, B. Olmedilla-Alonso, M.J. Llansola-Portoles, A.A. Pascal, and M. Rodriguez-Concepcion. 2021. An engineered extraplastidial pathway for carotenoid biofortification of leaves. *Plant Biotechnol. J.*, 19: 1008–1021.

Azmach, G., A. Menkir, C. Spillane, and M. Gedil. 2018. Genetic loci controlling carotenoid biosynthesis in diverse tropical maize lines. *G3*, 8: 1049–1065.

Badejo, A.A. 2018. Elevated carotenoids in staple crops: The biosynthesis, challenges, and measures for target delivery. *J. Genet. Eng. Biotechnol.*, 16: 553–562.

Baseggio, M., M. Murray, M. Magallanes-Lundback, N. Kaczmar, J. Chamness, E.S. Buckler, M.E. Smith, D. DellaPenna, W.F. Tracy, and M.A. Gore. 2020. Natural variation for carotenoids in fresh kernels is controlled by uncommon variants in sweet corn. *Plant Genome*, 13: e20008.

Berman, J., U. Zorrilla-López, G. Sandmann, T. Capell, P. Christou, and C. Zhu. 2017. The silencing of carotenoid β-hydroxylases by RNA interference in different maize genetic backgrounds increases the β-carotene content of the endosperm. *Int. J. Mol. Sci.*, 18: 2515.

Bouis, H.E., and A. Saltzman. 2017. Improving nutrition through biofortification: A review of evidence from HarvestPlus, 2003 through 2016. *Global Food Sec.*, 12: 49–58.

De Steur, H., S. Mehta, X. Gellynck, and J.L. Finkelstein. 2017. G.M. biofortified crops: Potential effects on targeting the micronutrient intake gap in human populations. *Curr. Op. Biotechn.*, 44: 181–188.

Diepenbrock, C.H., D.C. Ilut, M. Magallanes-Lundback, C.B. Kandianis, A.E. Lipka, P.J. Bradbury, J.B. Holland, J.P. Hamilton, E. Wooldridge, B. Vaillancourt, E. Góngora-Castillo, J.G. Wallace, J. Cepela, M. Mateos-Hernandez, B.F. Owens, T. Tiede, E.S. Buckler, T. Rocheford, C.B. Buell, M.A. Gore, and D. DellaPenna. 2021. Eleven biosynthetic genes explain the majority of natural variation in carotenoid levels in maize grain. *The Plant Cell,* 33: 882–900.

Diretto, G., S. Frusciante, C. Fabbri, N. Schauer, L. Busta, Z. Wang, A.J. Matas, A.K.C. Fiore, J. Rose, A.R. Fernie, R. Jetter, B. Mattei, J. Giovannoni, and G. Giuliano. 2020. Manipulation of b-carotene levels in tomato fruits results in increased ABA content and extended shelf life. *Plant Biotech. J.*, 18: 1185–1199.

Dong, O.X., S. Yu, R. Jain, N. Zhang, P.Q. Duong, C. Butler, Y. Li, A. Lipzen, J.A. Martin, K.W. Barry, J. Schmutz, L. Tian, and P.C. Ronald. 2020. Marker-free carotenoid-enriched rice generated through targeted gene insertion using CRISPR-Cas9. *Nat. Commun.*, 11: 1178.

Drapal, M., De Barros, E. Carvalho, T.M. Ovalle Rivera, L.A. Becerra Lopez-Lavalle, and P.D. Fraser. 2019. Capturing biochemical diversity in cassava (*Manihot esculenta* Crantz) through the application of metabolite profiling. *J. Agric. Food Chem.*, 67: 986–993.

Eggersdorfer, M., and A. Wyss. 2018. Carotenoids in human nutrition and health. *Arch. Biochem. Biophys.*, 652: 18–26.

Feder, A., N. Chayut, A. Gur, Z. Freiman, G. Tzuri, A. Meir, U. Saar, S. Ohali, F. Baumkoler, A. Gal-On, Y. Shnaider, D. Wolf, N. Katzir, A. Schaffer, J. Burger, L. Li, and Y. Tadmor. 2019. The role of carotenogenic metabolic flux in carotenoid accumulation and chromoplast differentiation: Lessons from the melon fruit. *Front. Plant Sci.*, 10: 1250.

Galani, Y.J.H., C. Orfila, and Y.Y. Gong. 2020. A review of micronutrient deficiencies and analysis of maize contribution to nutrient requirements of women and children in Eastern and Southern Africa. *Crit. Rev. Food Sci. Nutr.*, 62: 1568–1591.

Giuliano, G. 2017. Provitamin A biofortification of crop plants: A gold rush with many miners. *Cur. Op. Biotech.*, 30: 169–180.

Harrison, E.H., and R.W. Curley. 2016. Carotenoids and retinoids: Nomenclature, chemistry, and analysis. *Subcell. Biochem.*, 81: 1–19.

Harrison, E.H. 2022. Carotenoids, β-apocarotenoids, and retinoids: The long and the short of it. *Nutrients*, 14: 1411.

Jaramillo, A.M., S. Sierra, P. Chavarriaga Aguirre, D.K. Castillo, A. Gkanogiannis, L.A.B. Lopez-Lavalle, J.P. Arciniegas, T. Sun, L. Li, R. Welsch, E. Boy, and D. Alvarez. 2022. Characterization of cassava ORANGE proteins and their capability to increase provitamin A carotenoids accumulation. *PLoS One*, 17: e0262412.

Kaur, N., A. Alok, Shivani, P. Kumar, N. Kaur, P. Awasthi, S. Chaturvedi, P. Pandey, A. Pandey, A.K. Pandey, and S. Tiwari. 2020. CRISPR/Cas9 directed editing of lycopene epsilon-cyclase modulates metabolic flux for β-carotene biosynthesis in banana fruit. *Metab. Eng.*, 59: 76–86.

Kim, S.H., Y.O. Ahn, M-J. Ahn, J.C. Jeong, H.S. Lee, and S-S. Kwak. 2013. Cloning and characterization of an Orange gene that increases carotenoid accumulation and salt stress tolerance in transgenic sweet potato cultures. *Plant Physiol. Biochem.*, 70: 445–454.

Liu, L., D. Jeffers, Y.D. Zhang, M.L. Ding, W. Chen, M.S. Kang, and X. Fan. 2015. Introgression of the *crtRB1* gene into quality protein maize inbred lines using molecular markers. *Mol. Breed.*, 35: 154.

Low, J.W., R. Ortiz, E. Vandamme, M. Andrade, B. Biazin, and W.J. Grüneberg. 2020. Nutrient-dense orange-fleshed sweet potato: Advances in drought-tolerance breeding and understanding of management practices for sustainable next-generation cropping systems in Sub-Saharan Africa. *Front. Sustain. Food Syst.*, 4: 50.

Martínez-Cámara, S., A. Ibañez, S. Rubio, C. Barreiro, and J.L. Barredo. 2021. Main carotenoids produced by microorganisms. *Encyclopedia*, 1: 1223–1245.

Meléndez-Martínez, A.J., A.I. Mandić, F. Bantis, V. Böhm, G.I.A. Borge, M. Brnčić, A. Bysted, M. Pilar Cano, M. Graça Dias, A. Elgersma, M. Fikselová, J. García-Alonso, D. Giuffrida, V.S.S. Gonçalves, D. Hornero-Méndez, K. Kljak, V. Lavelli, G.A. Manganaris, P. Mapelli-Brahm, M. Marounek, B. Olmedilla-Alonso, M.J. Periago-Castón, A. Pintea, J.J. Sheehan, V.T. Šaponjac, M. Valšíková-Frey, L. Van Meulebroek, and N. O'Brien. 2022. A comprehensive review on carotenoids in foods and feeds: Status quo, applications, patents, and research needs. *Crit. Rev. Food Sci. Nutr.*, 62: 1999–2049.

Mi, J., J.G. Vallarino, I. Petrík, O. Novák, S.M. Correa, M. Chodasiewicz, M. Havaux, M. Rodriguez-Concepcion, S. Al-Babili, A.R. Fernie, A. Skirycz, and J.C. Moreno. 2022. A manipulation of carotenoid metabolism influence biomass partitioning and fitness in tomato. *Metab. Eng.*, 70: 166–180.

Milani, A., M. Basirnejad, S. Shahbazi, and A. Bolhassani. 2017. Carotenoids: Biochemistry, pharmacology, and treatment. *Brit. J. Pharm.*, 174: 1290–1324.

Morelli, L., S. Torres-Montilla, G. Glauser, V. Shanmugabalaji, F. Kessler, and M. Rodriguez-Concepcion. 2022. Novel insights on the contribution of plastoglobules and reactive oxygen species to chromoplast differentiation. *New Phytol.*, 237: 1696–1710.

Nassar, N.M.A. 2007. Cassava genetic resources and their utilization for breeding of the crop. *Gen. Mol. Res.*, 6: 1151–1168.

Njukwe, E., R. Hanna, H. Kirscht, and S. Araki. 2013. Farmers perception and criteria for cassava variety preference in Cameroon. *Afr. Study Monographs*, 34: 221–234.

Nkhata, S.G., S. Chilungo, A. Memba, and P. Mponela. 2020. Biofortification of maize and sweet potatoes with provitamin A carotenoids and implication on eradicating vitamin A deficiency in developing countries. *J. Agr. Food Res.*, 2: 100068.

Okello, J., L. Wanjohi, P. Makohka, J.W. Low, and N. Kwikiriza. 2019. Status of sweet potato in Sub-Saharan Africa: September 2019. *In:* Sweet Potato for Profit and Health Initiative. Nairobi: International Potato Center. *Front. Sustain. Food Syst.*, 4: 50.

Paine, J.A., C.A. Shipton, S. Chaggar, R.M. Howells, M.J. Kennedy, G. Vernon, S.Y. Wright, E. Hinchliffe, J.L. Adams, A.L. Silverstone, and R. Drake. 2005. Improving the nutritional value of Golden Rice through increased provitamin A content. *Nature Biotech.*, 23: 482–487.

Palmer, A.C., I. Darnton-Hill, and K.P. West. 2017. *Vitamin A Deficiency. Nutrition and Health in a Developing World*, 181–234. Cham: Springer International Publishing.

Pixley, K., R. Babu, J. Yan, and N. Palacios-Rojas. 2010. Challenges, progress, and state of breeding for provitamin A. *In: First Global Conference on Biofortification: From Discovery to Delivery*, 9–11 Nov. 2010. Washington, D.C.

Prasanna, B.M., N. Palacios-Rojas, F. Hossain, V. Muthusamy, A. Menkir, T. Dhliwayo, T. Ndhlela, F. San Vicente, S.K. Nair, B.S. Vivek, X. Zhang, M. Olsen, and X. Fan. 2020. Molecular breeding for nutritionally enriched maize: Status and prospects. *Front, Genet.*, 10: 1392.

Rodriguez-Bonilla, L., H.E. Cuevas, M. Montero-Rojas, F. Bird-Pico, D. Luciano-Rosario, and D. Sirtunga. 2014. Assessment of genetic diversity of sweet potato in Puerto Rico. *PLoS One*, 9: e116184.

Rodriguez-Concepcion, M., J. Avalos, M.L. Bonet, A. Boronat, L. Gomez-Gomez, D. Hornero-Mendez, M.C. Limon, A.J. Meléndez-Martínez, B. Olmedilla-Alonso, A. Palou, J. Ribot, M.J. Rodrigo, L. Zacarias, and C. Zhu. 2018. A global perspective on carotenoids: Metabolism, biotechnology, and benefits for nutrition and health. *Prog. Lipid Res.*, 70: 62–93.

Rodriguez-Amaya, D.B. 2016. *Food Carotenoids: Chemistry, Biology, and Technology*. Sussex: John Wiley and Sons Ltd.

Sadali, N.M., R.G. Sowden, Q. Ling, and R.P. Jarvis. 2019. Differentiation of chromoplasts and other plastids in plants. *Plant Cell Rep.*, 38: 803–818.

Schaub, P., F. Würst, J. Koschmieder, Q. Yu, P. Virk, J. Tohme, and P. Beyer. 2017. Nonenzymatic β-carotene degradation in provitamin A-biofortified crop plants. *J. Agric. Food Chem.*, 65: 6588–6598.

Simpungwe, E., T. Dhliwayo, M. Palenberg, V. Taleon, E. Birol, A. Oparinde, A. Saltzman, and M.T. Diressie. 2017. Orange maize in Zambia: Crop development and delivery experience. *Afr. J. Food Agric. Nutr. Dev.*, 17: 11973–11999.

Siwela, M., K. Pillay, L. Govender, S. Lottering, F.N. Mudau, A.T. Modi, and T. Mabhaudhi. 2020. Biofortified crops for combating hidden hunger in South Africa: Availability, acceptability, micronutrient retention and bioavailability. *Foods*, 9: 815.

Sprink, T., R. Wilhelm, and F. Hartung. 2022. Genome editing around the globe: An update on policies and perceptions. *Plant Phys.*, 190: 1579–1587.

Sun, T., Q. Zhu, Z. Wei, L.A. Owens, T. Fish, H. Kim, T.W. Thannhauser, E.B. Cahoon, and L. Li. 2021. Multi-strategy engineering greatly enhances provitamin A carotenoid accumulation and stability in Arabidopsis seeds. *Biotech.*, 2: 191–214.

Suwarno, W.B., K.V. Pixley, N. Palacios-Rojas, S.M. Kaeppler, and R. Babu. 2014. Formation of heterotic groups and understanding genetic effects in a provitamin A biofortified maize breeding program. *Crop Sci.*, 54: 14–24.

Tanumihardjo, S.A., R.M. Russell, C.B. Stephensen, B.M. Gannon, N.E. Craft, M.J. Haskell, G. Lietz, K. Schulze, and D.J. Raiten. 2016. Biomarkers of nutrition for development (BOND) – vitamin: A review. *J. Nutr.*, 146: 1816S–1848S.

Telengech, P.K., J.N. Maling'a, A.B. Nyende, S.T. Gichuki, and B.W. Wanjala. 2014. Gene expression of beta carotene genes in transgenic biofortified cassava. *Biotech.*, 5: 465–472.

Torres-Montilla, S., and M. Rodriguez-Concepcion. 2021. Making extra room for carotenoids in plant cells: New opportunities for biofortification. *Prog. Lipid Res.*, 84: 101128.

Watkins, J.L., and B.J. Pogson. 2020. Prospects for carotenoid biofortification targeting retention and catabolism. *Trends Plant Sci.*, 25: 501–512.

Welsch, R., and L. Li. 2022. Golden Rice – Lessons learned for inspiring future metabolic engineering strategies and synthetic biology solutions. *Methods Enzymol.*, 671: 1–29.

Wiseman, E.M., S. Bar-El Dadon, and R. Reifen. 2017. The vicious cycle of vitamin A deficiency: A review. *Crit. Rev. Food Sci. Nutr.*, 57: 3703–3714.

Wu, S., K.H. Lau, Q. Cao, J.P. Hamilton, H. Sun, and C. Zhou. 2018. Genome sequences of two diploid wild relatives of cultivated sweet potato reveal targets for genetic improvement. *Nat. Commun.*, 9: 4580.

Wurtzel, E.T. 2019. Changing form and function through carotenoids and synthetic biology. *Plant Phys.*, 179: 830–843.

Yang, J., M-H. Moeinzadeh, H. Kuhl, J. Helmuth, P. Xiao, and S. Haas. 2017 Haplotype-resolved sweet potato genome traces back its hexaploidization history. *Nat. Plants*, 3: 696–703.

Zheng, X., H.N.J. Kuijer, and S. Al-Babili. 2021. Carotenoid biofortification of crops in the CRISPR era. *Trends Biotech.*, 39: 857–860.

Zheng, X., G. Giuliano, and S. Al-Babili. 2020. Carotenoid biofortification in crop plants: Citius, altius, fortius. *Bioch. et Biophys. Acta (BBA) – Mol. Cell Biol. Lipids*, 1865: 158664.

Zhu, H., C. Li, and C. Gao. 2020. Applications of CRISPR–Cas in agriculture and plant biotechnology. *Nat. Rev. Mol. Cell Biol.*, 21: 661–677.

Zubair Khan., M., M. Takemura, T. Maoka, J. Hattan, M. Otani, and N. Misawa. 2022. *Molecular Breeding of Sweet Potato Carotenoids*. Intech Open.

Zuma, M.K., U. Kolanisi, and A.T. Modi. 2018. The potential of integrating provitamin A-biofortified maize in smallholder farming systems to reduce malnourishment in South Africa. *International J. Env. Res. Public Health*, 15: 805.

Zunjare, R.U., F. Hossain, V. Muthusamy, and A. Baveja. 2018. Development of biofortified maize hybrids through marker-assisted stacking of β-carotene hydroxylase, lycopene-ε-cyclase and opaque 2. *Genes*, 9: 1–12.

CHAPTER

8

Vitamin B: Eight Unsung Heroes, Longing for Attention

Bhawna Sheoran[a], Bhawna Babal[b], Anita Kumari[a], Reena Rani[c], Usman Ali[a], Sheetal Mehla[c], Simon Strobbe[d] and Teresa B. Fitzpatrick[d]*

[a] Department of Agri-Biotechnology, National Agri-Food Biotechnology Institute, Mohali, Punjab - 140306, India

[b] Department of Soil Science, CSK Himachal Pradesh Agriculture University, Palampur, Himachal Pradesh - 176061, India

[c] Department of Molecular Biology and Biotechnology, Chaudhary Charan Singh Haryana Agricultural University, Hisar, Haryana - 125001, India

[d] Department of Plant Sciences, University of Geneva, 1211 Geneva, Switzerland

1. Introduction

The micronutrient vitamin B-complex comprises eight water-soluble vitamins that play crucial and interconnected roles. Vitamins B_1 (thiamine), B_2 (riboflavin), B_3 (niacin), B_5 (pantothenic acid), B_6 (pyridoxine), B_7 (biotin), B_9 (folate), and B_{12} (cobalamin) are all included in the complex. Note that some chemical compounds, previously considered vitamins, have lost their vitamin status, thereby explaining the gap in numerical order of the different B-vitamins. The above-mentioned B-vitamins have very diverse roles, being involved in the metabolism of macronutrients such as proteins, carbohydrates and fats, as well as a plethora of specialized metabolites such as cholesterol, neurotransmitters, *S*-adenosylmethionine, and nucleotide bases, to name a few (Alina et al., 2019). Generally, they are essential for maintaining healthy eyes, nerves, muscles, liver, skin, gut mucous membranes; hormone production by the adrenal glands; absorbing and activating iron; and fetal development because they act as cofactors for various enzymes (Table 1). Each B vitamin comprises several forms known as vitamers (Table 1); these are chemically related compounds

*Corresponding author: theresa.fitzpatrick@unige.ch

that can be interconverted to alleviate the corresponding vitamin deficiency. The human body does not typically have long-term reserves of these water-soluble B-vitamins, B_{12} being an exception (Mahmood, 2014), and is, therefore, reliant on these vitamins as a necessary part of a daily diet. Their deficiency leads to a plethora of diseases and developmental disorders/impairments (Table 1) (Bailey et al., 2015), although the global prevalence is difficult to map due to a paucity of deficiency data. The effect of these B-vitamin deficiencies should, however, not be underestimated, as it was recently calculated that folate deficiency was responsible for over 200,000 neural tube defects (leading to many deaths) in 2019 alone (Kancherla et al., 2021). The artificial addition of micronutrients to dietary sources is impractical on a bigger scale, as it requires a sophisticated distribution infrastructure, government coordination, recurring investments and customer compliance. Furthermore, although the gut microbiome supplies B-vitamins, it is inadequate for dietary sufficiency. Biofortification is a pivotal tool in the fight against micronutrient malnutrition (Garg et al., 2018; Van Der Straeten et al., 2020). To increase the level of vitamins specifically in the edible portion of food crops, (molecular) breeding as well as engineering of metabolic pathways through transgenic technology is being applied (Roychoudhury, 2021; Bouis & Saltzman, 2017). This chapter is a compilation of extensive knowledge of B vitamins in the light of biofortification, therefore considering sources, metabolism, strategies, and their associated challenges.

1.1. Types of B Vitamins: Sources and Deficiency

1.1.1. Vitamin B_1 (Thiamine)

Historically, thiamine deficiency was linked to symptoms described in ancient Chinese texts, later connected to nutrition in the late 19th century. Japanese sailors eating only rice suffered diseases until a diversified diet improved their health. In 1897, a Dutch scientist found polished rice caused paralysis, while unpolished rice did not, leading to thiamine's discovery in rice's outer layers (Carpenter, 2012). Thiamine-rich sources include meats, fish, whole grains, fortified foods and legumes. Thiamine exists as various vitamers (Table 1), mainly thiamine itself and its phosphate esters like thiamine mono/diphosphates, crucial for carbohydrate metabolism, protein, and lipid digestion, impacting energy production and metabolism (Lonsdale, 2006). Thiamine deficiency affects the brain and heart, causing cognitive decline and congestive heart failure (DiNicolantonio et al., 2013; Gibson et al., 2016).

1.1.2. Vitamin B_2 (Riboflavin)

Riboflavin's discovery traces back to Alexander Wynter Blyth in 1872, later identified as riboflavin in the 1930s by György, Kuhn and Wagner-Jauregg (Northrop-Clewes & Thurnham, 2012). Riboflavin sources include dairy, meat, nuts, green vegetables, and fortified foods (Harvard, 2023). Three vitamer forms

Table 1: Compilation of B vitamin chemical structures, vitamers, coenzyme forms, governed processes, and deficiency diseases.

Vitamin	Structure	Vitamers	Coenzyme and Function	Deficiency Prevalence	Deficiency Disorders	Reference
B_1 (Thiamine)	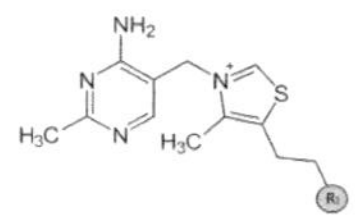	R_1: -OH, $-PO_3^{2-}$, $-OP_2O_6^{3-}$, $-OP_3O_9^{4-}$	**TDP** Decarboxylation reactions	Japan, Thailand	Beriberi, Wernicke-Korsakov syndrome	Strobbe & Van Der Straeten, 2018; Fitzpatrick & Chapman 2020.
B_2 (Riboflavin)	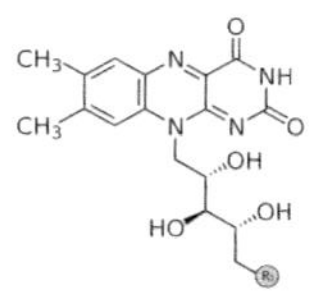	R_1: -OH, $-PO_3^{2-}$ (+/- adenine nucleotide)	**FMN and FAD** Oxidation-reduction reactions involving two hydrogen atoms	Developing countries	Dermatitis, cheilosis, glossitis, anemia, insomnia, conjunctivitis	Uebanso et al. 2020
B_3 (Niacin)	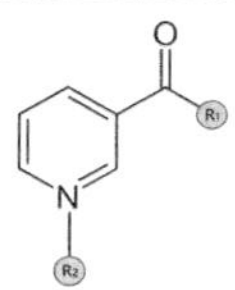	R_1: -OH, $-NH_2$ R_2: -ribose (+/- adenine nucleotide diphosphate: +/- O PO_3^{2-})	**NADH and NADPH** Oxidation-reduction reactions involving the hydride ions	Developing countries, famine conditions	Diarrhoea, dementia, dermatitis	Williams et al. 2017
B_5 (Pantothenic acid)	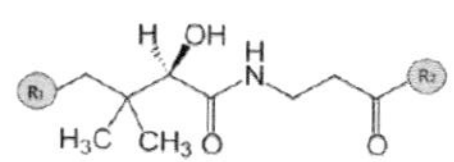	R_1: -OH, $-PO_3^{2}$, +/- adenine nucleotide diphosphate: +/- O PO_3^{2-} R_2:Cysteamine	**Coenzyme A** Acetyl group and other acyl group transfer	Insufficient data	Acne and paresthesia (rare)	Farah et al. 2022

B_6 (Pyridoxine)	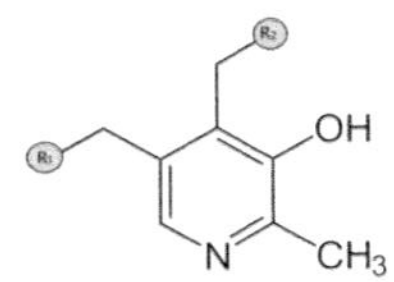	R_1: -OH, $-NH_2$, =O R_2: -OH, $-OPO_3^{2-}$	**Pyridoxal 5′-Phosphate** Amino and carboxyl group transfer	Developing countries	Neuropathy, seborrheic dermatitis, glossitis, cheilosis, depression, confusion, and seizures	Coburn, 2015
B_7 (Biotin)	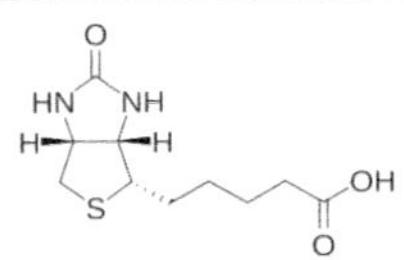	–	**Biotin** Carboxylation reactions	Saudi Arabia, Turkey and Brazil	Decreased nail and hair development, reduced development and nervous syndromes in children	Canda et al. 2020
B_9 (Folate)	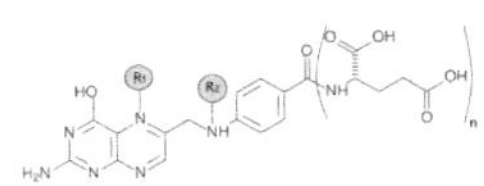	R_1: -H, -OH, $-CH_3$, $-CH_2$-, =CH- R_2: -H, -OH, $-CH_2$-, =CH-	**THF** Methyl, formyl, methylene and formamino group transfer	Low-income economies	Neural Tube Defects (NTDs) fatigue, anemia, and muscle weakness	Pique et al. 2021 Rogers et al. 2018
B_{12} (Cobalamin)	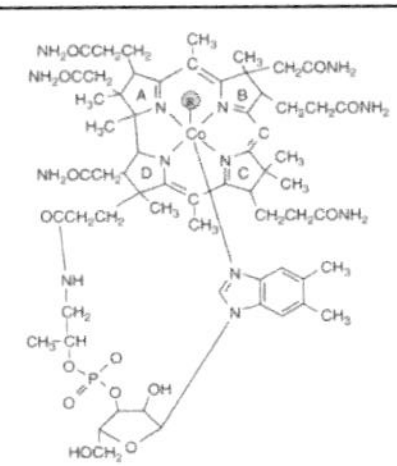	R: -CN, -5′-Deoxyadenosyl, -OH, $-CH_3$, $-H_2O$, $-NO_2$	Radical and methyl transfer	Vegans, elderly population, pregnant/ lactating women and preschool-age children	Megaloblastic anemia, atrophic gastritis, poor balance, memory trouble	Ata et al. 2020 Green et al. 2017

(Adopted from Fitzpatrick & Noordally, 2021; Rathee et al., 2022) exist: riboflavin, flavin mononucleotide (FMN), and flavin adenine dinucleotide (FAD), which play coenzyme roles (Table 1) in energy metabolism (USDHHS, 2023). Riboflavin aids in preventing DNA damage from carcinogens, converting tryptophan to niacin, and normal mitochondrial function (Thakur et al., 2017). It supports digestion, nutrient absorption, and tissue preservation; it is investigated for migraine prevention and is linked to lower colorectal and lung cancer risk (Condo et al., 2009). Its deficiency causes symptoms ranging from sore throat and mouth inflammation to anemia and impaired nerve functioning (Mosegaard et al., 2020; Thakur et al., 2017; Mahabadi et al., 2023).

1.1.3. Vitamin B_3 (Niacin)

Niacin was identified as a vital nutrient in the 20th century during investigations into pellagra, by American epidemiologist Joseph Goldberger through human and animal studies (Lanska, 2012). Sources of niacin encompass meats, fish, rice, nuts, and whole grains (Harvard, 2023). Nicotinic acid and nicotinamide are the main vitamers (Table 1), which can be converted to nicotinamide adenine dinucleotide (phosphate) (NAD(P)H) within the human body (Sauve, 2008). NAD(P)H act as coenzymes in over 400 enzyme-driven reactions, aiding energy conversion, cholesterol metabolism, fat production, DNA synthesis, and exerting antioxidant and anti-inflammatory effects (IM, 1998; USDHHS, 2023). Niacin itself has been used to treat dyslipidemia (the imbalance of lipids in the body) for decades, addressing cardiovascular risks (Kim & Park, 2022). A deficiency causes headaches, depression, memory loss, circulatory problems, to name a few ailments (Knip et al., 2000). Further disorders associated with niacin deficiency include digestive issues, mental disorientation, pellagra with diarrhoea, dermatitis, and dementia (Hegyi et al., 2004).

1.1.4. Vitamin B_5 (Pantothenate)

In the 1930s, Roger J. Williams was credited with discovering vitamin B_5, or pantothenic acid, by isolating it from liver and yeast extracts and recognizing its essential role in growth and development, coining the name "pantothenic" from the Greek word "pantos," signifying its widespread presence in various foods (Lanska, 2012). Pantothenate dietary sources are beef, chicken, mushrooms, avocados, nuts, dairy, potatoes, eggs, rice, oats, and broccoli (Harvard, 2023). Its absorption is affected by gut bacteria (Harvard, 2023). It is crucial for blood cell production, energy conversion and hormone formation. It enables coenzyme A (CoA) and acyl carrier protein formation, supporting fatty acid metabolism and other metabolic functions (IM, 1998). When investigated for cholesterol reduction in dyslipidemia, it could counteract high lipid levels (Chen et al., 2015). Its antioxidant potential might alleviate early heart disease inflammation (Jung et al., 2017). Furthermore, cerebral pantothenate deficiency is found to cause neurodegeneration in sporadic Alzheimer's disease patients (Xu et al., 2020). Deficiency manifests as fatigue and headaches (Jung et al., 2017).

1.1.5. Vitamin B_6 (Pyridoxine)

Pyridoxine was first discovered in the 1930s by the Hungarian biochemist Paul György, as the compound that could cure skin inflammation in rats (György 1964). Pyridoxine sources encompass liver, tuna, fortified cereals, chickpeas, poultry, dark leafy greens, bananas, papayas, oranges, and cantaloupe (Harvard, 2023). The vitamin has various forms, including pyridoxine (PN), pyridoxal (PL), pyridoxamine (PM), and phosphorylated esters, with pyridoxal 5′-phosphate (PLP) as a key coenzyme form (Table 1) (IM, 1998; McCormick, 2006). PLP aids over 200 enzymes in protein, carbohydrate, and lipid digestion, maintains homocysteine levels, and supports brain health (Stach et al., 2021). Deficiency leads to anemia, skin disorders, confusion, depression, nausea, heightened infection susceptibility, skin issues, swelling, white patches, reduced attention, and convulsions (NIH, 2023).

1.1.6. Vitamin B_7 (Biotin)

The identification of biotin as a vitamin stemmed from studies on microbial growth factors, with Paul György's work in the 1930s isolating a growth-promoting substance from egg yolk, which he named "biotin" after the Greek word "biotos," signifying "life" (Lanska, 2012). Biotin-rich sources include meats, eggs (cooked), baker's yeast, wheat germ, and oysters, while avidin in raw eggs can hinder its absorption (Harvard, 2023). Biotin aids in lipid, carbohydrate, and protein metabolism, gene activity, and cellular signaling (NIH, 2023). Deficiency symptoms include rash, hair loss, numbness, depression, and impaired immunity, which are more common in smokers, pregnant women, low-biotin breastfed infants, and certain medical conditions (Canda et al., 2020). Biotin deficiency is rare due to dietary availability (Dasgupta, 2019).

1.1.7. Vitamin B_9 (Folate)

Lucy Wills, a British hematologist, identified folate as a "new hemopoietic factor" in yeast and liver that cured tropical macrocytic anemia in the 1930s (Wills, 1931). In 1941, this compound got its name, folic acid, when scientists isolated it from spinach as a growth factor for *Streptococcus lactis* (Mitchell et al., 1941). Folate rich sources include liver, seafood, leafy greens, fruits, beans and nuts (Harvard, 2023). Folates consist of tri-partite structures with different vitamers differing in oxidation state, glutamate tail, and C_1-substituent nature, affecting stability (Scott, 2000). Tetrahydrofolates are active coenzymes, serving as C_1-donor and acceptor in many metabolic reactions, thereby playing an essential role in DNA (purine) formation, protein metabolism, and homocysteine breakdown, red blood cell production, fetal growth and brain health (Strobbe and Van Der Straeten, 2017). Deficiency symptoms are tiredness, irritability, (megaloblastic) anaemia, and tongue issues (NIH, 2023). Inadequate folic acid levels have been causally linked to the development of neural tube defects during early pregnancy (Blom et al., 2006).

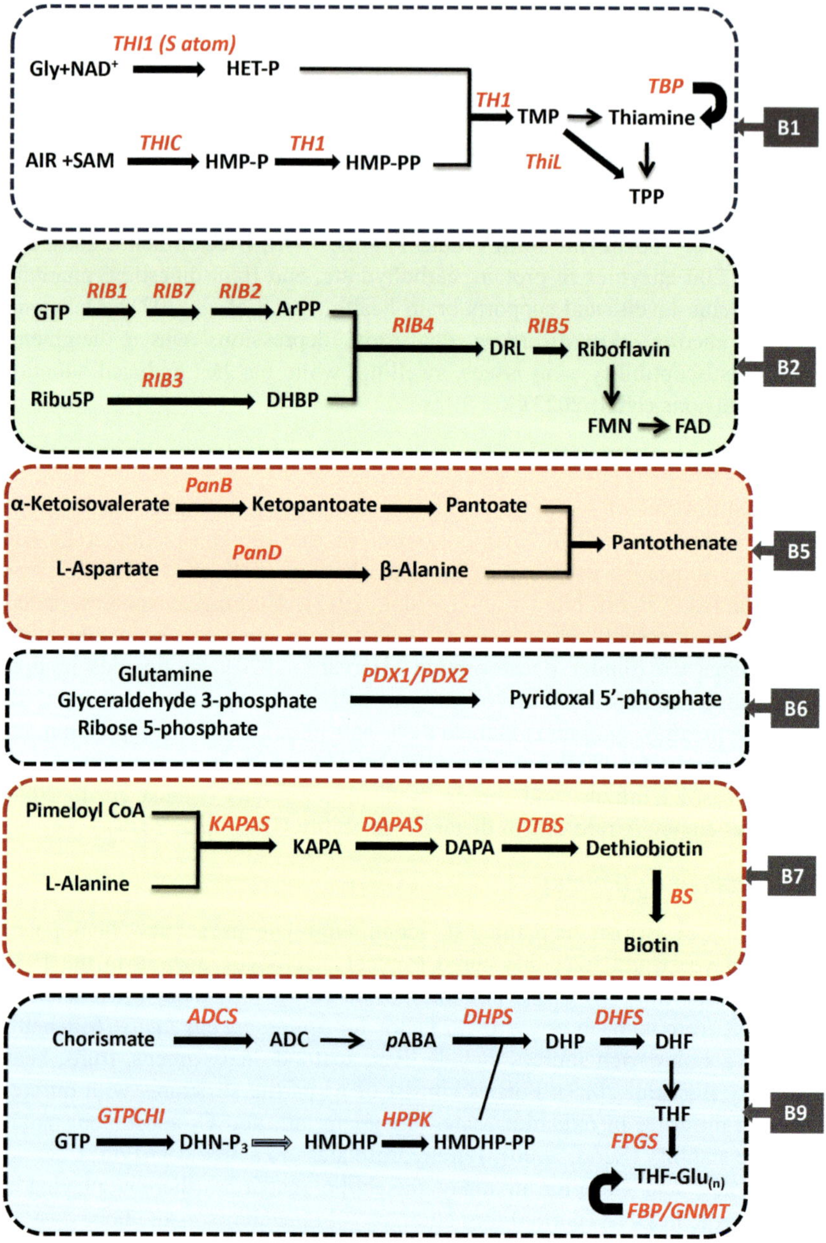

Figure 1: Brief outline of selected steps of the biosynthetic pathways of different B vitamins depicting genes that have been used for biofortification by genetic engineering as elaborated in the text. See Table 2 for gene sources.

(Contd.)

Figure 1 (*Contd.*)

Abbreviations: Genes are *THI1*: 4-methyl-5-hydroxyethylthiazole phosphate (HET-P) synthase; *THIC*: 4-amino-5-hydroxymethyl-2-methylpyrimidine phosphate (HMP-P) synthase; *TH1*: HMP-P kinase/thiamine monophosphate (TMP) phosphorylase; *ThiL*: TMP kinase; *TBP*: thiamine binding protein; *RIB1*: Guanosine triphosphate (GTP) cyclohydrolase II; *RIB7*: 2,5-diamino-6-ribosyl-amino-4(3H)pyrimidinedione 5′-phosphate (DARPP) reductase; *RIB2*: 2,5-diamino-6-ribityl-amino-2,4(1H,3H) pyrimidinedione 5′-phosphate (DArPP) deaminase; *RIB3*: 3,4-dihydroxy-2-butanone-4-phosphate (DHBP) synthase; *RIB4*: lumazine synthase; *RIB5*: riboflavin synthase; *PanB*: ketopantoate hydroxymethyltransferase; *PanD*: L-aspartate α-decarboxylase; *PDX1*: pyridoxal phosphate synthase; *PDX2*: pyridoxal phosphate glutaminase; *KAPAS*: 7-keto-8-amino-pelargonic acid synthase; *DAPAS*: 7-8-diamino- pelargonic acid synthase; *DTBS*: Dethiobiotin synthase; *BS*: Biotin synthase; *ADCS*: 4-amino-4-deoxychorismate synthase; *GTPCHI*: GTP cyclohydrolase I. *HPPK*: HMDHP pyrophosphokinase; *DHPS*: dihydropteroate synthase; *DHFS*: dihydrofolate synthetase; *FPGS*: folylpolyglutamate synthetase; *FBP*: folate binding protein; *GNMT*: glycine N-methyltransferase.

Metabolites: Gly: glycine; NAD^+: nicotinamide adenine dinucleotide; HET-P: 4-methyl-5-β-hydroxyethylthiazole phosphate; AIR: 5-aminoimidazole ribonucleotide; SAM: *S*-adenosylmethionine; HMP-P: 4-amino-2-methyl-5-hydroxymethylpyrimidine phosphate; HMP-PP: HMP-pyrophosphate; TMP: thiamine monophosphate; TPP: thiamine pyrophosphate; GTP: guanosine triphosphate; Ribu5P: ribulose 5-phosphate; ArPP: 5-amino-6-ribityl-amino-2,4(1H,3H)pyrimidinedione 5′-phosphate; DHBP: 3,4-dihydroxy-2-butanone-4-phosphate; DRL: 6,7-dimethyl-8-ribityllumazine; FMN: flavin mononucleotide; FAD: flavin adenine dinucleotide; KAPA: 7-keto-8-amino-pelargonic acid; DAPA: 7-8-diamino-pelargonic acid; ADC, aminodeoxychorismate; pABA: *para*-aminobenzoic acid; DHN-P_3: dihydroneopterin triphosphate; HMDHP: 6-hydroxymethyldihydropterin; HMDHP-PP: 6-hydroxymethyl-7,8-dihydroplerin pyrophosphate; DHP: dihydropteroate; DHF: dihydrofolate; THF: tetrahydrofolate; THF-$Glu_{(n)}$: tetrahydrofolate polyglutamate. The compound arrow between DHN-P_3 and HMDHP indicates that multiple steps are involved.

1.1.8. *Vitamin B_{12}* (Cobalamin)

Pioneering clinical studies in the 19th century led to Minot and Murphy's Nobel Prize-winning work showing the effectiveness of a liver-rich diet for pernicious anemia. Dorothy Hodgkin's research in 1964 resulted in the isolation and characterization of vitamin B_{12} earning her a Nobel Prize (Castle 1966). The dietary need for the vitamin B_{12} complex was discovered by Castle as a factor missing in anaemic people (Castle 1966). Bovine liver, fish, eggs, clams and dairy are the main sources of cobalamin, with options for non-meat eaters like dairy and some algae (Harvard, 2023). Given these typical sources of B_{12}, the vegan diet is particularly susceptible to inadequate intake and is, therefore, recommended to be fortified by vitamin B_{12} supplementation (Woo et al., 2014). Deficiency leads to megaloblastic anemia, cognitive issues, and irreversible neurological damage. Symptoms include mood changes, numbness and dizziness (Ata et al., 2020). Cobalamin aids neurological regulation, red blood cell formation, and DNA biosynthesis (NIH, 2023).

2. Biofortification of Vitamin B

Biofortified crops provide a long-term sustainable way for individuals who rely on staple crops for the majority of their caloric intake to enhance the amount of dietary B vitamins they consume. Efforts have been undertaken to boost crop B vitamin content, particularly for vitamin B_9, as well as vitamins B_1 and B_6, but efficacy within human diets has not yet been studied; consequently, further research is required to give sustainable and scalable solutions in many parts of the world (Titcomb and Tanumihardjo, 2019). Several genes have been targeted (Figure 1, Table 2) in efforts to increase selective B vitamins in staple crops mainly through genetic engineering as diversity is sometimes limited among cultivated crops and wild relatives for breeding purposes (Goyer and Sweek, 2011; Mangel et al., 2017, 2022).

2.1. Vitamin B_1 (Thiamine)

The biosynthesis of thiamine involves the independent biosynthesis of its pyrimidine and thiazole moieties and is diverse between microorganisms and plants. However, the process is well-characterized in the plant model *Arabidopsis* (Minhas et al., 2018). In *Arabidopsis*, biosynthesis starts in plastids, producing the heterocycle precursors (Fitzpatrick and Noordally, 2021). The enzyme 4-methyl-5-hydroxyethylthiazole phosphate (HET-P) synthase (THI1) transforms glycine and NAD^+ into the thiazole moiety (HET-P) as shown in Figure 1. In doing so the THI1 protein donates a sulfur atom from a backbone cysteine, thereby functioning as a single-turnover suicidal enzyme (Chatterjee et al., 2011). 4-amino-2-methyl-5-hydroxymethylpyrimidine phosphate (HMP-P) synthase (THIC) forms the pyrimidine intermediate HMP-P using *S*-adenosylmethionine (SAM) and an iron-sulfur cluster as coenzymes and 5-aminoimidazole ribonucleotide (AIR) as substrate. The bifunctional enzyme TH1 catalyzes the conversion of HMP-P to HMP-pyrophosphate (HMP-PP). TH1 also catalyzes HMP-PP and HET-P condensation into thiamine monophosphate (TMP) (Figure 1). The TH2 enzyme further processes TMP to thiamine (Hsieh et al., 2017; Mimura et al., 2016), which facilitates thiamine pyrophosphate (TPP) production by a kinase (TPK) apparently in the cytosol (Ajjawi et al., 2007). The expression of the biosynthesis gene *THIC* in higher plants follows circadian regulation, and is influenced by a riboswitch that binds TPP (Wachter et al., 2007; Bocobza et al., 2013; Noordally et al., 2020, 2023).

Breeding Approach Potential

Recent studies have delved into the natural genetic diversity of staple cereal kernels, revealing varying thiamine concentrations. Thiamine content fluctuations of 5.5–13.6 $\mu g\ g^{-1}$ in wheat (Shewry et al., 2011) and 1.08-26.54 $\mu g\ g^{-1}$ in maize (Safiul Azam et al., 2022) have been observed. Diversity was also evident in cassava and rice, indicating potential for thiamine content enhancement through introgressive hybridization (Mangel et al., 2017, 2022). In cassava, significant

Table 2: Biofortification examples in crops via breeding and genetic engineering approaches

Breeding Approach

Vitamin	Crop	Major Gene and its Source	Precursors	Fold Increase	Reference
B_1	Potato		Thiamine and its derivatives	2 (Tubers)	Goyer and Sweek 2011
B_3	Maize	*o2* (Maize *VQL1*)	Tryptophan	2 (Endosperm)	Hossain et al., 2018
B_9	Potato		Folate	10 (Tubers)	Robinson et al., 2015
	Wheat		Folate	9 (Grains)	Garg et al., 2018; Jha et al., 2015; Riaz et al., 2019
	Common bean, Lentil, Chickpea and Pea		Folate	~1.4, ~1.3, ~1.7 & ~1.3, respectively	Jha et al., 2015

Genetic Engineering Approach

Vitamin	Crop	Genes and their Source	Fold Increase	Reference
B_1	Rice	*THIC/THI1 (Rice)* *THIC/THI1/TH1* (*A. thaliana*), *TBP* (*S. indicum*), *ThiL* (*E. coli*)	~3 (Unpolished seeds), ~3 (Unpolished seeds), ~3-4 (Polished seeds), <0.3 (Bran, embryo)	Dong et al. 2015; Strobbe et al., 2021; Fitzpatrick et al., 2024; Chung et al., 2024
B_2	Rice	*RIB1/RIB7/RIB2/RIB3/RIB4/RIB5* (*S. cerevisiae*)	4.24 (Polished seeds)	Tian et al., 2021
B_5	*Brassica napus*	*PanB* (*E. coli*)	~2.5 (Whole seeds)	Chakauya et al., 2008
	Tobacco	*PanD* (*E. coli*)	3-4 (Whole seeds)	Fouad and Rathinasabapathi, 2006

(Contd.)

Table 2: (*Contd.*)

Genetic Engineering Approach				
Vitamin	**Crop**	**Genes and their Source**	**Fold Increase**	**Reference**
B_6	Rice	*PDX1/PDX2* (*A. thaliana*)	50 (Leaves)	Mangel et al., 2019
	Cassava	*PDX1/PDX2* (*A. thaliana*)	48 & 4 (Leaves & storage roots, respectively)	Li et al., 2015
B_9	Rice	*ADCS* (*A. thaliana*)	0.2 (Unpolished seeds)	Storozhenko et al., 2007
	Potato		Insignificant (Tubers)	Blancquaert et al., 2013
	Maize	*GTPCHI* (*E. coli*)	2 (Endosperm)	Naqvi et al., 2009
	Potato	*GTPCHI* (*A. thaliana*)	2 (Leaves)	Blancquaert et al., 2013
	Mexican common bean		3 (Whole seeds)	Ramírez Rivera et al., 2016
	Rice		Insignificant (Endosperm)	Storozhenko et al., 2007
	Rice		6.1 (Unpolished seeds)	Dong et al., 2014
	Tomato	*GTPCHI* (*M. musculus*)	2 (Fruit)	de la Garza et al., 2004
	Lettuce	*GTPCHI* (*G. gallus*)	9 (Leaves)	Nunes et al., 2009
	Rice	*HPPK/DHPS* (*T. aestivum*)	2, 1.4 (Leaves & unpolished seeds, respectively)	Gillies et al., 2008
	Rice	*HPPK/DHPS* (*A. thaliana*)	1.27 (Unpolished seeds)	Dong et al., 2014
	Rice	*DHFS* (*A. thaliana*)	6.1 (Unpolished seeds)	

Rice	*GTPCHI*/biosynthesis genes (*A. thaliana*)	6.1 (Unpolished seeds)	Dong et al., 2014
Rice	*GTPCHI* /*ADCS* (*A. thaliana*)	100 (Unpolished seeds)	Storozhenko et al., 2007
Tomato	*GTPCHI* (*M. musculus*) + *ADCS* (*A thaliana*)	25 (Fruit)	de la Garza et al., 2007
Potato	*GTPCHI*/*ADCS* (*A. thaliana*)	2 (Microtubers)	Blancquaert et al., 2013
Rice	*FPGS* (*A. thaliana*)	1.45 (Unpolished seeds)	Dong et al., 2014
Rice	*FPGS* (*O. sativa*)	4.7 (Unpolished seeds)	Abilgos Ramos, 2010
Rice	*GTPCHI* /*ADCS*/*FPGS* (*A. thaliana*)	100 (Polished seeds)	Blancquaert et al., 2015
Rice	*FBP* (*B. taurus*)	6.2 (Unpolished seeds)	Ramos, 2010
Rice	*GNMT* (*R. norvegicus*)	8.8 (Unpolished seeds)	Ramos, 2010
Rice	*GTPCHI* /*ADCS* (*A. thaliana*) + *FBP* (*B. taurus*)	150 (Polished seeds)	Blancquaert et al., 2015
Potato	*GTPCHI*/*ADCS* (*A. thaliana*) + *HPPK*/*DHPS* (*O. sativa*) +/- *FPGS* (*A. thaliana*)	12 (Tubers)	De Lepeleire et al., 2018
Maize	*GTPCHI*/*ADCS* (*G. max*)	4.2 (Whole kernal)	Liang et al., 2019
Wheat	*GTPCHI* (*G. max*) + *ADCS* (*S. lycopersicum*)	5.6 (Whole seeds)	Liang et al., 2019

Organism name abbreviations: *Arabidopsis* (*A.*) *thaliana*; *Bos* (*B.*) *taurus*; *Escherichia* (*E.*) *coli*; *Gallus* (*G.*) *gallus*; *Glycine* (*G.*) *max*; *Mus* (*M.*) *musculus*; *Oryza* (*O.*) *sativa*; *Rattus* (*R.*) *norvegicus*; *Saccharomyces* (*S.*) *cerevisiae*; *Sesamum* (*S.*) *indicum*; *Solanum* (*S.*) *lycopersicum*; *Triticum* (*T.*) *aestivum*. See text for gene names and annotations.

differences of thiamine content in leaves and storage roots, showing a 5.3 (± 3.6) and a 2.7 (± 0.6) fold difference between the 41 accessions, respectively. Contemporary wheat breeding has identified quantitative trait loci (QTL) associated with higher thiamine content (Li et al., 2018). Two genetic single nucleotide polymorphisms (SNP) were revealed to underly the thiamine content in wheat (Li et al., 2018). A breeding project in potato involving wild cultivars with over 2-fold thiamine content (Table 2) was shown to be possible based on the identified variations in vitamin content (Goyer and Sweek, 2011).

Genetic Engineering Approach

Genetic engineering approaches have been attempted for enhancing thiamine content in a few species. In Arabidopsis and rice, *THIC* and *THI1* overexpression led to elevated thiamine levels in leaves (up to 3.4 and 2.5-fold) and seeds (up to 2.6 and 5-fold) (Dong et al., 2015, 2016; Mangel, 2016). However, the enhanced levels failed to reach the endosperm of rice and therefore polished seeds, which is the target tissue for biofortification. This approach was later extended in both species to include the *TH1* gene (Strobbe et al., 2021), while ectopic expression of a bacterial thiamine monophosphate kinase (ThiL) in rice was also recently shown to enhance thiamine content in certain tissues (Chung et al., 2024), but neither succeeded in significant increases in the polished grain. Recently, an approach using a thiamine binding protein (TBP) has proven to successfully enhance thiamine content in the endosperm of rice up to 4-fold (potentially providing up to 30% of the recommended daily intake) and is stable in the field with no impact on agronomic traits (Fitzpatrick et al., 2024). While the latter is an important achievement, further improvement will require an understanding of regulation but as for all of the B vitamins its metabolism is complex (Fitzpatrick & Noordally, 2021) and yet to be completely unraveled. In particular transport, storage and regulation by a riboswitch and the circadian clock are key areas in the case of thiamine that need more exploration and a deeper understanding may assist in enhancing levels further in consumed tissues. Of note is *Arabidopsis* plants impaired in riboswitch function led to slight thiamine increases (Wachter et al., 2007; Bocobza et al., 2013; Rosada-Souza et al., 2019) but this also leads to impaired development (Bocobza et al., 2013, Noordally et al., 2023). Although a long-distance thiamine transporter is known in *Arabidopsis*, it also transports polyamines (Martinis et al., 2016). While identification of other transporters is necessary, it is complicated by the non-specificity of tissue transporters and influence on general metabolic pathways if manipulated.

2.2. Vitamin B_2 (Riboflavin)

The riboflavin biosynthetic pathway is largely conserved between prokaryotes and plants, with slight variations in fungi such as *Ashbya gossypii* (Abbas and Sibirny, 2011). The mention of *A. gossypii* is pertinent as it naturally produces riboflavin and is utilized in industry. This pathway is also followed by yeast, which has been harnessed for crop biofortification (Aguiar et al., 2015). The pathway comprises six genes i.e., *RIB1*, *RIB2*, *RIB3*, *RIB4*, *RIB5*, *RIB7* which are equivalent to the *ribA*, *ribC*, *ribB*, *ribD*, *ribE*, and *ribF* respectively in plants and microbes. Starting with guanosine triphosphate (GTP), GTP cyclohydrolase II catalyzes its conversion into 2,5-diamino-6-ribosyl-amino 4(3H) pyrimidinedione 5′-phosphate (DARPP) in the purine route (Liu et al., 2020), ultimately leading to the production of 5-amino 6-ribityl-amino-2,4(1H,3H) pyrimidinedione 5′-phosphate (ArPP) through the action of the RIB1, RIB7 and RIB2 enzymes (Revuelta et al., 2016). In the pentose phosphate route, the enzyme 3,4-dihydroxy-2-butanone-4-phosphate (DHBP) synthase (RIB3) transforms ribulose 5-phosphate (Ribu5P) into DHBP (Figure 1). ArPP and DHBP merge, forming 6,7-dimethyl-8-ribityllumazine (DRL), facilitated by RIB4. Riboflavin synthase (RIB5) yields riboflavin which is converted into flavin mononucleotide (FMN) and flavin adenine dinucleotide (FAD) through riboflavin kinase and FAD-synthase, respectively (Liu et al., 2020).

Breeding Approach Potential

Analysis of crop germplasm has demonstrated significant variability in riboflavin content among different wheat and maize cultivars. Davis et al. (1981) reported riboflavin levels ranging from 1.0 to 1.7 μg/g across 231 wheat varieties. Furthermore, Batifoulier et al. (2006) observed values of 0.5 to 1.1 μg/g in 49 wheat types, while Shewry et al. (2011) found levels spanning 0.8 to 1.4 μg/g in 24 wheat cultivars. In maize inbred lines, it varied between 1.19–37.37 μg/100 g, as shown by germplasm screening (Safiul Azam et al., 2022). Employing the QTL mapping and genome wide association studies (GWAS) approaches, seven loci and multiple SNP markers were found to be linked to riboflavin content in wheat in different studies (Li et al., 2018; Nurit et al., 2021). These findings hold promise for future research on riboflavin genetics (by unravelling the molecular mechanisms underlying the different QTLs) and their integration into wheat breeding efforts.

Genetic Engineering Approach

Rice is an important target crop for transgenic approaches due to its vulnerability to riboflavin loss during milling (38%), primarily occurring in the pericarp. By introducing six genes from yeast (*Saccharomyces cerevisiae, Sc*) (*ScRIB1*, *ScRIB7*, *ScRIB2*, *ScRIB3*, *ScRIB4*, and *ScRIB5*) into rice endosperm, transgenic rice displayed a noteworthy riboflavin content in milled seeds (0.1091 mg/100 g DW), exhibiting a 4.24-fold increase compared to wild type (Tian et al., 2021).

Alternative Approaches Used for Biofortification

Several bacteria, fungi and yeast are used for industrial riboflavin production (Levit et al., 2021). Currently, lactic acid-producing bacteria enhance riboflavin concentration in fermented foods (Levit et al., 2021).

2.3. Vitamin B_3 (Niacin)

Plant biosynthesis *de novo* of vitamin B_3 primarily occurs through aspartate precursors, contrasting with fungi where the tryptophan pathway dominates (Suo et al., 2022). Furthermore, while *Arabidopsis* relies on the aspartate pathway, rice employs both the aspartate and tryptophan pathways (Suo et al., 2022; Katoh et al., 2006). The aspartate pathway cascade begins with the conversion of L-aspartate into nicotinate mononucleotide (NaMN) through aspartate oxidase (AOX), quinolinate synthase (QS), and quinolinate phosphoribosyltransferase (QPT) (Noctor et al., 2011). NaMN is further transformed into nicotinic acid adenine dinucleotide, then to nicotinamide adenine dinucleotide (NAD) by NaMN adenyltransferase (NNAT) and NAD^+ synthase (NADS), respectively (Noctor et al., 2011, Gerdes et al., 2012). Ethylene treatment was found to induce niacin accumulation via enhanced expression of key genes *AOX* and *QS* (Suo et al., 2022). The information regarding regulation of niacin biosynthesis in plants is limited. However, in bacteria it is known that the transcriptional regulator NiaR binds to the promoter regions of niacin genes and represses their expression (Rodionov et al., 2008).

Breeding Approach Potential

Potential for increasing niacin content through plant breeding indicates differences among crops. Nurit et al. (2021) found a variation of 1.31-5.75 μg/g and 0.24-3.59 μg/g of nicotinic acid and nicotinamide, respectively, in 166 bread wheat accessions. However, genetic variation of niacin in wheat with low heritability suggests that increasing the niacin content through plant breeding might not be feasible (Shewry et al., 2013), whereas germplasm screening of maize inbred lines suggests that significant variation in niacin content, ranging from 19.60 to 213.7 μg/100 g, could be achieved through plant breeding methods (Safiul Azam et al., 2022). Efforts focused on improving amino acids, notably in maize through traditional plant breeding methods (Maqbool et al., 2021), might have potential for improving niacin content. Quality Protein Maize (QPM), a maize variety with elevated lysine and tryptophan levels was generated by combining natural *opaque-2* (*o2*) mutants through backcrossing and recurrent selection of maize kernels (Vasal, 2000). These varieties, however, did not depict adequate increases in niacin content (Hossain et al., 2019) (Table 2). Thus, more efforts are required for greater insights into biofortification of niacin through breeding approaches in crops.

Genetic Engineering Approach

Research on *Arabidopsis* underscores the pivotal roles of AOX, QS, NNAT, and NADS in the aspartate pathway of niacin biosynthesis (Hashida et al., 2007; Schippers et al., 2008; Suo et al., 2022). Mutations in these enzymes disrupt the biosynthetic process and significantly impact plant growth and development

(Hashida et al., 2007; Sinclair et al., 2000, Hao et al., 2018). Nonetheless, increases in niacin and nicotinamide were observed due to the overexpression of NADS in *Arabidopsis* (Shinnosuke et al., 2010). However, NAD^+ itself did not increase and the plants showed early flowering and senescence, negatively affecting seed development (Shinnosuke et al., 2010). Given the complexity of the biosynthetic process and our limited understanding of its genetic regulation, augmenting niacin synthesis through genetic alteration remains to be further explored. Moreover, NAD^+ and intermediates are known to be involved in other biological processes such as regulation of the circadian clock and responses to stress (De Flora, 2011; Dutilleul et al., 2005; Adams-Phillips et al., 2010; Gakière et al., 2018) adding further complexity.

2.4. Vitamin B_5 (Pantothenate)

The biosynthetic pathway of pantothenate in plants starts from α-ketoisovalerate, which in turn is derived from valine (Huang et al., 2018). This biosynthesis is considered to take place in the cytosol, mitochondria and likely plastids, therefore requiring intracellular transport of biosynthetic intermediates (Huang et al., 2018). First, in the mitochondria (and plastids), ketopantoate is formed from α-ketoisovalerate (Figure 1) by the transfer of a methyl group catalyzed by ketopantoate hydroxymethyltransferase (KPHMT; *PanB*). Subsequently, pantoate is formed following reduction by ketopantoate reductase (KPR; *PanE*) (Raman & Rathinasabapathi, 2004; Siddique et al., 2022). Finally, following its subcellular transport (Huang et al., 2018), pantoate is condensed with β-alanine (derived from L-aspartate by *PanD*) to yield pantothenate in an enzymatic step catalyzed by pantothenate synthetase (PS; *PanC*) in the cytosol (Genschel et al., 1999; Jonczyk et al., 2008). In bacteria, pantothenate biosynthesis is well understood and resembles the pathway as currently understood in plants (Leonardi & Jackowski, 2007). In bacteria, KPHMT exhibits end-product inhibition by ketopantoate and pantothenate. Controlling the size of the coenzyme A (CoA) pool is also primarily achieved by restricting its production through feedback inhibition, and secondarily by breaking it down during shortage of pantothenate (Leonardi and Jackowski, 2007). Plant metabolism regarding regulation of pantothenate biosynthesis needs to be focused for more biofortification possibilities.

Breeding Approach Potential

Given that pantothenate biosynthesis and its compartmentalization is not yet completely understood in plants (Chakauya at al., 2006; Huang et al., 2018; Siddique et al., 2022), mapping germplasm populations for pantothenate variations could pinpoint novel genetic factors playing a role in pantothenate accumulation, potentially revealing currently unknown enzymatic steps (Siddique et al., 2022). Interestingly, a GWAS found over 10-fold variation of pantothenate content in 167 wheat cultivars, allowing identification of SNPs and putative candidate genes, including multiple pantothenate kinase orthologs and uncharacterized proteins (Nurit et al., 2021).

Genetic Engineering Approach

Transgenic *Brassica napus* lines were developed to express *Escherichia coli* (*E. coli*) *PanC* or *PanB* (Figure 1, Table 2) (Chakauya et al., 2008). The *PanC* transgenic lines showed no significant change in vitamin B_5 content; however, those harboring *PanB* had up to 2.5-fold elevated pantothenate levels in leaves, flowers, siliques, and seeds, with the highest increase in seeds (Chakauya et al., 2008). Another study reported that the level of pantothenate in tobacco leaves increased by 3-4 fold with the expression of *E. coli PanD* (Fouad & Rathinasabapathi, 2006). Further exploration of the pantothenate pathway in plants, including its regulation, is required to facilitate enhanced levels in the consumed tissues of crops of interest.

2.5. Vitamin B_6 (Pyridoxine)

Vitamin B_6 is biosynthesized *de novo* in microorganisms and plants. While *E. coli* utilizes a pathway the requires condensation of deoxyxylulose-5-phosphate (DXP) and 4-phosphohydroxy-L-threonine via *PdxA* and *PdxJ* (Fitzpatrick et al., 2007); most other organisms, including higher plants, employ an alternate route that uses ribose 5-phosphate, glyceraldehyde 3-phosphate and ammonium (derived from glutamine) and the enzymes PDX1 and PDX2 (referred to as the DXP-independent pathway) (Tambasco-Studart et al., 2005, 2007; Titiz, 2008). PDX1 and PDX2 form a complex that catalyzes formation of the coenzyme pyridoxal 5′-phosphate directly (Tambasco-Studart et al., 2005, 2007; Robinson et al., 2016; Titiz, 2008). The DXP-independent route offers a more direct way for vitamin B_6 synthesis (two enzymes) compared to the DXP-dependent route that relies on seven enzymes (Fitzpatrick et al., 2007). Regulation of vitamin B_6 biosynthesis in plants is currently being unraveled and appears to at least be coupled to carbon/nitrogen balance and environmental scenarios (Steensma et al., 2023, Labella et al., 2020).

Breeding Approach Potential

Maize inbred lines have shown remarkable variations ranging from 138.59–1065.11 μg per 100 g in vitamin B_6 content amongst 156 lines, with some identified as good sources of vitamin B_6 and promising germplasms for maize breeding (Safiul Azam et al., 2022). Nurit et al. (2021) found a variation ranging from 0.47 to 2.58 in pyridoxine content among 167 bread wheat accessions. Mangel et al (2022) found vitamin B_6 variation ranging from 0.75 to 3.12 ng mg DW^{-1} in unpolished rice seeds and from 0.34 to 1.01 ng mg DW^{-1} in polished rice seeds of 59 genetically diverse accessions. More research, by further extending the screening of germplasm collections in different crops, will improve the potential in breeding towards higher vitamin B_6 content.

Genetic Engineering Approach

In plants, the DXP-independent pathway has been targeted to improve the accumulation of vitamin B_6. Initial studies illustrated that expression of the *Cercospora nicotianae PDX1* and *PDX2* genes in tobacco resulted in a 1.2-fold increase in total vitamin B_6 (Herrero & Daub, 2007). The overexpression of the *PDX1.1* paralog or *PDX2* or in combination in *Arabidopsis* resulted in increased content in various tissues (4 to 9-fold) (Chen & Xiong, 2009; Raschke et al., 2011). Expression of the same *Arabidopsis* genes in cassava led to a 48-fold increase in leaves and 4-fold increase in storage roots (Li et al., 2015) (Table 2). Importantly, these increases were observed in field grown plants and over two cycles of vegetative propagation (Li et al., 2015). Moreover, cooking and bioavailability studies indicated that the recommended daily intake for vitamin B_6 could be met in a single meal that includes storage roots and is thus of real significance for populations that rely on this crop for sustenance (Li et al., 2015). Although, while a similar approach in rice lead to enhanced levels in leaves (up to 50-fold), it did not lead to increases in the grain (Mangel et al., 2019). This suggests that crop-specific approaches may be required, at least in the case of vitamin B_6.

2.6. Vitamin B_7 (Biotin)

Biotin biosynthesis is similar in bacteria and plants and involves several enzymes: keto-8-aminopelargonic acid synthase (KAPAS encoded by *bioF, Bio4*), 7,8-diaminopelargonic acid synthase (DAPAS encoded by *bioA, Bio1*), dethiobiotin synthetase (DTBS encoded by *bioD, Bio3*), and biotin synthase (BS encoded by *bioB; Bio2*) (Alban, 2011; Muralla et al., 2008). KAPAS transforms pimeloyl-CoA into 7-keto-8-aminopelargonic acid (KAPA) using L-alanine, in a cytosolic reaction. DAPAS then converts KAPA into 7,8-diaminopelargonic acid (DAPA) with the help of *S*-adenosylmethionine (SAM). DTBS changes DAPA into dethiobiotin (DTB), needing CO_2 and ATP, in a mitochondrial reaction (Muralla et al., 2008). Lastly, BS closes the sulfur ring of biotin, requiring sulfur, electrons from flavodoxin, SAM, and NADPH. On an industrial scale, biotin is prepared by chemical synthesis, while microbiological routes are being explored and utilized (Laudert and Hohmann, 2011). Efficient and easy biotin fermentation is highly desirable for the supply of large quantities. Different strategies like genetic engineering of microorganisms, composition of the growth media, processing conditions and supplementation have significantly increased strains' efficiency (Xiao et al., 2020). Biotin biosynthesis was found to be upregulated during dessication stress in *Auricularia fibrillifera* (Guo et al., 2022). The implications of organelle-specific transporters seem to play a major role in maintenance of optimum biotin levels via biosynthesis regulation but more research is needed for a better understanding (Pinon et al., 2005).

Breeding and Genetic Engineering Approaches

No significant research has been performed regarding biofortification of biotin in plants.

2.7. Vitamin B_9 (Folate)

The folate biosynthetic *de novo* pathway is similar in microbes and plants and differs mainly due to the compartmentalization of the pathway in plants. It initiates with the conversion of chorismate into aminodeoxychorismate (ADC) by ADC synthase, which is carried out by a single protein in plants (Sahr et al., 2006), while requiring a heterodimer in *E. coli*. Then, ADC lyase converts ADC to *p*-aminobenzoate (*p*ABA). This generation of *p*ABA happens in plastids of higher plants. The other branch of folate biosynthesis occurs in the cytosol, where 6-hydroxymethyldihydropterin (HMDHP) is formed starting from GTP. First, GTP cyclohydrolase I (GTPCHI) generates dihydroneopterin triphosphate (DHN-P_3) from GTP (Basset et al., 2002). HMDHP is converted into DHN-P_3 in the cytosol through a series of enzymatic steps involving phosphatases and aldolases. The final steps of folate biosynthesis occur in the mitochondria, where *p*ABA and pyrophosphorylated HMDHP are condensed to form dihydropteroate (DHP). In plants, this is carried out by a bifunctional HMDHP pyrophophokinase/DHP synthase (HPPK/DHPS) enzyme. Still in the mitochondria, DHP is converted to dihydrofolate (DHF) by DHF synthetase (DHFS). Dihydrofolate reductase (DHFR) reduces DHF into tetrahydrofolate (THF), a glutamate tail is attached to the THF molecule by folylpolyglutamate synthetase (FPGS) to form polyglutamylated folate (Gorelova et al., 2017). That these steps take place across multiple cellular compartments might help in regulation (Pinon et al., 2005).

Breeding Approach Potential

Breeding efforts to develop elite crop varieties with increased edible folate content have demonstrated to be viable (Table 2). A study of 78 accessions of polished rice showed that the amount of folate content can vary by up to 7-fold (Dong et al., 2011). Another study of 150 rice accessions showed that the amount of folate content in brown unpolished rice can vary by up to 3-fold (Aiyswaraya et al., 2017). Similar screenings have been conducted on various other crops, including barley (Andersson et al., 2008), potato (Robinson et al., 2015), different beans (Jha et al., 2015), fruits and vegetables (Mezzetti et al., 2016) and breeding programs have been conducted for a few (Table 2). Recently, a major-effect QTL cluster (qFo-05) and a regulator (hemerythrin RING zinc finger) have been identified in soybean (Agyenim-Boateng et al., 2023) that may help in enhancing folate content.

Genetic Engineering Approach

Folate biosynthetic genes have been used for ectopic expression in several crops (compiled in Table 2). The generation of cis-genic *Arabidopsis* lines that over-express *GTPCHI* paved the way for this strategy in rice, tomato, maize, lettuce, and potato (Hossain et al., 2004). The highest (9-fold) boost in crops' edible parts, by this one-gene strategy, occurred in lettuce (Nunes et al., 2009). Given that *GTPCHI* expressing transgenic lines are depleted in the *p*ABA branch of folate biosynthesis, combined expression of *GTPCHI* and *ADCS* was much more effective resulting in a 25- and 100-fold increase in folate in tomato (de la Garza et al., 2007) and rice (Storozhenko et al., 2007), respectively. However, only 2-fold improvement was reported in potatoes (Blancquaert et al., 2013). Inclusion of *FPGS* in the previously *GTPCHI* and *ADCS*-engineered plants increased folate levels in rice endosperm (100-fold) and potato tubers (12-fold), respectively (Blancquaert et al., 2015; De Lepeleire et al., 2018). Moreover, enhancing the expression of *FPGS*, which lengthens the folate glutamate tail, has been demonstrated to increase the storage stability of folate in both rice (Blancquaert et al., 2015) and potato (De Lepeleire et al., 2018). Additionally, the introduction of mammalian folate-binding proteins has also been shown to be an effective strategy for improving folate storage stability (Blancquaert et al., 2015).

2.8. Vitamin B_{12} (Cobalamin)

While eukaryotes cannot make cobalamin (CBL), some prokaryotes can biosynthesize the vitamin *de novo* with up to 30 enzymatic steps under aerobic or anaerobic conditions or more efficient contracted salvage pathways (Balabanova et al., 2021). The *de novo* process involves three stages: generating uroporphyrinogen-III (UroIII), transforming UroIII into cobinamide (Cbi) through corrin ring formation and adenylation, followed by the assembly of a nucleotide loop, attaching a lower axial ligand (often 5,6-dimethylbenzimidazole) to the corrin ring. Pathways are designated as cob (oxygen-dependent) with *cob* genes (A to M) and cbi (oxygen-independent) with *cbi* genes (A-H, J). The B_{12}-element, a conserved RNA structural element is a riboswitch and binds to adenosylcobalamin (Ado-CBL), the active form of vitamin B_{12} to modulate gene expression (Vitreschak et al., 2003). In Gram-negative proteobacteria, Ado-CBL binding inhibits translation initiation, whereas in Gram-positive bacteria, it promotes transcriptional antitermination. The *cbiM*Cbl riboswitch, found in *Propionibacterium* strain UF1 controls the expression of the *cobA* operon (Li et al., 2020). Most plants neither produce nor require cobalamin (Watanabe and Bito, 2018). While cobalamin is made exclusively by certain prokaryotes, some of which are found in the intestinal flora of ruminants and human gut microbiota, it is not made in sufficient amounts in the latter (Degnan et al., 2014). Thus, its essentiality in the human diet has inspired various approaches to meet requirements. For example, agronomic fortification by seed priming of vitamin B_{12} or its nano-particles has been used to enhance vitamin B_{12} accumulation in kaiware daikon sprouts (Sato et al., 2004). However, it has

Bailey, R.L., K.P. West Jr., and R.E. Black. 2015. The epidemiology of global micronutrient deficiencies. *Ann. Nutr. Metab.*, 66: 22–33.

Balabanova, L., L. Averianova, M. Marchenok, O. Son, and L. Tekutyeva. 2021. Microbial and genetic resources for cobalamin (vitamin B_{12}) biosynthesis: From ecosystems to industrial biotechnology. *Int. J. Mol. Sci.*, 22(9): 4522.

Basset, G., E.P. Quinlivan, M.J. Ziemak, R. Díaz de la Garza, M. Fischer, S. Schiffmann, A. Bacher, J.F. Gregory III, and A.D. Hanson. 2002. Folate synthesis in plants: The first step of the pterin branch is mediated by a unique bimodular GTP cyclohydrolase I. *Proc. Natl. Acad. Sci. U.S.A.*, 99(19): 12489–12494.

Batifoulier, F., M.A. Verny, E. Chanliaud, C. Rémésy, and C. Demigné. 2006. Variability of B vitamin concentrations in wheat grain, milling fractions, and bread products. *Eur. J. Agron.*, 25(2): 163–169.

Blancquaert, D., S.J. Van Daele Strobbe, F. Kiekens, S. Storozhenko, H. De Steur, and D. Van Der Straeten. 2015. Improving folate (vitamin B9) stability in biofortified rice through metabolic engineering. *Nat. Biotechnol.*, 33(10): 1076–1078.

Blancquaert, D., S. Storozhenko, J. Van Daele, C. Stove, R.G. Visser, W. Lambert, and D. Van Der Straeten. 2013. Enhancing pterin and para-aminobenzoate content is not sufficient to successfully biofortify potato tubers and *Arabidopsis thaliana* plants with folate. *J. Exp. Bot.*, 64(12): 3899–3909.

Blom, H.J., G.M. Shaw, M. den Heijer, and R.H. Finnell. 2006. Neural tube defects and folate: Case far from closed. *Nat. Rev. Neurosci.*, 7(9): 724–731.

Bocobza, S.E., S. Malitsky, W.L. Araújo, A. Nunes-Nesi, S. Meir, M. Shapira, and A. Aharoni. 2013. Orchestration of thiamin biosynthesis and central metabolism by combined action of the thiamin pyrophosphate riboswitch and the circadian clock in Arabidopsis. *Plant Cell*, 25(1): 288–307.

Bouis, H.E., and A. Saltzman. 2017. Improving nutrition through biofortification: A review of evidence from HarvestPlus, 2003 through 2016. *Glob. Food Secur.*, 12: 49–58.

Canda, E., S. Kalkan Uçar, and M. Çoker 2020. Biotinidase deficiency: Prevalence, impact and management strategies. *Pediatric Health Med. Ther.*, 11: 127–133.

Carpenter, K.J. 2012. The discovery of thiamin. *Ann. Nutr. Metab.*, 61(3): 219–223.

Castle, W.B. (1966). Treatment of pernicious anemia: Historical aspects. *Clin. Pharmacol. Ther.*, 7(2), 147–161.

Chakauya, E., K.M. Coxon, M. Wei, M.V. MacDonald, T. Barsby, C. Abell, and A.G. Smith. 2008. Towards engineering increased pantothenate (vitamin B_5) levels in plants. *Plant. Mol. Biol.*, 68: 493–503.

Chakauya, E., K.M. Coxon, H.M. Whitney, J.L. Ashurst, C. Abell, and A.G. Smith. 2006. Pantothenate biosynthesis in higher plants: Advances and challenges. *Physiol. Plant.*, 126(3): 319–329.

Chatterjee, A., N.D. Abeydeera, S. Bale, P.J. Paj, P.C. Dorrestein, D.H. Russell, S.E. Ealick, and T.P. Begley. 2011. *Saccharomyces cerevisiae* THI1p is a suicide thiamine thiazole synthase. *Nature*, 487: 542–546.

Chen, H., and L. Xiong. 2009. Enhancement of vitamin B6 levels in seeds through metabolic engineering. *Plant Biotechnol. J.*, 7(7): 673–681.

Chen, Y.Q., S.P. Zhao, and Y.H. Zhao. 2015. Efficacy and tolerability of coenzyme A vs pantethine for the treatment of patients with hyperlipidemia: A randomized, double-blind, multicenter study. *J. Clin. Lipidol.*, 9(5): 692–697.

Chung, Y.H., T.C. Chen, W.J. Yang, S.Z. Chen, J.M. Chang, W.Y. Hsieh, and M.H. Hsieh. 2024. Ectopic expression of a bacterial thaimin monophosphate kinase enhances vitamin B_1 biosynthesis in plants. *Plant J.*, 117: 1330–1343.

Coburn, S.P. 2015. Vitamin B-6 metabolism and interactions with TNAP. *Subcell. Biochem.*, 76: 207–238.

Condo, M., A. Posar, A. Arbizzani, and A. Parmeggiani. 2009. Riboflavin prophylaxis in pediatric and adolescent migraine. *J. Headache. Pain*, 10: 361–365.

Dasgupta, A. 2019. *Biotin and Other Interferences in Immunoassays: A Concise Guide*. Elsevier.

Davis, K.R., and L. Tourneau. 1981. Evaluation of the nutrient composition of wheat. II: Proximate analysis, thiamin, riboflavin, niacin, and pyridoxine. *Cereal Chem.*, 58(2): 116–120.

De Flora, A. 2011. Intercellular and intracellular trafficking of NAD+ and NAD+ precursors, intermediate metabolites and signal metabolites. *J. Siena Acad. Sci.*, 3(1): 83–83.

de la Garza, R.D., E.P. Quinlivan, S.M. Klaus, G.J. Basset, J.F. Gregory III, and A.D. Hanson. 2004. Folate biofortification in tomatoes by engineering the pteridine branch of folate synthesis. *Proc. Natl. Acad. Sci. U.S.A.*, 101(38): 13720–13725.

De Lepeleire, J., S. Strobbe, J. Verstraete, D. Blancquaert, L. Ambach, R.G. Visser, C. Stove, and D. Van Der Straeten. 2018. Folate biofortification of potato by tuber-specific expression of four folate biosynthesis genes. *Mol. Plant*, 11(1): 175–188.

Degnan, P.H., M.E. Taga and A.L. Goodman. 2014. Vitamin B_{12} as a modulator of gut microbial ecology. *Cell. Metab.*, 20(5): 769–778.

Díaz de la Garza, R.I., J.F. Gregory III, and A.D. Hanson. 2007. Folate biofortification of tomato fruit. *Proc. Natl. Acad. Sci. U.S.A.*, 104(10): 4218–4222.

DiNicolantonio, J.J., C.J. Lavie, A.K. Niazi, J.H. O'Keefe, and T. Hu. 2013. Effects of thiamine on cardiac function in patients with systolic heart failure: Systematic review and metaanalysis of randomized, double-blind, placebo-controlled trials. *Ochsner J.*, 13: 495–499.

Dong, J., M. Tu, Y. Feng, A. Zdepski, F. Ge, D. Kumar, J.P. Slovin, and J. Messing. 2019. Candidate gene identification of existing or induced mutations with pipelines applicable to large genomes. *Plant J.*, 97(4): 673–682.

Dong, W., N. Thomas, P.C. Ronald, and A. Goyer. 2016. Overexpression of thiamin biosynthesis genes in rice increases leaf and unpolished grain thiamin content, but not resistance to *Xanthomonas oryzae* pv. oryzae. *Front. Plant Sci.*, 7: 616.

Dong, W., V.O. Stockwell, and A. Goyer. 2015. Enhancement of thiamin content in *Arabidopsis thaliana* by metabolic engineering. *Plant Cell Physiol.*, 56(12), 2285–2296.

Dong, W., Z.J. Cheng, C.L. Lei, X.L. Wang, J.L. Wang, J. Wang, and J.M. Wan. 2014. Overexpression of folate biosynthesis genes in rice (*Oryza sativa* L.) and evaluation of their impact on seed folate content. *Plant Foods Hum. Nutr.*, 69: 379–385.

Dong, W., Z. Cheng, X. Wang, B. Wang, H. Zhang, N. Su, and J. Wan. 2011. Determination of folate content in rice germplasm (*Oryza sativa* L.) using tri-enzyme extraction and microbiological assays. *Int. J. Food Sci. Nutr.*, 62(5): 537–543.

Dutilleul C., C. Lelarge, J.L. Prioul, R. De Paepe, C.H. Foyer, and G. Noctor. 2005. Mitochondria-driven changes in leaf NAD status exert a crucial influence on the control of nitrate assimilation and the integration of carbon and nitrogen metabolism. *Plant Physiol.*, 139: 64–78. doi: 10.1104/pp.105.066399.

Farah, N., V.K. Chin, P.P. Chong, W.F. Lim, C.W. Lim, R. Basir, and T.Y. Lee. 2022. Riboflavin as a promising antimicrobial agent? A multi-perspective review. *Curr. Res. Microb. Sci.*, 3: 100111.

Fitzpatrick, T.B., I. Dalvit, F.H. Chang, K. Wang, J.B. Fudge, S.H. Chang, B. Maillot and W. Gruissem. 2024. Vitamin B_1 enhancement in the endosperm of rice through thiamine sequestration. *Plant Biotechnol. J.* (in press) doi: 10.1111/pbi.14348.

Fitzpatrick, T.B., and Z. Noordally. 2021. Of clocks and coenzymes in plants: Intimately connected cycles guiding central metabolism. *New Phytol.*, 230(2): 416–432.

Fitzpatrick, T.B., and L. Chapman. 2020. The importance of thiamine (vitamin B_1) in plant health: From crop yield to biofortification. *J. Biol. Chem.*, 295: 12002–12013.

Fitzpatrick, T.B., N. Amrhein, B. Kappes, P. Macheroux, I. Tews, and T. Raschle. 2007. Two independent routes of *de novo* vitamin B_6 biosynthesis: Not that different after all. *Biochem. J.*, 407(1): 1–13.

Fouad, W.M., and B. Rathinasabapathi. 2006. Expression of bacterial L-aspartate-α-decarboxylase in tobacco increases β-alanine and pantothenate levels and improves thermotolerance. *Plant Mol. Biol.*, 60: 495–505.

Gakière, B., A.R. Fernie, and P. Pétriacq. 2018. More to NAD^+ than meets the eye: A regulator of metabolic pools and gene expression in *Arabidopsis*. *Free Rad. Biol. Med.*, 122: 86–95.

Garg, M., N. Sharma, S. Sharma, P. Kapoor, A. Kumar, V. Chunduri, and P. Arora. 2018. Biofortified crops generated by breeding, agronomy, and transgenic approaches are improving lives of millions of people around the world. *Front. Nutr.*, 5: 12.

Genschel, U., C.A. Powell, C. Abell, and A.G. Smith. 1999. The final step of pantothenate biosynthesis in higher plants: Cloning and characterization of pantothenate synthetase from *Lotus japonicus* and *Oryza sativum* (rice). *Biochem. J.*, 341(3): 669–678.

Gerdes, S., C. Lerma-Ortiz, O. Frelin, S.M. Seaver, C.S. Henry, V. de Crécy-Lagard, and A.D. Hanson. 2012. Plant B vitamin pathways and their compartmentation: A guide for the perplexed. *J. Exp. Bot.*, 63(15): 5379–5395.

Gibson, G.E., J.A. Hirsch, P. Fonzetti, B.D. Jordan, R.T. Cirio, and J. Elder. 2016. Vitamin B_1 (thiamine) and dementia. *Ann. N.Y. Acad. Sci.*, 1367(1): 21–30.

Gillies, S.A., S.R. McIntosh, and R. Henry. 2008. A transgenic cereal crop with enhanced folate: Rice expressing wheat *HPPK/DHPS*. Presentation to the 11th International Wheat Genetics Symposium, Brisbane, Australia.

Gorelova, V., J. De Lepeleire, J. Van Daele, D. Pluim, C. Meï, A. Cuypers, and D. Van Der Straeten. 2017. Dihydrofolate reductase/thymidylate synthase fine-tunes the folate status and controls redox homeostasis in plants. *Plant Cell*, 29(11): 2831–2853.

Goyer, A. 2017. Thiamin biofortification of crops. *Curr. Opin. Biotechnol.*, 44: 1–7.

Goyer A., and K. Sweek. 2011. Genetic diversity of thiamin and folate in primitive cultivated and wild potato (*Solanum*) species. *J. Agric. Food Chem.*, 59(24): 13072–13080.

Green, R., L.H. Allen, A.-L. Bjørke-Monsen, A. Brito, J.-L. Guéant, J.W. Miller, A.M. Molloy, E. Nexo, S. Stabler, and B.-H. Toh. 2017. Vitamin B_{12} deficiency. *Nat. Rev. Dis. Primers*, 3(1), 1–20.

Guo, H., X. Xiong, Y. Wang, H. Tian, S. Zhang, and G. Geng.2022. Integrative proteomic and physiological analyses of the molecular response to dessication-stress in *Auricularia fibrillifera*. *Front. Plant Sci.*, 13: 995810.

György, P. 1964. The history of vitamin B_6. Introductory remarks. *Vitamins and Hormones*, 22: 361–365. Academic Press.

Hao, J., P. Pétriacq, L. de Bont, M. Hodges, and B. Gakière. 2018. Characterization of L-aspartate oxidase from *Arabidopsis thaliana*. *Plant Sci.*, 271: 133–142.

Hashida, S.N., H. Takahashi, M. Kawai-Yamada, and H. Uchimiya. 2007. *Arabidopsis thaliana* nicotinate/nicotinamide mononucleotide adenyltransferase (AtNMNAT) is required for pollen tube growth. *Plant J.*, 49(4): 694–703.

Harvard T.H. Chan School of Public Health. *The Nutrition Source*. https://www.hsph.harvard.edu/nutritionsource/. Accessed 30.08.2023.

Hegyi, J., R.A. Schwartz, and V. Hegyi. 2004. Pellagra: Dermatitis, dementia, and diarrhea. *Int. J. Dermatol.*, 43(1): 1-5.

Herrero, S., and M.E. Daub. 2007. Genetic manipulation of vitamin B-6 biosynthesis in tobacco and fungi uncovers limitations to up-regulation of the pathway. *Plant Sci.*, 172(3): 609–620.

Hossain, F., K. Sarika, V. Muthusamy, R.U. Zunjare, and H.S. Gupta. 2019. Quality protein maize for nutritional security. In: A. Qureshi, Z. Dar and S. Wani (eds). *Quality Breeding in Field Crops*, 217–237. Springer Nature.

Hossain, F., V. Muthusamy, N. Pandey, A.K. Vishwakarma, A. Baveja, R.U. Zunjare, and H.S. Gupta. 2018. Marker-assisted introgression of opaque2 allele for rapid conversion of elite hybrids into quality protein maize. *J. Genet.*, 97: 287–298.

Hossain, T., I. Rosenberg, J. Selhub, G. Kishore, R. Beachy, and K. Schubert. 2004. Enhancement of folates in plants through metabolic engineering. *Proc. Natl. Acad. Sci. U.S.A.*, 101(14): 5158–5163.

Hsieh, W.Y., J.C. Liao, H.T. Wang, T.H. Hung, C.C. Tseng, T.Y. Chung, and M.H. Hsieh. 2017. The *Arabidopsis* thiamin-deficient mutant *pale green1* lacks thiamin monophosphate phosphatase of the vitamin B_1 biosynthesis pathway. *Plant J.*, 91(1): 145–157.

Huang, L., M. Pyc, S. Alseekh, D.R. McCarty, V. de Crécy-Lagard, J.F. Gregory III, C.S. Henry, A.R. Fernie, R.T. Mullen, and A.D. Hanson. 2018. A plastidial pantoate transporter with a potential role in pantothenate synthesis. *Biochem. J.*, 475(4): 813–825.

Institute of Medicine. 1998. Food and Nutrition Board. Dietary reference intakes: Thiamin, riboflavin, niacin, vitamin B_6, folate, vitamin B_{12}, pantothenic acid, biotin, and choline. Washington DC.: National Academy Press.

Jalilian, N., G.D. Najafpour, and M. Khajouei. 2019. Enhanced vitamin B_{12} production using *Chlorella vulgaris*. *Int. J. Eng.*, 32(1): 1–9.

Jha, A.B., K. Ashokkumar, M. Diapari, S.J. Ambrose, H.X. Zhang, and B. Taran. 2015. Genetic diversity of folate profiles in seeds of common bean, lentil, chickpea and pea. *J. Food Compost. Anal.*, 42: 134–140.

Jonczyk, R., S. Ronconi, M. Rychlik, and U. Genschel. 2008. Pantothenate synthetase is essential but not limiting for pantothenate biosynthesis in *Arabidopsis*. *Plant Mol. Biol.*, 66: 1–14.

Jung, S., M.K. Kim and B.Y. Choi. 2017. The long-term relationship between dietary pantothenic acid (vitamin B_5) intake and C-reactive protein concentration in adults aged 40 years and older. *Nutr. Metab. Cardiovasc. Dis.*, 27(9): 806–816.

Kancherla, V., K. Wagh, H. Pachón, and G.P. Oakley Jr. 2021. A 2019 global update on folic acid-preventable spina bifida and anencephaly. *Birth Defects Res.*, 113(1): 77–89.

Katoh, A., K. Uenohara, M. Akita, and T. Hashimoto. 2006. Early steps in the biosynthesis of NAD in *Arabidopsis* start with aspartate and occur in the plastid. *Plant Physiol.*, 141(3): 851–857.

Kim, C., and K. Park. 2022. Dietary niacin intake and risk of dyslipidemia: A pooled analysis of three prospective cohort studies. *Clin. Nutr.*, 41(12): 2749–2758.

Knip, M., I.F. Douek, W.P. Moore, H.A. Gillmor, A.E. McLean, P.J. Bingley, and E.A. Gale. 2000. Safety of high-dose nicotinamide: A review. *Diabetologia*, 43(11): 1337–1345.

Labella, J.I., R. Cantos, P. Salinas, J. Espinosa, and A. Contreras. 2020. Distinctive features of PipX, a unique signaling protein of cyanobacteria. *Life*, 10(6): 79.

Lanska, D.J. 2012. The discovery of niacin, biotin, and pantothenic acid. *Ann. Nutr. Metab.*, 61(3): 246–253.

Laudert, D., and Hohmann, H.P. 2011. Application of enzymes and microbes for the industrial production of vitamins and vitamin-like compounds. *In:* M. Moo-Young (ed.), *Comprehensive Biotechnology*, 2nd edn. 583–602. Burlington, MA, USA: Academic Press.

Lawrence, A.D., E. Nemoto-Smith, E. Deery, J.A. Baker, S. Schroeder, D.G. Brown, J.M. Tullet, M.J. Howard, I.R. Brown, and A.G. Smith. 2018. Construction of fluorescent analogs to follow the uptake and distribution of cobalamin (vitamin B_{12}) in bacteria, worms, and plants. *Cell Chem. Biol.*, 25: 941–951.

Leonardi, R. and S. Jackowski. 2007. Biosynthesis of pantothenic acid and coenzyme A. *EcoSal Plus*, 2, 10.1128/ecosalplus.

Levit, R., G. Savoy de Giori, A. De Moreno, De Leblanc, and J.G. LeBlanc. 2021. Recent update on lactic acid bacteria producing riboflavin and folates: Application for food fortification and treatment of intestinal inflammation. *J. Appl. Microbiol.*, 130(5): 1412–1424.

Li, J., Y. Ge, M. Zadeh, R. Curtiss III, and M. Mohamadzadeh. 2020. Regulating vitamin B_{12} biosynthesis via the cbiM Cbl riboswitch in *Propionibacterium* strain UF1. *Proc. Natl. Acad. Sci. U.S.A.*, 117(1): 602–609.

Li, J., J. Liu, P. Zhang, Y. Wan, X. Xia, Y. Zhang and Z. He. 2018. Genome-wide association mapping of vitamins B_1 and B_2 in common wheat. *Crop J.*, 6(3): 263–270.

Li, K.T., M. Moulin, N. Mangel, M. Albersen, N.M. Verhoeven-Duif, Q. Ma, P. Zhang, T.B. Fitzpatrick, W. Gruissem and H. Vanderschuren. 2015. Increased bioavailable vitamin B_6 in field-grown transgenic cassava for dietary sufficiency. *Nature Biotechnol.*, 33(10): 1029–1032.

Liang, Q., K. Wang, X. Liu, B. Riaz, L. Jiang, X. Wan, X. Ye, and C. Zhang. 2019. Improved folate accumulation in genetically modified maize and wheat. *J. Exp. Bot.*, 70(5): 1539–1551.

Liu, S., W. Hu, Z. Wang, and T. Chen. 2020. Production of riboflavin and related cofactors by biotechnological processes. *Microb. Cell Fact.*, 19(1): 1–16.

Lonsdale, D. 2006. A review of the biochemistry, metabolism, and clinical benefits of thiamin(e) and its derivatives. *eCAM*, 3: 49–59.

Mahabadi, N., A. Bhusal, and S.W. Banks. 2023. Riboflavin Deficiency. *In: StatPearls* [Internet]. Treasure Island (FL): StatPearls Publishing.

Mahmood, L. 2014. The metabolic processes of folic acid and Vitamin B_{12} deficiency. *JHRR* (In Developing Countries), 1(1): 5–9.

Mangel, N., J.B. Fudge, W. Gruissem, T.B. Fitzpatrick, and H. Vanderschuren. 2022. Natural variation in vitamin B_1 and vitamin B_6 contents in rice germplasm. *Front. Plant Sci.*, 13: 856880.

Mangel, N., J.B. Fudge, K.T. Li, T.Y. Wu, T. Tohge, A.R. Fernie, B. Szurek, T.B. Fitzpatrick, W. Gruissem, and H. Vanderschuren. 2019. Enhancement of vitamin B_6 levels in rice expressing Arabidopsis vitamin B_6 biosynthesis *de novo* genes. *Plant J.*, (6): 1047–1065.

Mangel, N., J.B. Fudge, T.B. Fitzpatrick, W. Gruissem, and H. Vanderschuren. 2017. Vitamin B_1 diversity and characterization of biosynthesis genes in cassava. *J. Exp. Bot.*, 68(13): 3351–3363.

Mangel, N. 2016. *Natural Variation, Molecular Determinants and Genetic Engineering of Vitamin B_1 and Vitamin B_6 Biosynthesis in Cassava and Rice.* Doctoral dissertation, ETH Zurich.

Maqbool, M.A., A. Beshir Issa, and E.S. Khokhar. 2021. Quality protein maize (QPM): Importance, genetics, timeline of different events, breeding strategies, and varietal adoption. *Plant Breed.*, 140(3): 375–399.

Martinis, J., E. Gas-Pascual, N. Szydlowski, M. Crèvecoeur, A. Gisler, L. Bürkle, and T.B. Fitzpatrick. 2016. Long-distance transport of thiamine (vitamin B_1) is concomitant with that of polyamines. *Plant Physiol.*, 171(1): 542–553.

McCormick, D. 2006. Vitamin B_6. *In*: B. Bowman, and R. Russell (Eds.), *Present Knowledge in Nutrition.* 9th Edn. Washington, D.C.: International Life Sciences Institute.

Mezzetti, B., F. Balducci, F. Capocasa, C.F. Zhong, R. Cappelletti, and L. Di Vittori. 2016. Breeding strawberry for higher phytochemicals content and claim it: Is it possible? *Int. J. Fruit Sci.*, 16: 194–206.

Mimura, M., R. Zallot, T.D. Niehaus, G. Hasnain, S.K. Gidda, and T.N.D. Nguyen. 2016. Arabidopsis TH2 encodes the orphan enzyme thiamin monophosphate phosphatase. *Plant Cell*, 28: 2683–2696.

Minhas, A.P., R. Tuli, and S. Puri. 2018. Pathway editing targets for thiamine biofortification in rice grains. *Front. Plant Sci.*, 9: 975.

Mitchell, H.K., E.E. Snell, and R.J. Williams. 1941. The concentration of "folic acid". *J. Amer. Chem. Soc.*, 63: 2284.

Mosegaard, S., G. Dipace, P. Bross, J. Carlsen, N. Gregersen, and R.K.J. Olsen. 2020. Riboflavin deficiency: Implications for general human health and inborn errors of metabolism. *Int. J. Mol. Sci.*, 21(11): 3847.

Muralla, R., E. Chen, C. Sweeney, J.A. Gray, A. Dickerman, B.J. Nikolau, and D. Meinke. 2008. A bifunctional locus (BIO3-BIO1) required for biotin biosynthesis in Arabidopsis. *Plant Physiol.*, 146(1): 60–73.

Naqvi, S., C.F.G. Zhu, K. Farre, L. Ramessar, Bassie, and J. Breitenbach. 2009. Transgenic multivitamin corn through biofortification of endosperm with three vitamins representing three distinct metabolic pathways. *Proc. Natl. Acad. Sci. U.S.A.*, 106: 7762–7767.

National Institutes of Health (NIH). https://www.nih.gov/factsheets. Accessed 30/08/2023.

Noctor, G., J. Hager, and S. Li. 2011. Biosynthesis of NAD and its manipulation in plants. *Adv. Bot. Res.*, 58: 153–201. Elsevier.

Noordally, Z., L. Land, C. Trichtinger, I. Dalvit, M. de Meyer, K. Wang, and T.B. Fitzpatrick. 2023. Clock and riboswitch control of THIC in tandem are essential for appropriate gauging of TDP levels under light/dark cycles in *Arabidopsis*. *iScience*, 26(3).

Noordally, Z.B., C. Trichtinger, I. Dalvit, M. Hofmann, C. Roux, N. Zamboni, and T.B. Fitzpatrick. 2020. The coenzyme thiamine diphosphate displays a daily rhythm in the *Arabidopsis* nucleus. *Commun. Biol.*, 3(1): 209.

Northrop-Clewes, C.A., and D.I. Thurnham. 2012. The discovery and characterization of riboflavin. *Ann. Nutr. Metab.*, 61(3): 224–230.

Nunes, A.C., D.C. Kalkmann, and F.J. Aragao. 2009. Folate biofortification of lettuce by expression of a codon optimized chicken *GTP cyclohydrolase I* gene. *Transgenic Res.*, 18: 661–667.

Nurit, E., J. Bordes, F. Balfourier, E. Paux, A. Piquet, D. Fossati, and G. Branlard. 2021. Association between SNP markers and 11 vitamin contents in grains of a worldwide bread wheat core collection. *J. Agri. Food Chem.*, 61(35): 8295–8303.

Pinon, V., S. Ravanel, R. Douce, and C. Alban. 2005. Biotin synthesis in plants. The first committed step of the pathway is catalyzed by a cytosolic 7-keto-8-aminopelargonic acid synthase. *Plant Physiol.*, 139(4): 1666–1676.

Pique, K., W. Taber, A. Thompson, and C.G. Maitland. 2021. Isolated optic neuropathy due to folate deficiency with associated iron overload. *BMJ Case Rep.*, 14(7): e242399.

Quadram Institute Newsletter, 2023. https://quadram.ac.uk/plant-powered-innovations-uk-researchers-boost-b12-in-pea-shoots-using-aeroponics/

Raman, S.B., and B. Rathinasabapathi. 2004. Pantothenate synthesis in plants. *Plant Sci.*, 167(5): 961–968.

Ramirez Rivera, N.G., C. Garcia-Salinas, F.J. Aragao, and R.I. Diaz de la Garza. 2016. Metabolic engineering of folate and its precursors in Mexican common bean (*Phaseolus vulgaris* L.). *Plant Biotechnol. J.*, 14: 2021–2032.

Ramos, R.A. 2010. *Folate profiling in rice.* Doctoral dissertation, University of Nottingham.

Raschke, M., S. Boycheva, M. Crèvecoeur, A. Nunes-Nesi, S. Witt, A.R. Fernie, N. Amrhein, and T.B. Fitzpatrick. 2011. Enhanced levels of vitamin B_6 increase aerial organ size and positively affect stress tolerance in *Arabidopsis*. *Plant J*, 66(3): 414–432.

Rathee, S., V. Nayak, K.R. Singh, and A. Ojha. 2022. Nanofortification of vitamin B-complex in food matrix: Need, regulations, and prospects. *Food Chem.: Mol. Sci.*, 4: 100100.

Revuelta, J.L., R. Ledesma-Amaro, and A. Jiménez. 2016. Industrial production of vitamin B_2 by microbial fermentation. Chapter 2. *Industrial Biotechnology of Vitamins, Biopigments, and Antioxidants*, 15–40. Weinheim, Germany: Wiley-VCH Verlag GmbH & Co. KGaA.

Riaz, B., Q. Liang, X. Wan, K. Wang, C. Zhang, and X. Ye. 2019. Folate content analysis of wheat cultivars developed in the North China Plain. *Food Chem.*, 289: 377–383.

Robinson, G.C., M. Kaufmann, C. Roux, and T.B. Fitzpatrick. 2016. Structural definition of the lysine swing in *Arabidopsis thaliana* PDX1: Intermediate channeling facilitating vitamin B_6 biosynthesis. *Proc. Natl. Acad. Sci. U.S.A.*, 113(40): E5821–E5829.

Robinson, B.R., V. Sathuvalli, J. Bamberg, and A. Goyer. 2015. Exploring folate diversity in wild and primitive potatoes for modern crop improvement. *Genes*, 6(4): 1300–1314.

Rodionov, D.A., X. Li, I.A. Rodionova, C. Yang, L. Sorci, E. Dervyn, D. Martynowski, H. Zhang, M.S. Gelfand, and A.L. Osterman. 2008. Transcriptional regulation of NAD metabolism in bacteria: Genomic reconstruction of NiaR (YrxA) regulon. *Nucleic Acids Res.*, 36(6): 2032–2046.

Rogers, L.M., A.M. Cordero, C.M. Pfeiffer, D.B. Hausman, B.L. Tsang, L.M. De-Regil, J. Rosenthal, H. Razzaghi, E.C. Wong, and A.P. Weakland. 2018. Global folate status in women of reproductive age: A systematic review with emphasis on methodological issues. *Ann. N. Y. Acad. Sci.*, 1431(1): 35–57.

Rosado-Souza, L., S. Proost, M. Moulin, S. Bergmann, S.E. Bocobza, A. Aharoni, T.B. Fitzpatrick, M. Mutwil, A.R. Fernie, and T. Obata. 2019. Appropriate thiamin pyrophosphate levels are required for acclimation to changes in photoperiod. *Plant Physiol.*, 180(1):185–197.

Roychoudhury, A. 2021. Metabolic engineering of water-soluble vitamins in plants for enhanced nutrition. *Int. J. Recent Adv. Biotechnol. Nanotechnol.*, 4(1): 1–10.

Safiul Azam, F.M., T. Lian, Q. Liang, W. Wang, C. Zhang, and L. Jiang. 2022. Variation of vitamin B contents in maize inbred lines: Potential genetic resources for biofortification. *Front. Nutr.*, 9: 1029119.

Sahr, T., S. Ravanel, G. Basset, B.P. Nichols, A.D. Hanson, and F.F. Rébeillé. 2006. Folate synthesis in plants: Purification, kinetic properties, and inhibition of aminodeoxychorismate synthase. *Biochem. J.*, 396(1): 157–162.

Sato, K., Y. Kudo, and K. Muramatsu. 2004. Incorporation of a high level of vitamin B_{12} into a vegetable, *kaiware daikon* (Japanese radish sprout), by the absorption from its seeds. *Biochim. Biophys. Acta*, 1672: 135–137.

Sauve, A.A. 2008. NAD^+ and vitamin B_3: From metabolism to therapies. *J. Pharmacol. Exp. Ther.*, 324(3): 883–893.

Schippers, J.H., A. Nunes-Nesi, R. Apetrei, J. Hille, A.R. Fernie, and P.P. Dijkwel. 2008. The Arabidopsis *onset of leaf death5* mutation of quinolinate synthase affects nicotinamide adenine dinucleotide biosynthesis and causes early ageing. *Plant Cell*, 20: 2909–2925.

Scott, J., F. Rébeillé, and J. Fletcher. 2000. Folic acid and folates: The feasibility for nutritional enhancement in plant foods. *J. Sci. Food Agri.*, 80(7): 795–824.

Shewry, P.R., M.J. Hawkesford, V. Piironen, A.M. Lampi, K. Gebruers, D. Boros, and J.L. Ward. 2013. Natural variation in grain composition of wheat and related cereals. *J. Agric. Food Chem.*, 61(35): 8295–8303.

Shewry, P.R., F. Van Schaik, C. Ravel, G. Charmet, M. Rakszegi, Z. Bedo, and J.L. Ward. 2011. Genotype and environment effects on the contents of vitamins B_1, B_2, B_3, and B_6 in wheat grain. *J. Agric. Food Chem.*, 59(19): 10564–10571.

Shinnosuke, H., T. Kentaro, T. Hirabayashi, M. Kawai-Yamada, K. Shoji, F. Goto, and H. Uchimiya. 2010. Metabolome analysis of Arabidopsis plants over-expressing NAD synthetase gene: Poster 8, Gene Regulation, and Molecular Biology. *In:* Plant Biology 2010, July 31-August 4, 2010, Montreal, Canada. American Society of Plant Biologists, Rockville, MD, no. 08021.

Siddique, S., Z.S. Radakovic, C. Hiltl, C. Pellegrin, T.J. Baum, H. Beasley, A.F. Bent, O. Chitambo, D. Chopra, and E.G. Danchin. 2022. The genome and lifestage-specific transcriptomes of a plant-parasitic nematode and its host reveal susceptibility genes involved in trans-kingdom synthesis of vitamin B_5. *Nature Commun.*, 13(1): 6190.

Sinclair, S.J., K.J. Murphy, C.D. Birch, and J.D. Hamill. 2000. Molecular characterization of quinolinate phosphoribosyltransferase (QPRTase) in Nicotiana. *Plant Mol. Biol.*, 44: 603–617.

Stach, K., W. Stach, and K. Augoff. 2021. Vitamin B_6 in health and disease. *Nutrients*, 13(9): 3229.

Steensma, P., M. Eisenhut, M. Colinas, L. Rosado-Souza, A.R. Fernie, A.P. Weber, and T.B. Fitzpatrick. 2023. PYRIDOX (AM) INE 5′-PHOSPHATE OXIDASE3 of *Arabidopsis thaliana* maintains carbon/nitrogen balance in distinct environmental conditions. *Plant Physiol.*, 193(2): 1433–1455.

Storozhenko, S., V. De Brouwer, M. Volckaert, O. Navarrete, D. Blancquaert, and G.F. Zhang. 2007. Folate fortification of rice by metabolic engineering. *Nature Biotechnol.*, 25: 1277–1279.

Strobbe, S., J. Verstraete, C. Stove, and D. Van Der Straeten. 2021. Metabolic engineering of rice endosperm towards higher vitamin B_1 accumulation. *Plant Biotechnol. J.*, 19(6): 1253–1267.

Strobbe, S., and D. Van Der Straeten. 2018. Toward eradication of B-vitamin deficiencies: Considerations for crop biofortification. *Front. Plant Sci.*, 9: 443.

Strobbe, S., and D. Van Der Straeten. 2017. Folate biofortification in food crops. *Curr. Opin. Biotechnol.*, 44: 202–211.

Suo, J., Y. Gao, H. Zhang, G. Wang, H. Cheng, Y. Hu, and J. Wu. 2022. New insights into the accumulation of vitamin B_3 in *Torreya grandis* nuts via ethylene induced key gene expression. *Food Chem.*, 371: 131050.

Suzuki, M., S. Wu, M. Mimura, S. Alseekh, A.R. Fernie, A.D. Hanson, D.R. McCarty. 2020. Construction and applications of a B vitamin genetic resource for investigation of vitamin-dependent metabolism in maize. *Plant J.*, 101(2): 442–454.

Tambasco-Studart, M., I. Tews, N. Amrhein, and T.B. Fitzpatrick. 2007. Functional analysis of PDX2 from Arabidopsis, a glutaminase involved in vitamin B_6 biosynthesis. *Plant Physiol.*, 144(2): 915–925.

Tambasco-Studart, M., O. Titiz, T. Raschle, G. Forster, N. Amrhein, and T.B. Fitzpatrick. 2005. Vitamin B_6 biosynthesis in higher plants. *Proc. Natl. Acad. Sci. U.S.A.*, 102(38): 13687–13692.

Thakur, K., S.K. Tomar, A.K. Singh, S. Mandal, and S. Arora. 2017. Riboflavin and health: A review of recent human research. *Crit. Rev. Food Sci. Nutr.*, 57(17): 3650–3660.

Tian, Y.S., J. Xu, B. Wang, X.Y. Fu, J.J. Gao, H.J. Han, and Q.H. Yao. 2021. Riboflavin fortification of rice endosperm by metabolic engineering. *Plant Biotechnol. J.*, 19(8): 1483.

Titcomb, T.J., and S.A. Tanumihardjo. 2019. Global concerns with B vitamin statuses: Biofortification, fortification, hidden hunger, interactions, and toxicity. *Compr. Rev. Food Sci. Food Saf.*, 18(6): 1968–1984.

Titiz, O. 2008. *Investigation of PDX1 and its role* in *a novel vitamin B_6 biosynthesis pathway in Arabidopsis thaliana*. Doctoral dissertation, ETH Zurich.

Uebanso, T., A. Yoshimoto, S. Aizawa, M. Nakamura, R. Masuda, T. Shimohata, and A. Takahashi. 2020. Glycolate is a novel marker of vitamin B_2 deficiency involved in gut microbe metabolism in mice. *Nutrients*, 12(3): 736.

U.S. Department of Health and Human Services (USDHHS). *Vitamin B_2 Fact Sheet for Health Professionals*. https://ods.od.nih.gov/factsheets/Riboflavin-HealthProfessional/. Accessed 30/08/2023.

Van Der Straeten, D., N.K. Bhullar, H. De Steur, W. Gruissem, D. MacKenzie, W. Pfeiffer, M. Qaim, I. Slamet-Loedin, S. Strobbe, and J. Tohme. 2020. Multiplying the efficiency and impact of biofortification through metabolic engineering. *Nature Commun.*, 11(1): 5203.

Vasal, S.K. 2000. High quality protein corn. *In: Specialty Corns*, 97–142. Boca Raton, Florida, US: CRC Press.

Vitreschak, A.G., D.A. Rodionov, A.A. Mironov, and M.S. Gelfand. 2003. Regulation of the vitamin B_{12} metabolism and transport in bacteria by a conserved RNA structural element. *RNA*, 9(9): 1084–1097.

Wachter, A., M. Tunc-Ozdemir, B.C. Grove, P.J. Green, D.K. Shintani, and R.R. Breaker. 2007. Riboswitch control of gene expression in plants by splicing and alternative 3′ end processing of mRNAs. *Plant Cell*, 19(11): 3437–3450.

Watanabe, F., and T. Bito. 2018. Vitamin B_{12} sources and microbial interaction. *Exp. Biol. Med.*, 243(2): 148–158.

Williams, P.A., J.M. Harder, N.E. Foxworth, K.E. Cochran, V.M. Philip, V. Porciatti, and S.W.M. John. 2017. Vitamin B_3 modulates mitochondrial vulnerability and prevents glaucoma in aged mice. *Science*, 355(6326): 756–760.

Wills, L. 1931. Treatment of "pernicious anaemia of pregnancy" and "tropical anaemia" with special reference to yeast extract as curative agent. *Br. Med. J.*, 1: 1059–1064.

Woo, K.S., T.C. Kwok, and D.S. Celermajer. 2014. Vegan diet, subnormal vitamin B-12 status, and cardiovascular health. *Nutrients*, 6(8): 3259–3273.

Xiao, F., H. Wang, Z. Shi, Q. Huang, L. Huang, J. Lian, and Z. Xu. 2020. Multi-level metabolic engineering of *Pseudomonas mutabilis* ATCC31014 for efficient production of biotin. *Metab. Eng.*, 61: 406–415.

Xu, J., S. Patassini, P. Begley, S. Church, H.J. Waldvogel, R.L. Faull, R.D. Unwin, and G.J. Cooper. 2020. Cerebral deficiency of vitamin B_5 (d-pantothenic acid; pantothenate) as a potentially-reversible cause of neurodegeneration and dementia in sporadic Alzheimer's disease. *Biochem. Biophys. Res. Comm.*, 527(3): 676–681.

CHAPTER
9

Vitamin C: Vital, Let's Achieve More Together

Shivani Kathi[a[0000-0003-2406-3447]] **and Catherine Simpson**[b*[0000-0002-8164-4275]]

[a] Department of Horticulture and Landscape Architecture, Oklahoma State University, Stillwater, OK, United States

[b] Department of Plant and Soil Science, Texas Tech University, Lubbock, TX, United States

1. Introduction

1.1. Vitamin C

Vitamin C is an essential antioxidant for humans and plants. However, humans do not have the ability to biosynthesize vitamin C and must acquire it through diet (Linster and Van Schaftingen, 2007). Vitamin C was not recognized as an essential micronutrient until the early 20th century, when deficiencies of the compound caused scurvy (Paciolla et al., 2019). The recommended daily intake of vitamin C is greater than other vitamins due to its important role in these diverse functions in the human body. The primary sources of vitamin C for humans are fruits like citrus, berries, kiwi, and mango, and vegetables like tomatoes, peppers, spinach, lettuce, and so on (Simkin, 2019; Strobbe et al., 2018). Consumption of grain/cereal-rich diets, excess processed foods, and insufficient fresh fruits and vegetables results in vitamin C deficiencies. Furthermore, vitamin C deficiencies are commonly seen in low-income countries and less frequently in high-income countries (Rowe and Carr, 2020). Still, because vitamin C deficiencies are rarely tested, this could be higher than current numbers indicate. Severe vitamin C deficiency causes scurvy, a multisystemic disorder that compromises human immunity (Magiorkinis et al., 2011). Alternatively, mild vitamin C deficiencies result in non-specific symptoms such as apathy, fatigue, depression, and anxiety

*Corresponding author: Catherine.simpson@ttu.edu

(Baradhi et al., 2018). Thus, the prevalence of vitamin C deficiencies is difficult to identify and quantify and may be misdiagnosed, leading to underreporting of deficiencies (Strobbe et al., 2018). So, it is important to consume enough amounts of vitamin C to meet daily dietary recommendations.

1.2. Vitamin C Contribution to Human Health

A minimum of 10 mg of vitamin C was first recommended for daily dietary intake for adults to avoid scurvy (Padayatty and Levine, 2016). However, it was later discovered that vitamin C plays a major role in enzyme activation, collagen production, immunity function, and oxidation stress reduction, which led to increasing this recommendation to 75 and 90 mg/day for adult women and men in the U.S., respectively; but recommended amounts vary by life stage, gender, and pregnancy (Frei et al., 2012). Vitamin C in plants exists naturally in two forms: predominantly ascorbic acid (AA) and occasionally its oxidized form, dehydroascorbic acid (DHA). The combined amount of AA and DHA is considered to be total vitamin C. DHA can also be absorbed and converted back into AA by humans under favorable circumstances (Carr and Vissers, 2013; Zaoui et al., 2016). Therefore, AA and DHA contribute to overall vitamin C intake and biological activity. It should be noted that AA, DHA, and vitamin C are often used interchangeably in the literature, and this review will reflect the language used in articles represented herein.

Vitamin C from fruits and vegetables is vital to human health. While supplements can meet dietary guidelines, plant-derived AA is more bioavailable and reduces chronic disease risk (Carr and Vissers, 2013). High-dose vitamin C is being studied to reduce cardiovascular disease and cancer risk (Paciolla et al., 2019). Therefore, consuming more vitamin C-rich foods is important. Many factors affect plant vitamin C concentrations, but several methods have been used to boost them. Among those techniques, biofortification increases crop nutritional quality by adding vitamins, minerals, and other beneficial components (Garg et al., 2018).

2. Vitamin C Biofortification Approaches

Biofortification of AA offers a sustainable solution to improve quantities of consumed vitamin C. Different approaches have been explored to achieve AA biofortification. Of the few notable attempts, AA has been biofortified in different crops through three major techniques, i.e., agronomically, breeding, and bioengineering. Furthermore, some studies have explored the biofortification of vitamin C, considering both AA and DHA are bioavailable (Carr and Vissers, 2013; Zaoui et al., 2016). This approach recognizes the importance of both forms of vitamin C in human nutrition and aims to increase the overall content of this nutrient in biofortified crops. By increasing the levels of AA and/or total vitamin C in crops, biofortification efforts can help address vitamin C deficiencies in populations that heavily rely on restrictive diets. These initiatives offer a sustainable and cost-effective approach to improve communities' nutritional status and overall health particularly in regions where access to diverse diets and vitamin C-rich foods may be limited (Martin and Li, 2017).

2.1. Agronomic Approaches

Vitamin C has been increased agronomically by adding with other plant vital nutrients through fertilization, foliar sprays or supplemented nutrient solutions (Table 1). However, the majority of vitamin C biofortification has been through the application of selenium (Se). Applying Se increased AA in various crops such as cherry tomato (Sabatino et al., 2021), tomato, lettuce, and melon (de los Santos-Vázquez et al., 2016). Application of Se was shown to increase vitamin C, ranging between 14.9–257.7% among various crops as seen in Table 1. Moreover, Se applied in combination with iodine (I) (up to 23.1%) or AA (up to 121.4%) also resulted in increased vitamin C. Additionally, other mineral applications that have shown promising results in vitamin C biofortification are silica (Si) and I. Briefly, foliar sprays of Si improved AA content in grains like rice (Das et al., 2018), quinoa (Lata-Tenesaca et al., 2022), and horticultural crops like tomato (dos Santos et al., 2022). The increase in AA through Si application ranged between 12–58% depending on the rates applied and crops (Table 1). Applying I through foliar sprays or supplemented nutrient solution has increased AA mostly in horticultural crops like strawberry (Macías et al., 2021), lettuce (Leyva et al., 2011), tomato (Mageshen et al., 2022), and sweet corn (Grzanka et al., 2021). When I is applied in the form of potassium iodide (KI) or potassium iodate (KIO_3), the increase in AA ranged between 6 to 33% in different crops (Table 1). Other mineral nutrients application that improved AA content are potassium (K), zinc (Zn), calcium (Ca), and magnesium (Mg) as seen in Table 1. Alternatively, the application of different engineered nanomaterials (NM's) has also been shown to increase ascorbate antioxidant activity in plants. Silica NM's increased radish ascorbate peroxidase activity (Taqdees et al., 2022) and Se NM's increased AA content in bok choy (Wang et al., 2022) and tomato (Neysanian et al., 2020).

Another way to agronomically biofortify crops for increased vitamin C content is through exogenous application of AA. Noreen et al. (2021) found that barley leaves synthesized more AA after foliar application. Se and AA foliar application increased sweet basil leaf ascorbate (Ardebili et al., 2015). Arugula and broccoli microgreens with supplemented nutrient solution had higher vitamin C and AA content (Kathi et al., 2022, 2023). Adding AA to the nutrient solution increased vitamin C up to 289.4%, especially in microgreens (Table 1).

2.2. Bioengineering Approaches

Bioengineering is a popular vitamin C biofortification concept due to its minimal time and resource requirements. Bioengineered foods encounter certain challenges, and variables among consumers. Vitamin C in plants is produced through two major biosynthetic pathways such as the L-galactose/Smirnoff-Wheeler pathway

Table 1: Studies that demonstrated increased vitamin C through the application other nutrients like selenium (Se), silica (Si) or iodine (I), potassium (K), zinc (Zn), calcium (Ca), magnesium (Mg), and ascorbic acid (AA)

Element Applied	Crop	Plant Part Biofortified	Application Method	Application Form	Rate	% Increase in Ascorbic Acid/ Vitamin C	Reference
Se	Cherry tomato	Fruits	Fertigation	Selenium dioxide	4.0 μmol L^{-1}	Ungrafted: 14.9% Grafted: 16.0%	Sabatino et al., 2021
Se	Lettuce	Leaves	Supplemented nutrient solution	Sodium selenate	40 mmol L^{-1}	257.7%	Ríos et al., 2008
Se nanomaterials	Bok choy	Shoots	Fertilization	Negatively charged gum arabic-Se (Se (-))	0.05 mg kg^{-1}	44.4%	Wang et al., 2022
Se or Se nanoparticles	Tomato	Leaves	Foliar	Se nanoparticles-red elemental Se- sodium selenate	3 mg L^{-1}	36.0%	Neysanian et al., 2020
Se and I	Indian mustard	Leaves and stem	Foliar	Sodium selenate or potassium iodide	50 mg Se L^{-1} or 100 mg I L^{-1}	Leaves- 10.2 Stem: 19.2 to 23.1%	Golubkina et al., 2018
Se and AA	Sweet basil	Leaf tissues	Foliar	NA	30 or 60 mg L^{-1} of Se with 200 mg L^{-1} AA	Se_{30}+AA- 100 Se_{60}+AA- 121.4	Ardebili et al., 2015
Si	Quinoa	Grain	Foliar and root	Sodium silicate	3 mmol L^{-1}	58.0	Lata-Tenesaca et al., 2022

(*Contd.*)

Table 1: *(Contd.)*

Element Applied	Crop	Plant Part Biofortified	Application Method	Application Form	Rate	% Increase in Ascorbic Acid/ Vitamin C	Reference
Si	Rice	Roots and Shoots	Foliar	Silicon	2 mM	Roots: 12 to 58% Shoots: 3 to 25%	Das et al., 2018
I	Strawberry	Leaves	Foliar	KIO_3	100 μM	33%	Macías et al., 2021
I or Vanadium and I	Sweet corn	grains	Foliar	V: Ammonium metavanadate I: KIO_3 or KI	V: 1.0 μM I: 10 or 100 μM	6 to 25%	Grzanka et al., 2021
Several nutrients	Cabbage and rocket microgreens	Microgreens	Supplemented nutrient solution	Modified Hoagland formulation	2.0 mM nitrate, 0.25 mM S, 0.20 mM P, 0.62 mM K, 0.75 mM Ca, 0.17 mM Mg, 0.25 mM ammonium, 20 μM Fe, 9 μM Mn, 0.3 μM Cu, 1.6 μM Zn, 20 μM boron, and 0.3 μM molybdenum	Cabbage: 30% Rocket: 58%	El-Nakhel et al., 2021

K	Cherry tomato	Fruits	Fertilization	KCl	15 mM	2 to 22%	Constán-Aguilar et al., 2014
Zn, Ca, and Mg	Mustard	Leaves	Supplemented nutrient solution	NA	Zn- 500 µM, or Ca- 40 mM, or Mg- 20 mM	Zn: 5.3% Ca: 7.6% Mg: 2.6%	Blasco et al., 2015
AA	Arugula microgreens	Microgreens	Supplemented nutrient solution	L-(+)-ascorbic acid	0.25% AA	289.4%	Kathi et al., 2022
AA	Broccoli microgreens	Microgreens	Supplemented nutrient solution	L-(+)-ascorbic acid	0.5% AA	53.9 to 222%	Kathi et al., 2023

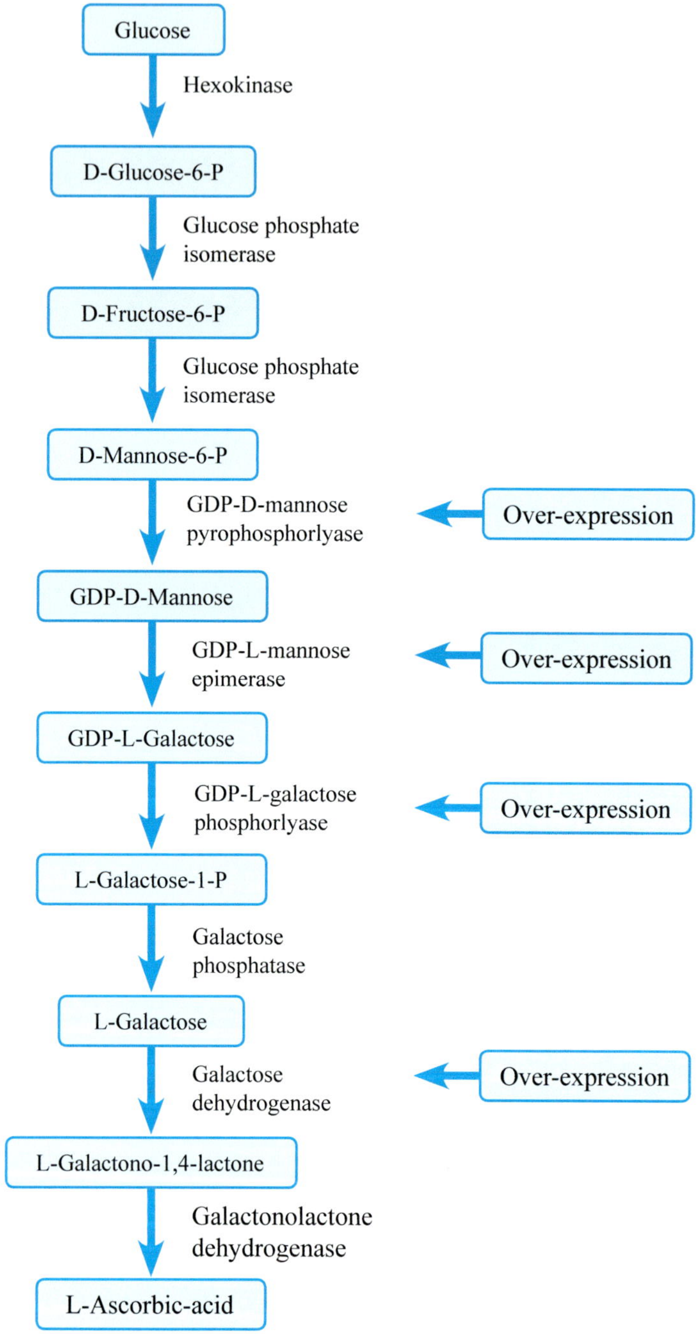

Figure 1: Biosynthetic pathway of vitamin C with key biofortification genes utilized.

(primary pathway), and the D-galacturonate pathway, extensively discussed in the review by Mellidou et al. (2021). To summarize, the L-galactose pathway begins with D-fructose-6-P derived from D-glucose converted to AA in eight enzymatic steps (Figure 1). Alternatively, the D-galacturonate pathway uses cell wall polysaccharides that ultimately results in the production of AA (Mellidou et al., 2021). However, AA production through the second pathway is rather unclear and seems to vary among different plant species and stages (Mellidou et al., 2021). Biofortification of vitamin C has been achieved through overexpression of genes such as GDP-D-mannose pyro-phosphorylase (GMP), GDP-D-mannose-3,5-epimerase (GME), GDP-L-galactose-phosphorylase (GGP), L-galactono-1,4-lactone dehydrogenase (GLDH) and D-galacturonate reductase (GalUR) involved in L-galactose pathway and D-galacturonate pathway, respectively encoding the enzymes in biosynthesis of AA. Some of the successful attempts of biofortifying vitamin C in different crops through overexpression of genes associated with enzymes involved in L-galactose and D-galacturonate pathways are shown in Table 2. This method of overexpression of genes involved in biosynthesis was able to increase AA content in various food crops such as rice (Broad et al., 2020), and tomato, tobacco, strawberry, lettuce, maize and potato as reviewed by Cruz-Rus et al. (2012).

Similarly, GDP-mannose pyrophosphatase aids in producing GDP-D-mannose during the first step of the L-galactose pathway. Thus, overexpression of tomato

Table 2: Biofortification of vitamin C through overexpression of genes associated with enzymes involved in vitamin C biosynthesis pathways

Crop	Target Enzyme	Process	Result	Reference
Tobacco	GDP-D-mannose pyro-phosphorylase	Ectopic expression	Increase in AA in leaves by 2-fold	Badejo et al., 2008
Tomato	GDP-D-mannose-3,5-epimerase	Overexpression	Increased AA ranging between 1.19 to 1.60-fold in leaves and fruits	Zhang et al., 2011
Tomato, strawberry, and potato	GDP-L-galactose-phosphorylase	Overexpression	2-to-6-fold increase in AA in fruits/ tubers	Bulley et al., 2012
Tobacco	L-galactono-1,4-lactone dehydrogenase	Overexpression	Up to 2-fold increase in AA in transgenic cells	Tokunaga et al., 2005
Arabidopsis and potato	D-galacturonate reductase	Ectopic expression	Increase in AA by 2 to 3-fold in tobacco leaves and potato tubers	Agius et al., 2003; Hemavathi et al., 2009

fruit-specific bacterial pyrophosphatase increased vitamin C content in ripe tomato fruits (Osorio et al., 2013). Alternatively, pyramiding of several genes involved in biosynthesis (such as GME × GMP, or GME × GMP × GGP × L-Gal-1-P phosphatase (GPP)), in tomatoes also resulted in increased ascorbate content in both leaves and red fruits (Li et al., 2019).

In contrast, another way to biofortify vitamin C is through silencing genes such as monodehydroascorbate reductase (MDHAR) and dehydroascorbate reductase (DHAR) that aid in ascorbate recycling (Jiang et al., 2017). However, this method resulted in lower increases in vitamin C in maize, potato, and tomato than that of overexpression methods (Gest et al., 2013; Jiang et al., 2017).

2.3. Breeding Approaches

Traditional breeding has greater potential in the vitamin C biofortification of plants (Strobbe et al., 2018). However, the conventional breeding process is more time-consuming than agronomic efforts, which can be a substantial drawback in biofortification efforts. Yet, the heritability of vitamin C concentrations from parent to progeny is high (Geleta and Labuschagne, 2006). Making it more important to determine suitable parents with high vitamin C concentrations to use in breeding programs. For example, among seven parental genotypes in pepper, "Mareko Shote" and "'PBC 142A" had higher concentrations of vitamin C (Geleta and Labuschagne, 2006). Similarly, Adalid et al. (2012) found that CDP4777 accession in orange-brownish wild cherry tomato can be a potential donor parent to improve AA content as its F_1 hybrids showed 130% of the AA content compared to that of the female parent. Roselló et al. (2011) also found that CDP4777 in small, orange-brownish cherry tomatoes, showed greater genotypic donor potential by accumulating more than 50 times AA than controls. For potatoes, the cultivar "Teton Russet" a hybrid between "Blazer Russet" and "Classic Russet" (Novy et al., 2014), "Targhee Russet" a hybridization product of A92303-7 and A96004-8 (Whitworth et al., 2016) and "Sage Russet" a cross between A89384–10 and A91194–4 (Yilma et al., 2017) showed higher amounts of AA than other hybrids after breeding attempts (Table 3).

Furthermore, Topuz and Ozdemir (2007) found that pepper cultivars 730 F1 and 1245 F1 had greater vitamin C which correlated with capsaicinoid content. This pattern of differences found between cultivars within a species has also been a common occurrence outside of the Solanaceae family. The strawberry cultivar 'Sahara' was found to be the best suited candidate for vitamin C biofortification breeding (Ali and Serçe, 2022). And, out of 79 apple cultivars, the greatest amount of vitamin C at harvest was found in the pulp of five cultivars, namely "Marie Joseph d'Othée", 'Braeburn', 'Coupette', 'Nicogreen', and 'Ontario' (Lemmens et al., 2020). Additionally, Shameh et al. (2019) found that *Rosa moschata* had greater AA content out of five different rose species. Consequently, the data from the evaluation of genetic sources could help screen genotypes containing traits to produce greater AA in breeding programs (Shameh et al., 2019). These correlations in different cultivars can be used to breed new cultivars with higher AA content. For instance, Karimi et al., 2020 mapped out the correlation between different biochemical characteristics such as vitamin C, anthocyanin, total phenolic content, titratable acidity, and total soluble carbohydrate in 20 genotypes of pomegranate (Karimi et al., 2020). Genotypes with higher anthocyanin, vitamin C and total phenolic content in leaves had higher vitamin C in arils, these results can help select genotypes with characteristics that produce greater AA content in the juvenile phase of plants.

Table 3: Crops with improved vitamin C content due to breeding

Crop	Cutivars or Hybrids with Greater Vitamin C Content	Results	Reference
Pepper	Mareko Shote and PBC 142A	Genotypes showed higher concentrations of vitamin C	Geleta and Labuschagne, 2006
Orange-brownish wild cherry tomato	CDP4777	F_1 hybrids showed 130% AA content	Adalid et al., 2012
Potato	Hybrids: Teton Russet, Targhee Russet, Sage Russet	Had increased amounts of AA than other cultivars and hybrids tested	Novy et al., 2014; Whitworth et al., 2016; Yilma et al., 2017
Pepper	Cultivars: 730 F1 and 1245 F1	Greater vitamin C content	Topuz and Ozdemir, 2007
Strawberry	Cultivar: Sahara	Best suited candidate for increased vitamin C content through breeding	Ali and Serçe, 2022
Apple	Cultivars: Marie Joseph d'Othée, Braeburn, Coupette, Nicogreen, and Ontario	Pulp had greatest amount of vitamin C	Lemmens et al., 2020
Rose	G10	Had greater AA content	Shameh et al., 2019

3. Bioavailability of Biofortified Vitamin C

Bioavailability refers to the amount of a nutrient absorbed into a body and utilized therein (Davidsson and Tanumihardjo, 2013). Therefore, it is essential to consider the bioavailability of biofortified vitamins and minerals for the functionality and practicality of crops that have been enriched. Vitamin C cannot be stored in the body and is absorbed through Na-dependent transporters or GLUT transporters, further underscoring the need for consistent dietary intake (Sanahuja et al., 2013). Bioavailability is affected by plant or food matrix, processing, nutrient interactions, and individual physiology. Both forms of vitamin C, AA, and DHA, are interconvertible and have activity in cells, therefore, both forms must be considered in biofortification and bioavailability studies. There is some evidence that the bioavailability of vitamin C is higher in plant-derived forms rather than synthetic, but this may be due to interaction with other micronutrients and phytonutrients that are present in plants and not in synthetic supplements (Beck et al., 2011; Carr and Vissers, 2013). For example, Vissers et al. (2011) found that, in mice, ascorbate levels in blood and organs were higher when fed a kiwifruit gel than synthetic ascorbate in water.

Bioavailability can also be affected by other mineral nutrients or plant matrix components. Vitamin E reduces vitamin C absorption in rats, and fruits rich in vitamin E seem to have higher absorption of vitamin C (Celik et al., 2007; Vissers et al., 2011, 2013). Vitamin C is essential for iron absorption in the intestine and metabolism (Lane and Richardson, 2014; Paciolla et al., 2019). Absorption and bioavailability of vitamin C may also be affected by anti-nutrients such as glucosinolates, tannins, fiber, phytates, polyphenols, and oxalates (Petroski and Minich, 2020). Furthermore, Koh et al. (2012) found that nitrates and oxalate were inversely related to vitamin C and flavonoids in spinach. Flavonoids may also reduce the bioavailability of vitamin C, but the exact effects are unclear (Melse-Boonstra, 2020). However, these effects vary as there are indications that vitamin C may counteract the negative impacts of phytates on iron absorption (López-Moreno et al., 2022). Despite abundant vitamin C research in plants and humans, biofortified crop bioavailability has not been extensively studied. Most studies have focused on improving iron bioavailability, but vitamin C bioavailability was not examined (Lockyer et al., 2018).

4. Stability of Biofortified Vitamin C in Foods

Processed plant products can also affect bioavailability. Vitamin C is a water-soluble vitamin and is not as stable as most mineral nutrients (Sanahuja et al., 2013). Vitamin C is easily oxidized by exposure to heat, oxygen, and some metal ions (Vissers et al., 2013). Extreme temperatures, plant or fruit maturity, water loss, bruising, mechanical injuries, and trimming affect vitamin C stability after harvest (Lee and Kader, 2000). Yet, studies indicate that storage in reduced oxygen or 10% CO_2 conditions can reduce vitamin C losses (Lee and Kader, 2000). Storage has also been shown to affect concentrations of vitamin C in human subjects. After strawberries were stored for four days at 4°C, vitamin C in plasma was higher even though concentrations of analyzed strawberries were identical to non-stored strawberries (Azzini et al., 2010). Conversely, Johnston and Hale (2005) found that vitamin C was reduced in human plasma after consuming reconstituted frozen orange juice stored for 8 days at 4°C compared to day 1. This indicates that storage in addition to cooking and post-harvest processes can affect vitamin C's bioavailability. Treatments have been developed to increase the retention of vitamin C, such as surface coating, heat treatments, irradiation, and ethylene degreening in addition to managing temperatures and water loss during storage (Mditshwa et al., 2017). However, further research is needed to elucidate vitamin C retention treatments for many plant products, especially when the plants are biofortified.

After harvest and post-harvest storage, vitamin C levels are subject to further changes if not consumed raw. While many fruits and vegetables rich in vitamin C are commonly consumed fresh, increasing availability and absorption, they are also consumed as food products that have been cooked or processed. Cooking can degrade vitamin C in foods, but the cooking method influences the magnitude of degradation (Sanahuja et al., 2013). For example, wet cooking methods reduce concentrations of vitamin C to a greater extent, while dry cooking methods reduce concentrations to a lesser extent (Lee et al., 2018). In several studies, vitamin C was degraded to a greater degree when potatoes were boiled or fried, but to a lesser degree when they were braised, sauteed, or microwaved (Sanahuja et al.,

Table 4: Summary of decreases in vitamin C due to processing storage methods

Product	Processing Method	% Retention	Reference
Vegetables	Microwaved	92.7	Lee et al., 2018
	Steamed	89.2	
Potato	Boiling	16	Sanahuja et al., 2013
	Microwave	70.6	
	Pressure cooking	35	
	Frying	81	
	Baking	48	
Tomato Puree	Thermal processing	20	Mieszczakowska-Frąc et al., 2021
Strawberry Juice	Pasteurization	65	
Blackcurrant Juice	Pectinase enzyme treatment	69-74	
Broccoli	Boiling	44-53	
Carrot	Blanching	1-85	
Vegetables	Boiling	0-74	
	Blanching	58-89	
	Steaming	0-89	
	Microwave	67-112	
Vegetables	Boiling	31-73	Popova, 2018
	Microwave	32-68	
	Steamed	35-72	

2013). However, this can be affected by crop species and cultivar. Of the four cooking methods studied by Lee et al. (2018) vitamin C was degraded the most in boiled samples and was lower in chard and retained in higher concentrations in sweet potatoes. Yet, microwaving had higher retentions of vitamin C (>90%) in spinach, carrots, sweet potato, and broccoli (Lee et al., 2018) (Table 4).

Post-harvest processing also affects in vitro vitamin C bioavailability. Castro et al. (2020) found that cold plasma processing increased AA bioavailability from camu-camu juice. These findings are important when biofortifying crops because post-harvest processing and cooking can reverse nutrient quality gains. This may be circumvented by concentrating biofortification efforts on crops that are primarily consumed fresh, such as fruits, leafy greens, and microgreens.

5. Challenges in Biofortifying Vitamin C

Vitamin C biofortification faces several key challenges and constraints. Crop variability, driven by differing roles of AA in various plant organs and species, presents a hurdle (Cruz-Rus et al., 2012). Horticultural crops like fruits and vegetables have more AA than grains. Thus, biofortification has focused on these crops and a few agronomic varieties like corn, quinoa, and sweet corn (Table 2). Environmental factors greatly impact vitamin C biofortification. Researchers have found that light, drought, salinity, storage, processing, and temperature affect AA content (Locato et al., 2013; Suekawa, 2017). For instance, exposing sweet corn seeds to high-light conditions during germination increases AA (Liu et al., 2017). Salt stress can enhance AA content in crops like strawberries and lettuce with application of iodine (Leyva et al., 2011; Medrano Macías et al., 2021). Proper storage conditions like refrigeration, freezing, and controlled CO_2 levels can maintain vitamin C post-harvest. Furthermore, vitamin C's bioavailability and absorption in the human body can vary (Carr and Vissers, 2013). Bioavailability is influenced by the presence of micronutrients, dietary fiber, and phytochemicals. These challenges underscore the complexity of vitamin C biofortification and the need for multidisciplinary efforts to overcome them.

6. Conclusion

Prospects for vitamin C biofortification are promising, as it offers benefits for both plants and humans. In addition to its essential role in plant metabolism, vitamin C has been recognized for its contribution to plant stress response and the improvement of shelf life in produce. This highlights the potential of vitamin C biofortification in enhancing crops' nutritional value, promoting plant resilience, and reducing post-harvest losses. Combining multiple biofortification approaches mentioned above while considering genetic variability, bioavailability, and stability of vitamin C can lead to developing methods and crops with improved vitamin C content. Such advancements can contribute to promoting plant and human health while addressing nutritional challenges and ensuring food security in the future.

References

Adalid, A.M., S. Roselló, M. Valcárcel, and F. Nuez. 2012. Analysis of the genetic control of β-carotene and L-ascorbic acid accumulation in an orange-brownish wild cherry tomato accession. *Euphytica*, 184(2): 251–263.

Agius, F., R. González-Lamothe, J.L. Caballero, J. Muñoz-Blanco, M.A. Botella, and V. Valpuesta. 2003. Engineering increased vitamin C levels in plants by overexpression of a D-galacturonic acid reductase. *Nature Biotechnol.*, 21(2): 177–181.

Ali, M.N., and S. Serçe. 2022. Vitamin C and fruit quality consensus in breeding elite European strawberry under multiple interactions of environment. *Mol. Biol. Rep.*, 49(12): 11573–11586.

Ardebili, Z.O., N.O. Ardebili, S. Jalili, and S. Safiallah. 2015. The modified qualities of basil plants by selenium and/or ascorbic acid. *Turk. J. Bot.*, 39(3): 401–407.

Azzini, E., P. Vitaglione, F. Intorre, A. Napolitano, A. Durazzo, M.S. Foddai, A. Fumagalli, G. Catasta, L. Rossi, E. Venneria, A. Raguzzini, L. Palomba, V. Fogliano, and G. Maiani. 2010. Bioavailability of strawberry antioxidants in human subjects. *The Br. J. Nutr.*, 104(8): 1165–1173.

Badejo, A.A., N. Tanaka, and M. Esaka. 2008. Analysis of GDP-D-mannose pyrophosphorylase gene promoter from acerola (*Malpighia glabra*) and increase in ascorbate content of transgenic tobacco expressing the acerola gene. *Plant Cell. Physiol.*, 49(1): 126–132.

Baradhi, K.M., S. Vallabhaneni, and S. Koya. 2018. Scurvy in 2017 in the USA. *Baylor University Medical Center Proceedings*. 31(2): 227–228.

Beck, K., C.A. Conlon, R. Kruger, J. Coad, and W. Stonehouse. 2011. Gold kiwifruit consumed with an iron-fortified breakfast cereal meal improves iron status in women with low iron stores: A 16-week randomized controlled trial. *Br. J. Nutr.*, 105(1): 101–109.

Blasco, B., N.S. Graham, and M.R. Broadley. 2015. Antioxidant response and carboxylate metabolism in Brassica rapa exposed to different external Zn, Ca, and Mg supply. *J. Plant. Physiol.*, 176: 16–24.

Broad, R.C., J.P. Bonneau, J.T. Beasley, S. Roden, P. Sadowski, N. Jewell, C. Brien, B. Berger, E. Tako, R.P. Glahn, R.P. Hellens, and A.A.T. Johnson. 2020. Effect of rice GDP-L-galactose phosphorylase constitutive overexpression on ascorbate concentration, stress tolerance, and iron bioavailability in rice. *Front. Plant Sci.*, 11: 1953.

Bulley, S., M. Wright, C. Rommens, H. Yan, M. Rassam, K. Lin-Wang, and W. A. Laing. 2012. Enhancing ascorbate in fruits and tubers through over-expression of the l-galactose pathway gene GDP-l-galactose phosphorylase. *Plant Biotechnol. J.*, 10(4): 390–397.

Carr, A.C., and S. Rowe. 2020. The emerging role of vitamin C in the prevention and treatment of COVID-19. *Nutrients*, 12(11): 3286.

Carr, A.C., and M.C.M. Vissers. 2013. Synthetic or food-derived vitamin C: Are they equally bioavailable? *Nutrients*, 5(11): 4284–4304.

Castro, D.R.G., J.M. Mar, L.S. da Silva, K.A. da Silva, E.A. Sanches, J. de Araújo Bezerra, S. Rodrigues, F.A.N. Fernandes, and P.H. Campelo. 2020. Improvement of the bioavailability of Amazonian juices rich in bioactive compounds using glow plasma technique. *Food Bioproc. Tech.*, 13(4): 670–679.

Celik, A., S. Ercisli, and N. Turgut. 2007. Some physical, pomological, and nutritional properties of kiwifruit cv. Hayward. *Int. J. Food Sci. Nutr.*, 58(6): 411–418.

Constán-Aguilar, C., R. Leyva, B. Blasco, E. Sánchez-Rodríguez, T. Soriano, and J.M. Ruiz. 2014. Biofortification with potassium: Antioxidant responses during post-harvest of cherry tomato fruits in cold storage. *Acta Physiol. Plant*, 36: 283–293.

Cruz-Rus, E., I. Amaya, and V. Valpuesta 2012. The challenge of increasing vitamin C content in plant foods. *Biotechnol. J.*, 7: 1110–1121.

Das, P., I. Manna, A.K. Biswas, and M. Bandyopadhyay. 2018. Exogenous silicon alters ascorbate-glutathione cycle in two salt-stressed indica rice cultivars (MTU 1010 and Nonabokra). *Environ. Sci. Pollut. Res.*, 25(26): 26625–26642.

Davidsson, L., and S.A. Tanumihardjo. 2013. Bioavailability. *Encyclopedia of Hum. Nutr.*, 1–4: 149–155.

de los Santos-Vázquez, M.E., A. Benavides-Mendoza, N.A. Ruiz-Torres, M. Cabrera-de la Fuente, and Á. Morelos-Moreno. 2016. Sodium selenite treatment of vegetable seeds and seedlings and the effect on antioxidant status. *Emir. J. Food Agric.*, 28(8): 589–593.

dos Santos, M.M.M., G.P. da Silva, R. de M. Prado, J.S. Pinsetta Junior, B.H. Mattiuz, and H. Braun 2022. Biofortification of tomato with stabilized alkaline silicate and silicic acid, nanosilica, and potassium silicate via leaf increased ascorbic acid content and fruit firmness. *J. Plant. Nutr.*, 45(6): 896–903.

El-Nakhel, C., A. Pannico, G. Graziani, Kyriacou, A. Gaspari, A. Ritieni, and Y. Rouphael. 2021. Nutrient supplementation configures the bioactive profile and production characteristics of three *Brassica* L. microgreens species grown in peat-based media. *Agron.*, 11(2): 346.

Frei, B., I. Birlouez-Aragon, and J. Lykkesfeldt. 2012. Authors' perspective: What is the optimum intake of vitamin C in humans? *Crit. Rev. Food Sci. Nutr.*, 52: 815–829.

Garg, M., N. Sharma, S. Sharma, P. Kapoor, A. Kumar, V. Chunduri, and P. Arora. 2018. Biofortified crops generated by breeding, agronomy, and transgenic approaches are improving lives of millions of people around the world. *Front. Nutr.*, 5: 12.

Geleta, L.F., and M.T. Labuschagne. 2006. Combining ability and heritability for vitamin C and total soluble solids in pepper (*Capsicum annuum* L.). *J. Sci. Food Agric.*, 86(9): 1317–1320.

Gest, N., C. Garchery, H. Gautier, A. Jiménez, and R. Stevens. 2013. Light-dependent regulation of ascorbate in tomato by a monodehydroascorbate reductase localized in peroxisomes and the cytosol. *Plant Biotechnol. J.*, 11(3): 344–354.

Golubkina, N., H. Kekina, and G. Caruso. 2018. Yield, quality, and antioxidant properties of Indian mustard (*Brassica juncea* L.) in response to foliar biofortification with selenium and iodine. *Plants*, 7(4): 80.

Grzanka, M., S. Smoleń, Ł. Skoczylas, and D. Grzanka. 2021. Biofortification of sweetcorn with iodine: Interaction of organic and inorganic forms of iodine combined with vanadium. *Agron.*, 11(9).

Hemavathi, Upadhyaya, C.P., N. Akula, K.E. Young, S.C. Chun, D.H. Kim, and S.W. Park.

2010. Enhanced ascorbic acid accumulation in transgenic potato confers tolerance to various abiotic stresses. *Biotechnol. Letr.*, 32: 321–330.

Jiang, L., W. Wang, T. Lian, and C. Zhang. 2017. Manipulation of metabolic pathways to develop vitamin-enriched crops for human health. *Front. Plant Sci.*, 8: 937.

Johnston, C.S., and J.C. Hale. 2005. Oxidation of ascorbic acid in stored orange juice is associated with reduced plasma vitamin C concentrations and elevated lipid peroxides. *J. Am. Diet. Assoc.*, 105(1): 106–109.

Karimi, H.R., F. Zarei, S.H. Mirdehghan, A.A. Mohammadi Mirik, and A. Sarkhosh. 2020. Correlation among some biochemical compounds in fruit, leaf, and shoot bark of pomegranate trees for breeding programs. *Int. J. Fruit Sci.*, 20(4): 805–824.

Kathi, S., H. Laza, S. Singh, L. Thompson, W. Li, and C. Simpson. 2023. Vitamin C biofortification of broccoli microgreens and resulting effects on nutrient composition. *Front. Plant Sci.*, 14: 778.

Kathi, S., H. Laza, S. Singh, L. Thompson, W. Li, and C. Simpson. 2022. Increasing vitamin C through agronomic biofortification of arugula microgreens. *Sci. Rep.*, 12(1): 1–10.

Koh, E., S. Charoenprasert, and A.E. Mitchell. 2012. Effect of organic and conventional cropping systems on ascorbic acid, vitamin C, flavonoids, nitrate, and oxalate in 27 varieties of spinach (*Spinacia oleracea* L.). *J. Agri. Food Chem.*, 60(12): 3144–3150.

Lane, D.J.R., and D.R. Richardson. 2014. The active role of vitamin C in mammalian iron metabolism: Much more than just enhanced iron absorption! *Free Radic. Biol. Med.*, 75: 69–83.

Lata-Tenesaca, L.F., R. de Mello Prado, M. de Cássia Piccolo, D.L. da Silva, J.L. F. da Silva, and G.E. Ajila-Celi. 2022. Forms of application of silicon in quinoa and benefits involved in the association between productivity with grain biofortification. *Sci. Rep.*, 12(1): 1–9.

Lee, S.K., and A.A. Kader. 2000. Preharvest and postharvest factors influencing vitamin C content of horticultural crops. *Postharvest Biol. Technol.*, 20(3): 207–220.

Lee, S., Y. Choi, H.S. Jeong, J. Lee, and J. Sung. 2018. Effect of different cooking methods on the content of vitamins and true retention in selected vegetables. *Food Sci. Biotechnol.*, 27(2): 333–342.

Lemmens, E., E. Alós, M. Rymenants, N. De Storme, and W. Keulemans (Johan). 2020. Dynamics of ascorbic acid content in apple (*Malus x domestica*) during fruit development and storage. *Plant Physiol. Biochem.*, 151: 47–59.

Leyva, R., E. Sánchez-Rodríguez, J.J. Ríos, M.M. Rubio-Wilhelmi, L. Romero, J.M. Ruiz, and B. Blasco. 2011. Beneficial effects of exogenous iodine in lettuce plants subjected to salinity stress. *Plant Sci.*, 181(2): 195–202.

Li, X., J. Ye, S. Munir, T. Yang, W. Chen, G. Liu, W. Zheng, and Y. Zhang. 2019. Biosynthetic gene pyramiding leads to ascorbate accumulation with enhanced oxidative stress tolerance in tomato. *Int. J. Mol. Sci.*, 20(7): 1558.

Linster, C., and E. Van Schaftingen. 2007. Vitamin C: Biosynthesis, recycling and degradation in mammals. *FEBS J.*, 274: 1–22.

Liu, W., M. Tian, Y. Kong, J. Lu, N. Li, and J. Han. 2017. Multilayered vitamin C nanoliposomes by self-assembly of alginate and chitosan: Long-term stability and feasibility application in mandarin juice. *LWT.*, 75: 608–615.

Locato, V., S. Cimini, and L.D. Gara. 2013. Strategies to increase vitamin C in plants: From plant defense perspective to food biofortification. *Front. Plant Sci.*, 4: 152.

Lockyer, S., A. White, and J.L. Buttriss. 2018. Biofortified crops for tackling micronutrient deficiencies: What impact are these having in developing countries and could they be of relevance within Europe? *Nutr. Bull.*, 43: 319–357.

López-Moreno, M., M. Garcés-Rimón, and M. Miguel. 2022. Antinutrients: Lectins, goitrogens, phytates, and oxalates, friends or foe? *JFF*, 89: 104938.

Macías, J.M., M.G.L. Caltzontzit, E.N.R. Martínez, W.A.N. Ortiz, A.B. Mendoza, and P.M. Lagunes. 2021. Enhancement to salt stress tolerance in strawberry plants by iodine products application. *Agron.*, 11(3): 602.

Mageshen, V.R., P. Santhy, S. Meena, M.R. Latha, A. Senthil, T. Saraswathi, and P. Janaki. 2022. Influence of potassium iodate and chitosan iodate complex on growth, yield, quality and iodine uptake in 'shivam' hybrid of tomato (*Solanum lycopersicum* L.). *J. Appl. Nat. Sci.*, 14(3): 784–795.

Magiorkinis, E., A. Beloukas, and A. Diamantis. 2011. Scurvy: Past, present and future. *Eur. J. Intern. Med.*, 2: 147–152.

Martin, C., and J. Li. 2017. Medicine is not health care, food is health care: Plant metabolic engineering, diet, and human health. *New Phytol.*, 216: 699–719.

Mditshwa, A., L.S. Magwaza, S.Z. Tesfay, and U.L. Opara. 2017. Postharvest factors affecting vitamin C content of citrus fruits: A review. *Sci. Hortic.*, 218: 95–104.

Medrano Macias, J., M.G. Lopez Caltzontzit, E.N. Rivas Martinez, W.A. Narvaez Ortiz, A. Benavides Mendoza and P. Martinez Lagunes. 2021. Enhancement to salt stress tolerance in strawberry plants by iodine products application. *Agron.*, 11(3): 602.

Mellidou, I., A. Koukounaras, S. Kostas, E. Patelou, and A.K. Kanellis. 2021. Regulation of vitamin C accumulation for improved tomato fruit quality and alleviation of abiotic stress. *Genes*, 12(5): 694.

Melse-Boonstra, A. 2020. Bioavailability of micronutrients from nutrient-dense whole foods: Zooming in on dairy, vegetables, and fruits. *Front. Nutr.*, 7: 101.

Mieszczakowska-Frąc., M., K. Celejewska, and W. Płocharski. 2021. Impact of innovative technologies on the content of vitamin C and its bioavailability from processed fruit and vegetable products. *Antioxid.*, (Basel). 10(1): 54.

Neysanian, M., A. Iranbakhsh, R. Ahmadvand, Z.O. Ardebili, and M. Ebadi. 2020. Comparative efficacy of selenate and selenium nanoparticles for improving growth, productivity, fruit quality, and postharvest longevity through modifying nutrition, metabolism, and gene expression in tomato; potential benefits and risk assessment. *PLoS One*, 16(4): e0250192.

Noreen, S., M. Sultan, M.S. Akhter, K.H. Shah, U. Ummara, H. Manzoor, M. Ulfat, M.N. Alyemeni, and P. Ahmad. 2021. Foliar fertigation of ascorbic acid and zinc improves growth, antioxidant enzyme activity and harvest index in barley (*Hordeum vulgare* L.) grown under salt stress. *Plant Physiol. Biochem.*, 158: 244–254.

Novy, R.G., J.L. Whitworth, J.C. Stark, B.A. Charlton, S. Yilma, N.R. Knowles, M.J. Pavek, R.R. Spear, T.L. Brandt, N. Olsen, M. Thornton, and C.R. Brown. 2014. Teton Russet: An early-maturing, dual-purpose potato cultivar having higher protein and vitamin C content, low asparagine, and resistances to common scab and fusarium dry rot. *Am. J. Potato Res.*, 91(4): 380–393.

Osorio, S., A. Nunes-Nesi, M. Stratmann, and A.R. Fernie. 2013. Pyrophosphate levels strongly influence ascorbate and starch content in tomato fruit. *Front. Plant Sci.*, 4: 308.

Paciolla, C., S. Fortunato, N. Dipierro, A. Paradiso, S. De Leonardis, L. Mastropasqua and M.C. de Pinto. 2019. Vitamin C in plants: From functions to biofortification. *Antioxid.*, 8: 519.

Padayatty, S.J., and M. Levine. 2016. Vitamin C: The known and the unknown and Goldilocks. *Oral Dis.*, 22: 463–493.

Petroski, W., and D.M. Minich. 2020. Is there such a thing as "anti-nutrients"? A narrative review of perceived problematic plant compounds. *Nutr.*, 12(10): 1–32.

Popova, A.T. 2019. The effect of heating on the vitamin C content of selected vegetables. *World J. Adv. Res. Rev.*, 3(3): 27–32.

Ríos, J.J., M.A. Rosales, B. Blasco, L.M. Cervilla, L. Romero, and J.M. Ruiz 2008. Biofortification of Se and induction of the antioxidant capacity in lettuce plants. *Sci. Hortic.*, 116(3): 248–255.

Roselló, S., A.M. Adalid, J. Cebolla-Cornejo, and F. Nuez. 2011. Evaluation of the genotype, environment, and their interaction on carotenoid and ascorbic acid accumulation in tomato germplasm. *J. Sci. Food Agri.*, 91(6): 1014–1021.

Sabatino, L., S. La Bella, G. Ntatsi, G. Iapichino, F. D'Anna, C. De Pasquale, B.B. Consentino, and Y. Rouphael. 2021. Selenium biofortification and grafting modulate plant performance and functional features of cherry tomato grown in a soilless system. *Sci. Hortic.*, 285: 110095.

Sanahuja, G., G. Farré, J. Berman, U. Zorrilla-López, R.M. Twyman, T. Capell, P. Christou, and C. Zhu. 2013. A question of balance: Achieving appropriate nutrient levels in biofortified staple crops. *Nutr. Res. Rev.*, 26(2): 235–245.

Shameh, S., A. Alirezalu, B. Hosseini, and R. Maleki. 2019. Fruit phytochemical composition and color parameters of 21 accessions of five Rosa species grown in North West Iran. *J. Sci. Food Agric.*, 99(13): 5740–5751.

Simkin, A.J. (2019). Genetic engineering for global food security: Photosynthesis and biofortification. *Plants*, 8(12): 586.

Strobbe, S., J. De Lepeleire, and D. Van Der Straeten. 2018. From in planta function to vitamin-rich food crops: The ACE of biofortification. *Frontiers in Plant Science*, 871: 1862.

Suekawa, M., T. Kondo, Y. Fujikawa, and M. Esaka. 2017. Regulation of ascorbic acid biosynthesis in plants. *Ascorbic Acid in Plant Growth, Development, and Stress Tolerance*, 157–176.

Taqdees, Z., J. Khan, W.U.D Khan, S. Kausar, M. Afzaal, and I. Akhtar. 2022. Silicon and zinc nanoparticles-enriched miscanthus biochar enhanced seed germination, antioxidant defense system, and nutrient status of radish under NaCl stress. *Crop Pasture Sci.*, 73(5): 556–572.

Tokunaga, T., K. Miyahara, K. Tabata, and M. Esaka. 2005. Generation and properties of ascorbic acid-overproducing transgenic tobacco cells expressing sense RNA for L-galactono-1, 4-lactone dehydrogenase. *Planta*, 220: 854–863.

Topuz, A., and F. Ozdemir. 2007. Assessment of carotenoids, capsaicinoids, and ascorbic acid composition of some selected pepper cultivars (*Capsicum annuum* L.) grown in Turkey. *J. Food Compos. Anal.*, 20(7): 596–602.

Vissers, M.C.M., A.C. Carr, J.M. Pullar, and S.M. Bozonet. 2013. The bioavailability of vitamin C from kiwifruit. *Adv. Food Nutr.*, 68: 125–147.

Vissers, M.C.M., S.M. Bozonet, J.F. Pearson, and L.J. Braithwaite. 2011. Dietary ascorbate intake affects steady state tissue concentrations in vitamin C-deficient mice: Tissue deficiency after suboptimal intake and superior bioavailability from a food source (kiwifruit). *Am. J. Clin. Nutr.*, 93(2): 292–301.

Wang, C., X. Liu, F. Chen, L. Yue, X. Cao, J. Li, B. Cheng, Z. Wang, and B. Xing. 2022. Selenium content and nutritional quality of *Brassica chinensis* L. enhanced by selenium engineered nanomaterials: The role of surface charge. *EnviroN. Pollut.*, 308.

Whitworth, J.L., R.G. Novy, J.C. Stark, M. Thornton, N.R. Knowles, M.J. Pavek, R.R. Spear, C.R. Brown, B.A. Charlton, V. Sathuvalli, S. Yilma, T.L. Brandt, and N. Olsen.

2016. Targhee russet: A high yielding, dual purpose potato variety with high protein and vitamin C content and resistance to tuber soft rot. *Am. J. Potato Res.*, 93(2): 189–201.

Yilma, S., B. Charlton, C.C. Shock, D.C. Hane, S.R. James, A.R. Mosley, K.A. Rykbost, E.B.G. Feibert, N.R. Knowles, M.J. Pavek, J.C. Stark, R.G. Novy, J.L. Whitworth, J.J. Pavek, D.L. Corsini, T.L. Brandt, N. Olsen, C.R. Brown, M.I. Vales, and V. Sathuvalli. 2017. Sage russet: A new high yielding russet potato variety with cold-sweetening resistance, high vitamin C and protein contents, and excellent fresh pack and processing potential. *Am. J. Potato Res.*, 94(4): 379–389.

Zaoui, S., H. Gautier, D. Bancel, G. Chaabani, H. Wasli, M. Lachaâl, and N. Karray-Bouraoui. 2016. Antioxidant pool optimization in *Carthamus tinctorius* L. leaves under different NaCl levels and treatment durations. *Acta Physiol. Plant*, 38(8): 1–11.

Zhang, C., J. Liu, Y. Zhang, X. Cai, P. Gong, J. Zhang, and Z. Ye. 2011. Overexpression of SlGMEs leads to ascorbate accumulation with enhanced oxidative stress, cold, and salt tolerance in tomato. *Plant Cell Rep.*, 30: 389–398.

CHAPTER
10

Vitamin E: Pioneering the Shift from Stability to Elevated Oil Quality

Firoz Hossain*[0000-0001-6662-7752], Vignesh Muthusamy, Rajkumar U. Zunjare, Rashmi Chhabra, Ashvinkumar Katral and Gulab Chand

ICAR-Indian Agricultural Research Institute, New Delhi

1. Introduction

Diet quality is a crucial link between food security and nutrition, with poor diet quality contributing to micronutrient deficiencies. Micronutrients, such as vitamin E, while necessary in small amounts, play vital roles in human growth and development (Hossain et al., 2023). Vitamin E, known as 'tocochromanols', constitutes a group of potent, fat-soluble antioxidants that prevents the propagation of free radical reactions (Wu et al., 2022). This group encompasses tocopherols, tocotrienols, and plastochromanol-8 (PC-8), with saturated side chains referred to as 'tocopherols' and unsaturated side chains as 'tocotrienols'. These components owe their antioxidant capacity to the polar chromanol head and apolar polyprenyl side chain (Wu et al., 2022). Tocopherols and tocotrienols have four biosynthetically related isoforms (α, β, γ, and δ) varying in methyl group positions on their chromanol rings (Wu et al., 2022).

Vitamin-E deficiency (VED) contributes to various diseases/disorders like macular degeneration, neurological issues, cancer, cataracts, Alzheimer's, cardiovascular, and inflammatory diseases (Bramley et al., 2000; Das et al., 2019). Initial approaches to increase dietary vitamin E included food fortification, supplementation, and diet diversification. However, these strategies faced limited success in resource-limited regions, prompting the emergence of the 'crop biofortification' strategy (Hossain et al., 2023).

Enhancing vitamin E content in staple crops, such as cereals, leafy vegetables, oilseeds, and fruits, has been a target of crop breeding due to its health benefits. Genetic variation studies have been conducted in various crops to identify genomic

*Corresponding author: fh_gpb@yahoo.com

regions governing vitamin E accumulation (Fritsche et al., 2012). The vitamin E biosynthesis pathway has been elucidated, and genes have been transformed and overexpressed in various plants to enhance vitamin-E content and composition (Garg et al., 2021). However, the tocopherol biosynthesis pathway's complexity suggests the need for multiple strategies, including natural variations, gene editing, and transgenic approaches, to combat vitamin E deficiency effectively.

2. Vitamin E Biosynthesis Pathway

Vitamin E biosynthesis predominantly occurs in plastids, with enzymes found in plastoglobules or inner envelope (Figure 1). Here's an overview of vitamin E biosynthesis process, highlighting key genes and enzymes:

Step 1: Formation of Homogentisic acid (HGA)

The pathway initiates with the formation of HGA containing an aromatic head group. This reaction is catalyzed by p-hydroxyphenylpyruvate dioxygenase (HPPD), encoded by the *PDS1* gene, a crucial player in tocopherol biosynthesis (Norris et al., 1995). HGA serves as a substrate not only for tocochromanols' synthesis but also for other secondary metabolites like plastoquinones. Co-expression of *PDS1* with downstream genes in the pathway leads to elevation in tocopherol concentrations (Naqvi et al., 2011).

Step 2: Polyprenyl side chain synthesis

The tocopherols' polyprenyl side chain can derive from either (i) direct reduction of geranylgeranyl diphosphate (GGDP) by geranylgeranyl pyrophosphate synthase (GGPPS 11), produced through the methyl erythritol phosphate (MEP) pathway, or (ii) from free phytol, a product of chlorophyll degradation (Vom Dorp et al., 2015). Phytol is converted to phytyl phosphate (PP) and phytyl diphosphate (PDP) involving phytol kinase and phytyl-P-kinase, respectively, encoded by *Vte5* and *Vte6* genes (Vom Dorp et al., 2015).

Step 3: Condensation of HGA and GGDP

The condensation of HGA and GGDP leads to formation of 2-methyl-6-geranylgeranyl-1,4-benzoquinol (MGGBQ), forming tocotrienols through homogentisic acid geranylgeranyl transferase (HGGT). Alternatively, HGA reacts with solanesyl pyrophosphate (SPP) in the presence of homogentisate solanesyl transferase (HST) to yield PC-8, a homologue of γ-tocotrienol with a C-40 prenyl side chain.

Step 4: Condensation of HGA and PDP

The condensation of HGA and PDP results in the formation of 2-methyl-6-phytyl-1,4-benzoquinol (MPBQ) catalyzed by homogentisic acid phytyl transferase (HPT), encoded by *Vte2* gene. This step is a limiting factor in tocopherol biosynthesis. Cyclization reactions involving tocopherol cyclase

(TC) results in the formation of δ-tocopherol (Sattler et al., 2003) and methylation involving 2-methyl-6-phytyl-1,4-benzoquinol/2-methyl-6-solenesy l-1,4-benzoquinol methyl transferase (MPBQ/MSBQ MT) generate uncyclized precursor 2,3-dimethyl-6-phytyl-1,4-benzoquinone (DMPBQ). This enzyme also transforms MSBQ into PQ-9 (Mene-Saffrane and Dellapenna, 2010). The DMPBQ is transformed to γ-tocopherol through cyclization in the presence of TC encoded by *Vte1* gene. Both δ-tocopherol and γ-tocopherol are then methylated by γ-tocopherol methyl cyclase encoded by *Vte4* gene, yielding β-tocopherol and α-tocopherol, respectively. The *Vte2* mutant plants' phenotypes are counteracted by PC-8 produced in Step 2. Interestingly, γ-tocopherol can replace α-tocopherol functionally. The *Vte4* gene is crucial for metabolic engineering to enhance α-tocopherol levels in crop plants.

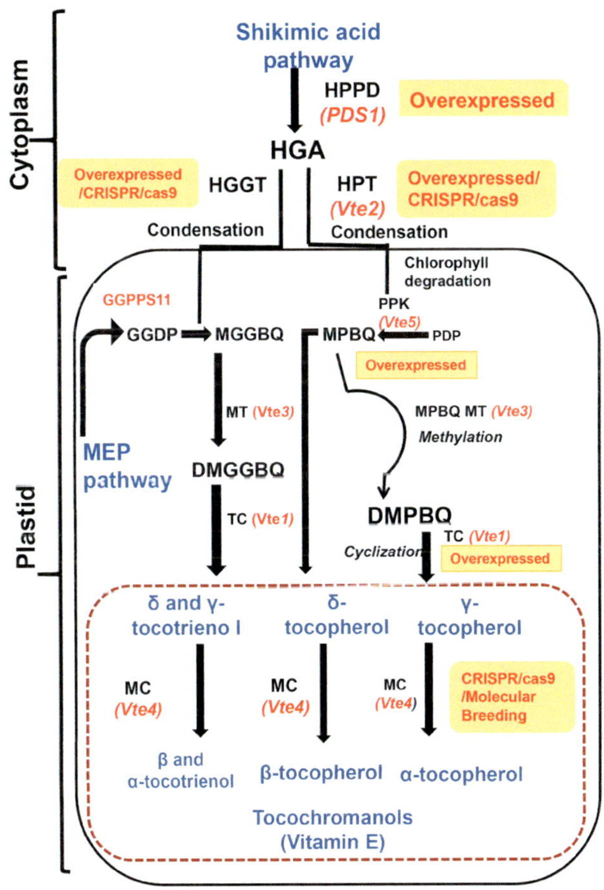

Figure 1: Vitamin E biosynthesis pathway with key genes and biofortification strategies.

Abbreviations: HPPD 4-hydroxyphenylpyruvate dioxygenase; HGA: Homogentisic acid; HGGT: Homogentisic acid transferase; HPT: Homogentisate phytyltransferase; GGPPS: geranylgeranyl diphosphate synthase; MEP: Methylerythritol phosphate; GGDP: geranylgeranyl diphosphate; MGGBQ: 2-methyl-6-geranylgeranyl-1,4-benzoquinol; DMGGBQ: 2,3-dimethyl-5-geranylgeranyl- benzoquinone; TC: Tocopherol cyclase; MC: Methyl cyclase; PP: Phytyl phosphate; PDP: Phytyl diphosphate; PK: Phytol kinase; PPK: Phytyl-P-kinase; MPBQ: 2-methyl-6-phytylbenzoquinone; MPBQ MT: 2-methyl-6-phytylbenzoquinone methyl transferase; DMPBQ: 2,3-dimethyl-5-phytyl-1,4-benzoquinone.

3. Genetic Variation for Tocopherol across the Food Crops

Mammals, photosynthetic bacteria, fungi, algae, and plants require dietary vitamin-E as an antioxidant. Only plastids and some green photosynthetic bacteria can synthesize it. α-tocopherol is the most active form, though γ-tocopherol predominates with ~80% proportion, while α-tocopherol constitutes ~20% (Hossain et al., 2023).

Studies on genetic variation in vitamin E content encompass key crop groups including cereals (wheat, barley, maize, rice, and sorghum), vegetables (potato, tomato, spinach, and lettuce), oilseed crops (soybean, rapeseed, and canola), and fruits (avocado and olive). Across the crops, α-tocopherol varied from 0.10–59.0 ppm (Table 1).

Table 1: Variations in tocopherols among diverse food crops (unit: ppm)

S. No.	Cultivars	α-T	β-T	δ-T	γ-T	Reference(s)
			Oil Producing Crops			
1.	Almond	0.50-49.24	0.14-1.42	0.50-3.19	16.06-60.0	Wang et al., 2019
2.	Apricot (Bitter)	1.36-3.10	-	1.67-2.50	66.42-81.0	Aksoz et al., 2020
3.	Argan	2.66-4.93	-	3.72-5.95	54.59-70.11	Hilali et al., 2020
4.	Cottonseed	30.5-57.3	0.04-0.30	-	10.5-31.7	Shahidi and De Camargo, 2016
5.	Linseed	0.54-1.20	-	0.75-0.95	52.0-57.3	Schwartz et al., 2008
6.	Olive	11.9-17.0	00-0.27	-	0.89-1.34	Shahidi and De Camargo, 2016
7.	Rapeseed	9.0-24.0	-	0.60-1.90	22-51	Schwartz et al., 2008; Wu et al., 2019

8.	Safflower	36.7-47.7	0.0-1.20	0.0-0.65	0.0-2.56	Syvaoja et al., 1986
9.	Soybean	9.53-12.0	1.00-1.31	23.9-26.0	61.0-69.9	Warner and Mounts, 1990
10.	Sunflower	32.7-59.0	00-2.40	0.27-0.50	1.40-4.50	Schwartz et al., 2008
11.	Palm	6.05-42.0	0.0-0.42	0.0-0.02	0.0-0.02	Ng et al., 2004; Shahidi and De Camargo, 2016
12.	Peanut	8.86-30.4	0.0-0.38	0.85-3.10	3.50-19.2	Syvaoja et al., 1986; Kornsteiner et al. (2006)
			Fruits and Vegetables			
13.	Avocado	3.22-5.06	-	-	-	Peraza-Magallanes et al., 2017
14.	Olives	26.7	-	-	2.10	Tekaya et al., 2014
15.	Tomato	0.10-0.26	0.007-0.014	0.035-0.119	0.001-0.007	Petropoulos et al., 2019
16.	Broccoli	3.33-3.77	0.02-0.03	-	0.29-0.40	Zaaboul and Liu, 2022
17.	Spinach	3.97	-	-	0.13	Zaaboul and Liu, 2022
			Cereals			
18.	Sorghum	43.17-44.88	70.66-103.93	34.94-35.73	39.58-41.49	Chung et al., 2013
19.	Oats	5.53-9.56	0.72-1.26	-	0.52-1.25	Peterson and Qureshi, 1992
20.	Wheat	15.2	7.8	-	-	Nielsen and Hansen, 2008
21.	Barley	7.15-12.7	0.40-1.7	0.39-1.40	0.37-8.5	Schuy et al., 2019; Panfili et al., 2008
22.	Maize	10.08-32.74	-	1.06-11.89	4.80-53.09	Das et al., 2018
23.	Rice	2.40-49.14	0.08-7.60	0.03-4.69	1.40-27.50	Goufo and Trindade, 2014

α-T: α-tocopherol, γ-T: γ-tocopherol, β-T: β-tocopherol, δ-T: δ-tocopherol

Malaysian rice cultivars displayed 2.56–16.47 ppm vitamin E in their kernels (Loganathan and Kim-Tiu, 2022). Rapeseed exhibited broad variation in total tocopherols (197.5–460.1 ppm) across 229 accessions (Fritsche et al., 2012). Likewise, maize inbreds showed 0.7–31.4 ppm α-tocopherol content (Das et al., 2019). Significant variations were noted in a set of 24 maize inbreds with favorable alleles of γ-tocopherol methyl transferase (Das et al., 2020). Higher α-tocopherol (3.3–4.3 ppm) was present in soybean cultivars from Hungry, Romania, and Yugoslavia compared to checks (0.49–0.96 ppm) (Ujiie et al., 2005).

Spinach, olives, and avocados exhibited elevated α-tocopherol (3.22–26.7 ppm) compared to other commonly used fruits and vegetables (0.10–0.26 ppm). Safflower, cottonseed, and soybean oils consistently had the highest α-tocopherol concentrations among oil-producing crops (Zaaboul and Liu, 2022). However, oil seeds' limited digestibility hampers their vitamin-E bioavailability (Stahl et al., 2002). Cereals, consumed widely, provide tocopherols in areas with scarce palm oil (Eitenmiller and Lee, 2004). Factors like variety, genotype, crop year, growth, processing, and storage influence tocopherol levels in plant-based foods (Zaaboul and Liu, 2022).

4. Molecular Approaches Employed for Enhancement of Vitamin-E

Genetic engineering with overexpression of single gene and overexpression in combination of multiple genes (co-expression) has been successfully achieved for enrichment of α-tocopherol content in arabidopsis, maize, soybean, sorghum, rice, barley, tomato, tobacco, lettuce, potato, and canola. Chloroplast engineering has been used by numerous researchers to increase vitamin E levels (Yabuta et al., 2013). Further, several QTLs and association mapping investigations on vitamin-E have been conducted in important crops, such as rapeseed, maize, soybean, rice, barley, and tomato for conventional and molecular breeding (Table 2). Main strategies for biofortification of vitamin E include:

(i) *Alteration in composition of the plant's tocochromanols into forms with higher biological efficacy:* This method seeks to increase the flux upstream of the tocopherol biosynthesis pathway (from the formation of HGA until they diverge into different tocopherols), which helps in converting other forms of tocopherols into α-tocopherol, and downregulate competing pathways that redirected flux away from tocopherol biosynthesis, such as deletion of homogentisate dioxygenase, that metabolizes HGA to central metabolic intermediates (Stacey et al., 2016). Elevating γ-tocopherol, α-tocopherol's precursor, via γ-TMT overexpression, considerably enhances vitamin E activity (Shintani and Dellapenna, 1998). This strategy yielded 2 to 26-fold vitamin E across diverse crops (Jiang, 2017).

(ii) *Raising the overall composition of tocochromanols in order to enhance crop vitamin-E activity:* To achieve this, genes that produce the aromatic head and/or the isoprenoid side chain of prenylquinols were overexpressed.

The enhancement in total composition of tocochromanols have been reported by several researchers by targeting *HPT* (Savidge et al., 2002; Collakova and Dellapenna, 2003; Karunanandaa et al., 2005), *HGGT* (Cahoon et al., 2003; Dolde and Wang, 2011; Tanaka et al., 2015), *GGR* (Patent US 6,624,342 B1), *TyrA* or *Tyr1* and *HPPD* genes (Karunanandaa et al., 2005).

Table 2: Crops targeted for vitamin E biofortification.

S. No.	Crop	Target Gene	Strategy	Reference(s)
1.	*Arabidopsis thaliana*	*Vte2*	Overexpression	Collakova and Dellapenna, 2003
2.		*Vte2*	Overexpression	Savidge et al., 2002
3.		*ZmTMT*	Overexpression	Zhang et al., 2020
4.		*Vte6*	Overexpression	Vom Dorp et al., 2015
5.		*Vte2 + Vte4*	Co-expression	Fritsche et al., 2017
6.		*BnVte4*	Overexpression	Endrigkeit et al., 2009
7.		*GmTMT2a*	Overexpression	Zhang et al., 2013b
8.		*Vte4*	Overexpression	Li et al., 2010
9.	Maize	*TMT*	Overexpression	Zhang et al., 2020
10.		*GmTMT2a*	Overexpression	Zhang et al., 2013a
11.		*Vte4*	Molecular Breeding	Das et al., 2018
12.		*Vte4*		Xiao et al., 2020
13.		*Vte4*		Feng et al., 2015
14.		*Vte4*		Chauhan et al., 2022
15.	Barley	*HvHPT + HvHGGT*	CRISPR/*Cas9*	Zeng et al., 2020
16.		*HvHGGT*	Overexpression	Chen et al., 2017
17.	*Brassica*	*PDS1*	Overexpression	Raclaru et al., 2006
18.		*Vte4*	CRISR/Cas9	Zhang et al., 2022
19.		*AtVte1, GmVte1*	Overexpression	Kumar et al., 2005
20.		*Vte1*	Overexpression	Kanwischer et al., 2005
21.		*BnCO2.TMTa*	Overexpression	Guo et al., 2022
22.	Soybean	*Pfy-TMT*	Overexpression	Tavva et al., 2007
23.		*AtVte3 and AtVte4*	Coexpression	Van Eenennaam et al., 2003
24.		*HvGGT, Gmγ-TMT*	Coexpression	Konda et al., 2020

(*Contd.*)

Table 2: (*Contd.*)

S. No.	Crop	Target Gene	Strategy	Reference(s)
25.	Rice	*HPPD*	Overexpression	Farre et al., 2012
		γ-TMT	Overexpression	Zhang et al., 2013b
26.		*Vte 1*	Overexpression	Liu et al., 2008
27.		*TC, HPT*	Overexpression	Sundararajan et al., 2021
28.		*Atγ-TMT*	Overexpression	Zhang et al., 2013b
29.	Tobacco	*γ-TMT*	Overexpression	Yabuta et al., 2013
30.		*Atγ-TMT*	Overexpression	Jin and Daniell, 2014
31.		*HPT, TC, TMT*	Overexpression	Lu et al., 2013
32.	Lettuce	*LSHPT homologue*	Overexpression	Ren et al., 2011
33.		*γ-TMT*	Overexpression	Yabuta et al., 2013
34.		*Atγ-TMT*	Overexpression	Cho et al., 2005
35.	Sorghum	*HvHGGT*	Overexpression	Che et al., 2016

5. Major Crops Targeted for Vitamin-E Biofortification

5.1. Arabidopsis

Arabidopsis, a model plant, has been extensively utilized to evaluate strategies for enhancement of vitamin E (Table 2). Overexpressing the *Vte2* gene boosted tocopherol content 2-fold in seeds 4.4-fold in leaves (Collakova and Dellapenna, 2003). Exclusive *Vte4* overexpression in arabidopsis leaves elevated α-tocopherol and total tocopherol content by up to 30% (Li et al., 2010). *Vte6* exhibited overexpression increased Phytyl-PP (PDP) and tocopherol in seeds (Vom Dorp et al., 2015). Concurrent *Vte2* and *Vte4* overexpression in seeds and leaves escalated total tocopherol 12-fold, converting γ-tocopherol to α-tocopherol (Frtische et al., 2017). Maize *ZmTMT* overexpression raised α-/γ-tocopherol ratio up to 15-fold and α-tocopherol up to 4–5 times in arabidopsis seeds (Zhang et al., 2020). A *Brassica napus Vte4* homolog elevated α-tocopherol 50-fold in arabidopsis seeds (Endrigkeit et al., 2009). Soybean *GmTMT2a* overexpression in arabidopsis boosted 4–6 times α-tocopherol (Zhang et al., 2013a).

5.2. Maize

In transgenic maize, Zhang et al. (2020) achieved a 17-fold increase in α-/γ-tocopherol and a 6.5-fold in α-tocopherol *via ZmTMT* overexpression. *GmTMT2a* gene raised α-tocopherol 3–4.5-fold in maize seeds (Zhang et al., 2013a). Co-expressing *PDS1* with downstream tocopherol biosynthesis genes in maize kernels increased tocopherol 3-fold (Naqvi et al., 2011). *Vte4*, was identified by Shutu et al. (2012) using SNP markers and segregating populations. (Das et al., 2021) used MAS to introgress the *Vte4* allele for improved vitamin E in four elite hybrids. Xiao et al. (2020) and Feng et al. (2015) improved sweet corn by introducing the *Vte4* allele.

5.3. Barley

The functional mutants of barley *HvHPT* and *HvHGGT* genes were produced and characterized by Zeng et al. (2020) using CRISPR. Chen et al. (2017) produced transgenic barley overexpressing *HvHGGT* under endogenous *D-hordein* promoter (*proHor*) which led to an increase in tocotrienol content by 10–15% and radical scavenging activity by 17–18% in seeds.

5.4. Brassica

Utilizing CRISPR/Cas9 technology, diverse mutations of four *B. napus Vte4* homologs were generated, causing significant shifts in α-tocopherol and α-/γ-tocopherol ratios compared to wild-type controls (Zhang et al., 2022). Rapeseed tocopherol levels doubled when downstream genes and *PDS1* were co-expressed (Raclaru et al., 2006). Transgenic rapeseed expressing arabidopsis and maize *Vte1* genes exhibited elevated tocopherol forms in seed oil (Kumar et al., 2005). *Vte1* overexpression in leaves shifted tocopherol composition from α- to γ-tocopherol and increased total tocopherol up to 7-fold (Kanwischer et al., 2005). Overexpressing *B. napus γ-TMT* paralogue *'BnaC02.TMT.a'* led to higher α- and total tocopherol content (Guo et al., 2022).

5.5. Soybean

Overexpressing the *γ-TMT* gene from *Perilla frutescens* in soybean seeds led to a 10.4-fold rise in α-tocopherol and a 14.9-fold increase in β-tocopherol content (Tavva et al., 2007). Co-expressing *At-Vte3* and *At-Vte4* in soybean seeds lead to more than an 8-fold increase in α-tocopherol and up to a 5-fold rise in seed vitamin E activity (Van Eenennaam et al., 2003). Soybean seeds expressing barley *HGGT* alone or with a *γ-TMT* gene exhibited 8–10-fold increase in total tocochromanols, mainly tocotrienols, with minimal impact on seed oil or protein levels (Konda et al., 2020).

5.6. Rice

Overexpressing the *HPPD* gene shifted synthesis from γ- to α- without raising total-tocopherol levels (Farre et al., 2012). *γ-TMT* overexpression increased transcript levels of *MPBQMT2* compared to wild-type plants (Zhang et al., 2013b). *Vte1* overexpression in rice enhanced salt stress tolerance and tocopherol content (Ouyang et al., 2011). Indica rice transformed with *TC* and *HPT* genes exhibited a 4-fold higher α-tocopherol (Sundararajan et al., 2021). Arabidopsis γ-TMT overexpression in Japonica rice raised α-tocotrienol levels 5-fold (Zhang et al., 2013b).

5.7. Tobacco

Overexpression and plastid transformation of *TMT* increased total tocopherol and α-tocopherol content (Yabuta et al., 2013). Similarly, arabidopsis *γ-TMT* overexpression in tobacco under high salinity and metal stress elevated α-tocopherol upto 9.6-fold (Jin and Daniell, 2014). Tobacco plastids with multigene engineering with *HGPT, TC*, and *TMT* achieved a 10-fold tocochromanol increase through synthetic operon development (Lu et al., 2013).

5.8. Lettuce

Overexpression of *LsHPT* elevated γ-tocopherol (4-fold) and α-tocopherol (2.6-fold) (Ren et al., 2011). Plastid-transformed γ-TMT overexpression increased α- and total tocopherol (Yabuta et al., 2013). Introduction of arabidopsis γ-TMT gene into lettuce led to conversion of γ-tocopherol to α-tocopherol (Cho et al., 2005).

5.9. Sorghum

Transfer of barley *HGGT* gene increased total vitamin E as well as all-trans β-carotene in sorghum (Che et al., 2016).

6. Optimization of Bioprocesses for *in vitro* α-tocopherol Production

The bioprocess optimization is used to enhance α-tocopherol through *in vitro* cultures. This approach has been applied to various crops including sunflower, safflower, grapevine, rose, argan, amaranth, arabidopsis, carrot, and cucumber. These optimizations involve media, photomixotrophic conditions, elicitor/precursor addition, light treatment, and aeration rate (Muthulakshmi et al., 2023) as discussed below:

(a) *Choice of suitable plant and explant:* Optimal selection of plant and explant is crucial for efficient vitamin E production *in vitro*. (Muresan et al., 2021) reported an 8.5-fold increase in γ-tocopherol in callus compared to seeds and leaves. Green callus exhibited higher α-tocopherol (Furuya et al., 1987). Green callus established under 7000 lx light and a 16/8 light-dark cycle yielded 3.3 times more α-tocopherol than in darkness. α-tocopherol was prominent in photosynthetic tissues and less in non-green callus (Fachechi et al., 2007). Continuous white light and lower sucrose (3 gL^{-1}) increased tocopherol (up to 4-fold), as sucrose hampers photosynthetic growth (Husemann and Barz, 1977).

(b) *Photomixotropic conditions:* Light conditions, along with medium nutrients, are crucial for α-tocopherol formation. Photomixotropic conditions in safflower cell cultures led to a 230% α-tocopherol increase compared to heterotrophic conditions (Geipel et al., 2013). Conversely, adding methyl jasmonate as an elicitor and phenylalanine as a precursor increased α-tocopherol content in dark-grown *Rosa damascena* callus cultures, contrasting light-grown cultures (Olgunsoy et al., 2017).

(c) *Different combinations of plant growth regulators:* Optimal plant growth regulator combinations impact secondary metabolite production in cell suspension cultures (Muthulakshmi et al., 2023). A blend of inositol and casamino acid outperformed inositol alone, enhancing safflower callus growth by 5-fold (Furuya et al., 1987). Sunflower callus cultures with added casamino acid exhibited a 3.6-fold α-tocopherol increase (Caretto et al., 2004).

(d) *Elicitors:* Elicitors like titanium dioxide and silicon dioxide nanoparticles significantly enhance α-tocopherol levels. Their addition to *Argania spinosa* cell suspension culture increased α-tocopherol from 500 mg/g to around 2800 mg/g dry weight (DW) (Hegazi et al., 2020). Sodium chloride boosted α-tocopherol content in safflower, sunflower and cucumber suspension cultures (Srinivasan et al., 2022).

(e) *Age of culture:* Longer subculture intervals positively influence α-tocopherol synthesis, possibly due to stress-induced production. Extended light exposure increases α-tocopherol in older sunflower cultures more than in younger ones (Muthulakshmi et al., 2023).

6.1. Status of Biofortified Cultivars with Vitamin E

In recent years, agricultural researchers have made strides in developing vitamin E enriched cultivars using metabolic engineering and molecular breeding across various crops. These advancements have been embraced in countries like Malaysia, China, India, and the USA (Das et al., 2021) (Table 2).

Noteworthy achievements include enhancing total vitamin E content in soybean cultivar A3244, with transgenic strategies introducing genes *Eh-TYRA, At-HPPD, At-Vte2,* and *At-GGH*, yielding an impressive nearly 15-fold increase (320 ppm in WT seed to 4800 ppm in transgenic) (Raclaru et al., 2006). Arabidopsis, too, showed success, with γ-tocopherol methyltransferase (γ-TMT) introduction resulting in an 80-fold α-tocopherol accumulation boost in pDC3 seeds (Mene-Saffrane and Pellaud, 2010). In elite rice cultivar Wuyujing 3, arabidopsis γ-TMT introduction led to a 36–57-fold increase in α-/γ-tocopherol (Zhang et al., 2013b). Maize transgenic version Zheng58 exhibited 6.5-fold higher α-tocopherol and a 17-fold α-/γ-tocopherol increase (Zhang et al., 2020). The transgenic potato Kufrijyoti and *B. juncea* cv. Varuna displayed 173–258% and 6-fold higher vitamin E content respectively with introduced arabidopsis γ-TMT (Upadhyaya et al., 2021).

Beyond transgenics, countries are focusing on vitamin E enrichment through marker-assisted backcross breeding. India introduced the *vte4* gene into provitamin A, lysine, tryptophan enriched maize hybrids (APQH1, APQH4, APQH5 and APQH7) and provitamin A enriched sweetcorn hybrids (PSSC-2 and ASKH2) (Chauhan et al., 2022). In the USA, Soybean germplasm-P1 was developed for higher vitamin E through molecular-assisted breeding for *γ-TMT3* gene (Hagely et al., 2021). Marked success was observed in sweet corn inbreds, where *vte4* gene introgression increased α-tocopherol and α-/γ-tocopherol levels (Feng et al., 2015). Chinese sweet corn cultivars Yuetian9, Yuetian22, and Yuetian28 also showed enhanced α- and γ-tocopherols (Xiao et al., 2020). China even achieved rice "Shangshida No. 5" cultivar (α-tocopherol-49.14 ppm) through a single-point mutation of the giant embryo gene (ge) in 'Chao2-10' (21.85 ppm) (Wang et al., 2013).

6.2. Future Prospects

Vitamin-E is one of the important micronutrients for human growth and development. The research in the area of vitamin E biofortification is at increasing pace using different strategies such as molecular breeding, genetic engineering, and tissue cultures. Although crops such as maize, rice, wheat, sorghum, barley, and brassica have been targeted for tocopherol enrichment, analogous initiatives in vegetable and fruit crops remain inadequate. The following research topics must be addressed for future Vitamin-E-rich crop breeding: (i) Expansion to vegetable and fruit crops for vitamin E biofortification; (ii) Integration of multiple nutritional traits alongside vitamin E, such as provitamin A, lysine, tryptophan, methionine, and anthocyanins; (iii) Broadening the genetic base of targeted crops to develop cultivars with greater adaptability to biotic and abiotic stresses; (iv) Incorporation of resilience traits to enhance tolerance to drought, heat, pests, and diseases; (v) Consideration of consumer acceptance and sensory attributes in biofortified crops.

7. Conclusion

Biofortified crops hold promise in combatting malnutrition, but exploiting genetic variation remains crucial. Studies across diverse crops underline the potential for enhancing vitamin E content. Several crop cultivars with its enhanced content have been developed and released. QTL mapping and GWAS studies have identified genetic determinants for vitamin E, yet introgression of these regions into elite crops, except maize, requires further attention. In crops like rice, wheat, sorghum, barley, and brassica, untapped genetic variations remain. By combining conventional and molecular breeding, transgenics, and cis-genics, potential biofortified crops can be developed, alleviating vitamin E malnutrition. Raising awareness, forging industry partnerships, and robust policies are pivotal in popularizing these cultivars to address malnutrition effectively.

References

Aksoz, E., O. Korkut, D. Aksit, and C. Gokbulut. 2020. Vitamin E (α-, β+γ- and δ-tocopherol) levels in plant oils. *Flavour Fragr. J.*, 35(5): 504–510.

Bramley, P.M., I. Elmadfa, A. Kafatos, F.J. Kelly, Y.R. Manios, and H.E. Roxborough, P.J.A. Sheehy, and K.H. Wagner, 2000. Vitamin E. *J. Sci. Food Agric.*, 80: 913–938.

Caretto, S., E. Bray Speth, C. Fachechi, R. Gala, G. Zacheo, and G. Giovinazzo. 2004. Enhancement of vitamin E production in sunflower cell cultures. *Plant Cell Rep.*, 23: 174–179.

Chauhan, H.S., R. Chhabra, T. Rashmi, V. Muthusamy, R.U. Zunjare, S.J. Mishra, N. Gain, B.K. Mehta, A.K. Singh, H.S. Gupta, and F. Hossain. 2022. Impact of vte4 and crtRB1 genes on composition of vitamin E and provitamin A carotenoids during kernel-stages in sweet corn. *J. Food Compos. Anal.*, 105: 104264.

Che, P., Z.Y. Zhao, K. Glassman, D. Dolde, T.X. Hu, T.J. Jones, D.F. Gruis, S. Obukosia, F. Wambugu, and M.C. Albertsen. 2016. Elevated vitamin E content improves all-trans β-carotene accumulation and stability in biofortified sorghum. *PNAS*, 113(39): 11040–11045.

Chen, J., C. Liu, B. Shi, Y. Chai, N. Han, M. Zhu, and H. Bian. 2017. Overexpression of HvHGGT enhances tocotrienol levels and antioxidant activity in barley. *J. Agric. Food Chem.*, 65(25): 5181–5187.

Cho, E.A., C.A. Lee, Y.S. Kim, S.H. Baek, B.G. Reyes, and S.J. Yun. 2005. Expression of γ-tocopherol methyltransferase transgene improves tocopherol composition in lettuce (*Latuca sativa* L.). *Mol. Cells* (Springer Science & Business Media BV): 19(1).

Chung, I.M., S.J. Yong, J. Lee, and S.H. Kim. 2013. Effect of genotype and cultivation location on β-sitosterol and α-, β-, γ-, and δ-tocopherols in sorghum. *Food Res. Int.*, 51(2): 971–976.

Collakova E., and D. DellaPenna. 2003. Homogentisate phytyltransferase activity is limiting for tocopherol biosynthesis in *Arabidopsis*. *Plant Physiol.*, 131: 632–642.

Das, A.K., M.M. Gowda, V. Muthusamy, R.U. Zunjare, H.S. Chauhan, A. Baveja, V. Bhatt, G. Chand, J.S. Bhat, S.K. Guleria, and S. Saha. 2021. Development of maize hybrids with enhanced vitamin E, vitamin A, lysine, and tryptophan through molecular breeding. *Front. Plant Sci.*, 12: 659381.

Das, A.K., V. Muthusamy, R.U. Zunjare, A. Baveja, H.S. Chauhan, J.S. Bhat, S.K. Guleria, B. Kumar, S. Saha and F. Hossain. 2020. Genetic variability for kernel tocopherols and haplotype analysis of γ-tocopherol methyl transferase (vte4) gene among exotic- and indigenous-maize inbreds. *J. Food Compos. Anal.*, 88: 103446.

Das, A.K., S.K. Jaiswal, V. Muthusamy, R.U. Zunjare, H.S. Chauhan, G. Chand, S. Saha, and F. Hossain. 2019. Molecular diversity and genetic variability of kernel tocopherols among maize inbreds possessing favorable haplotypes of γ-tocopherol methyl transferase (ZmVTE4). *J. Plant Biochem. Biotechnol.*, 28: 253–262.

Das, A.K., V. Muthusamy, R.U. Zunjare, H.S. Chauhan, P.K. Sharma, J.S. Bhat, S.K. Guleria, S. Saha, and F. Hossain. 2019. Genetic variability, genotype × environment interactions- and combining ability-analyses of kernel tocopherols among maize genotypes possessing novel allele of γ-tocopherol methyl transferase (ZmVTE4). *J. Cereal Sci.*, 86: 1–8.

Diepenbrock, C.H., C.B. Kandianis, A.E. Lipka, M. Magallanes-Lundback, B. Vaillancourt, E. Góngora-Castillo, J.G. Wallace, J. Cepela, A. Mesberg, P.J. Bradbury, and D.C. Ilut. 2017. Novel loci underlie natural variation in vitamin E levels in maize grain. *Plant Cell*, 29(10): 2374–2392.

Dolde, D., and T. Wang. 2011. Oxidation of crude corn oil with and without elevated tocotrienols. *J. Am. Oil Chem. Soc.*, 88(9): 1367–1372.

Eitenmiller, R.R., and J. Lee. 2004. *Vitamin E: Food Chemistry, Composition, and Analysis.* Marcel Dekker Inc., New York: CRC Press.

Endrigkeit, J., X. Wang, D. Cai, C. Zhang, Y. Long, J. Meng, and C. Jung. 2009. Genetic mapping, cloning, and functional characterization of the BnaX. VTE4 gene encoding a γ-tocopherol methyltransferase from oilseed rape. *Theor. Appl. Genet.*, 119: 567–575.

Espinoza, A., A. San Martín, M. López-Climent, S. Ruiz-Lara, A. Gómez-Cadenas, and J.A. Casaretto. 2013. Engineered drought-induced biosynthesis of α-tocopherol alleviates stress-induced leaf damage in tobacco. *J. Plant Physiol.*, 170(14): 1285–1294.

Fachechi, C., R. Nisi, R. Gala, A. Leone, and S. Caretto. 2007. Tocopherol biosynthesis is enhanced in photomixotrophic sunflower cell cultures. *Plant Cell Rep.*, 26: 525–530.

Farré, G., D. Sudhakar, S. Naqvi, G. Sandmann, P. Christou, T. Capell, and C. Zhu. 2012. Transgenic rice grains expressing a heterologous ρ-hydroxyphenylpyruvate dioxygenase shift tocopherol synthesis from the γ to the α isoform without increasing absolute tocopherol levels. *Transgenic Res.*, 21:1093–1097.

Feng, F., Q. Wang, C. Liang, R. Yang, and X. Li. 2015. Enhancement of tocopherols in sweet corn by marker-assisted backcrossing of ZmVTE4. *Euphytica*, 206: 513–521.

Fritsche, S., X. Wang, and C. Jung. 2017. Recent advances in our understanding of tocopherol biosynthesis in plants: An overview of key genes, functions, and breeding of vitamin E improved crops. *Antioxid.*, 6(4): 99.

Fritsche, S., X. Wang, J. Li, B. Stich, F.J. Kopisch-Obuch, J. Endrigkeit, G. Leckband, F. Dreyer, W. Friedt, J. Meng, and C. Jung. 2012. A candidate gene-based association study of tocopherol content and composition in rapeseed (*Brassica napus*). *Front. Plant Sci.*, 3: 129.

Furuya, T. Yoshikawa, T. Kimura, and H. Kaneko.1987. Production of tocopherols by cell culture of safflower. *Phytochem.*, 26(10): 2741–2747.

Garg, M., A. Sharma, S. Vats, V. Tiwari, A. Kumari, V. Mishra, and M. Krishania. 2021. Vitamins in cereals: A critical review of content, health effects, processing losses, bioaccessibility, fortification, and biofortification strategies for their improvement. *Front. Nutr.*, 8: 586815.

Geipel, K., M.L. Socher, C. Haas, T. Bley, and J. Steingroewer. 2013. Growth kinetics of a *Helianthus annuus* and a *Salvia fruticosa* suspension cell line: Shake flask cultivations with online monitoring system. *Eng. Life Sci.*, 13(6): 593–602.

Goufo, P., and H. Trindade. 2014. Rice antioxidants: Phenolic acids, flavonoids, anthocyanins, proanthocyanidins, tocopherols, tocotrienols, γ-oryzanol, and phytic acid. *Food Sci. Nutr.*, 2(2): 75–104.

Guo, Y., D. Li, T. Liu, M. Liao, Y. Li, W. Zhang, Z. Liu, and M. Chen. 2022. Effect of overexpression of γ-tocopherol methyltransferase on α-tocopherol and fatty acid accumulation and tolerance to salt stress during seed germination in *Brassica napus* L. *Int. J. Mol. Sci.*, 23(24): 15933.

Hagely, K., A.R. Konda, J.H. Kim, E.B. Cahoon, and K. Bilyeu. 2021. Molecular-assisted breeding for soybean with high oleic/low linolenic acid and elevated vitamin E in the seed oil. *Mol. Breed.*, 41(1): 3.

Hegazi, G.A., W.M. Ibrahim, M.H. Hendawy, H.M. Salem, and H.E. Ghareb. 2020. Improving α-tocopherol accumulation in Argania spinosa suspension cultures by precursor and nanoparticles feeding. *Plant Arch.*, 20: 2431–2437.

Hilali, M., H.E. Monfalouti, and B.E. Kartah. 2020. Evaluation of the chemical composition of Argan (*Argania spinosa* L.) oil according to its extraction method, origin of production and altitude. *J. Anim. Feed Sci.*, 10(3): 111–118.

Hossain, F., R.U. Zunjare, V. Muthusamy, A. Kumar, J. Madhavan, G. Ikkurti, A. Katral, Z.A. Talukder, R. Chhabra, G. Chand, and V. Bhatt. 2023. Genetic improvement of specialty corn for nutritional quality traits. *Maize Improvement: Current Advances in Yield, Quality, and Stress Tolerance under Changing Climatic Scenarios*, 235–257. Cham: Springer International Publishing.

Hossain, F., R.U. Zunjare, V. Muthusamy, J.S. Bhat, B.K. Mehta, D. Sharma, Z.A. Talukder, R. Chhabra, A. Katral, S. Dutta, and G. Chand. 2022. Biofortification of maize for nutritional security. *Biofortification of Staple Crops*, 147–174. Singapore: Springer Singapore.

Husemann, W., and W. Barz. 1977. Photoautotrophic growth and photosynthesis in cell suspension cultures of *Chenopodium rubrum*. *Physiol Plant*, 40(2): 77–81.

Jiang, Q. 2017. Natural forms of vitamin E as effective agents for cancer prevention and therapy. *Adv. Nutr.*, 8(6): 850–867.

Jin, S., and H. Daniell. 2014. Expression of gamma-tocopherol methyltransferase in chloroplasts results in massive proliferation of the inner envelope membrane and decreases susceptibility to salt and metal-induced oxidative stresses by reducing reactive oxygen species. *Plant Biotechnol. J.*, 12: 1274–1285.

Kanwischer, M., S. Porfirova, E. Bergmuller, and P. Dörmann. 2005. Alterations in tocopherol cyclase activity in transgenic and mutant plants of Arabidopsis affect tocopherol content, tocopherol composition, and oxidative stress. *Plant Physiol.*, 137(2): 713–723.

Karunanandaa, B., Q. Qi, M. Hao, S.R. Baszis, P.K. Jensen, Y.H.H. Wong, J. Jiang, M. Venkatramesh, K.J. Gruys, F. Moshiri, and D. Post-Beittenmiller. 2005. Metabolically engineered oilseed crops with enhanced seed tocopherol. *Metab. Eng.*, 7(5–6): 384–400.

Konda, A.R., T.J. Nazarenus, H. Nguyen, J. Yang, Ma. Gelli, S. Swenson, J.M. Shipp, M.A. Schmidt, R.E. Cahoon, O.N. Ciftci, C. Zhang, T.E. Clemente, and E.B. Cahoon. 2020. Metabolic engineering of soybean seeds for enhanced vitamin E tocochromanol content and effects on oil antioxidant properties in polyunsaturated fatty acid-rich germplasm. *Metab. Eng.*, 57: 63–73.

Kornsteiner, M., K.H. Wagner, and I. Elmadfa. 2006. Tocopherols and total phenolics in 10 different nut types. *Food Chem.*, 98(2): 381–387.

Kumar, R., M. Raclaru, T. Schüßeler, J. Gruber, R. Sadre, W. Lühs, K.M. Zarhloul, W. Friedt, D. Enders, M. Frentzen, and D. Weier. 2005. Characterization of plant tocopherol cyclases and their overexpression in transgenic *Brassica napus* seeds. *FEBS Lett.*, 579(6): 1357–1364.

Li, Yin, Y. Zhou, Z. Wang, X. Sun, and K. Tang. 2010. Engineering tocopherol biosynthetic pathway in Arabidopsis leaves and its effect on antioxidant metabolism. *Plant Sci.*, 178(3): 312–320.

Liu, X., X. Hua, J. Guo, Qi, D.L. Wang, Z. Liu, Z. Jin, S. Chen, and G. Liu. 2008. Enhanced tolerance to drought stress in transgenic tobacco plants overexpressing VTE1 for increased tocopherol production from Arabidopsis thaliana. *Biotechnol. Letters*, 30: 1275–1280.

Loganathan, R., and T. Kim-Tiu. 2022. Assessment of fatty acid and vitamin E profiles in edible oils in Malaysia. *J. Oil Palm Res.*, 34(4): 741–751.

Lu, Y., H. Rijzaani, D. Karcher, S. Ruf, and R. Bock. 2013. Efficient metabolic pathway engineering in transgenic tobacco and tomato plastids with synthetic multigene operons. *PNAS*, 110(8): E623–E632.

Luby, C.H., H.A. Maeda, and I.L. Goldman. 2014. Genetic and phenological variation of tocochromanol (vitamin E) content in wild (*Daucus carota* L. var. *carota*) and domesticated carrot (*D. carota* L. var. *sativa*). *Hortic. Res.*, 1: 14015.

Mene-Saffrane, L., and D. DellaPenna. 2010. Biosynthesis, regulation, and functions of tocochromanols in plants. *Plant Physiol. Biochem.*, 48: 301–309.

Mureşan, L., D. Clapa, T. Rusu, T.T. Wang, and J.B. Park. 2021. Soybean callus: A potential source of tocopherols. *Plants*, 10(12): 2571.

Muthulakshmi, V.M., A, Srinivasan, and S. Srivastava. 2023. Antioxidant green factories: Toward sustainable production of vitamin E in plant *in vitro* cultures. *ACS Omega*, 8(4): 3586–3605.

Naqvi, S., G. Farré, C. Zhu, G. Sandmann, T. Capell, and P. Christou. 2011. Simultaneous expression of Arabidopsis ρ-hydroxyphenylpyruvate dioxygenase and MPBQ methyltransferase in transgenic corn kernels triples the tocopherol content. *Transgenic Res.*, 20: 177–181.

Nielsen, M.M.L., and A. Hansen. 2008. Stability of vitamin E in wheat flour and whole wheat flour during storage. *Cereal Chem.*, 85(6): 716–720.

Norris, S.R., T.R. Barrette, and D. DellaPenna. 1995. Genetic dissection of carotenoid synthesis in arabidopsis defines plastoquinone as an essential component of phytoene desaturation. *Plant Cell*, 7: 2139–2149.

Olgunsoy, P., Ulusoy, S., and Akcay, U.C. 2017. Metabolite production and antibacterial activities of callus cultures from *Rosa damascena* Mill. petals. *Marmara Pharm. J.*, 21(3): 590–597.

Ouyang, S., S. He, P. Liu, W. Zhang, J. Zhang, and S. Chen. 2011. The role of tocopherol cyclase in salt stress tolerance of rice (*Oryza sativa*). *Sci. China Life Sci.*, 54: 181–188.

Panfili, G., A. Fratianni, T. Di Criscio, and E. Marconi. 2008. Tocol and β-glucan levels in barley varieties and in pearling by-products. *Food Chem.*, 107(1): 84–91.

Peraza-Magallanes, A.Y., M.A. Pereyra-Camacho, E. Sandoval-Castro, S. Medina-Godoy, M. Valdez-Morales, M.T. Clegg, and C.L. Calderón-Vázquez. 2017. Exploring genetic variation, oil and α-tocopherol content in avocado (*Persea americana*) from northwestern Mexico. *Genet. Resour. Crop Evol.*, 64: 443–449.

Peterson D.M., and A.A. Qureshi. 1992. Genotype and environment effects on tocols of barley and oats. *Cereal Chem.*, 70: 157e162.

Petropoulos, S.A., Â. Fernandes, N. Katsoulas, L. Barros, and I.C. Ferreira. 2019. The effect of covering material on the yield, quality, and chemical composition of greenhouse-grown tomato fruit. *J. Sci. Food Agric.*, 99(6): 3057–3068.

Raclaru, M., J. Gruber, and R. Kumar. 2006. Increase of the tocochromanol content in transgenic *Brassica napus* seeds by overexpression of key enzymes involved in prenylquinone biosynthesis. *Mol. Breed.*, 18: 93–107.

Ren, W., L. Zhao, L. Zhang, Y. Wang, L. Cui, Y. Tang, X. Sun, and K. Tang. 2011. Molecular analysis of a homogentisate phytyltransferase gene from *Lactuca sativa* L. *Mol. Bio. Rep.*, 38: 1813–1819.

Savidge, B., J.D. Weiss, Y.H.H. Wong, M.W. Lassner, T.A. Mitsky, C.K. Shewmaker, D. Post-Beittenmiller, and H.E. Valentin. 2002. Isolation and characterization of homogentisate phytyltransferase genes from Synechocystis sp. PCC 6803 and Arabidopsis. *Plant Physiol.*, 129(1): 321–332.

Schuy, C., J. Groth, A. Ammon, J. Eydam, S. Baier, G. Schweizer, A. Hanemann, M. Herz, L.M. Voll, and U. Sonnewald. 2019. Deciphering the genetic basis for vitamin E accumulation in leaves and grains of different barley accessions. *Sci. Rep.*, 9(1): 9470.

Schwartz, H., V. Ollilainen, V. Piironen, and A.M. Lampi. 2008. Tocopherol, tocotrienol and plant sterol contents of vegetable oils and industrial fats. *J. Food Compos. Anal.*, 21(2): 152–161.

Shahidi, F., and A.C. De Camargo. 2016. Tocopherols and tocotrienols in common and emerging dietary sources: Occurrence, applications, and health benefits. *Int. J. Mol. Sci.*, 17(10): 1745.

Shintani, D., and D. DellaPenna. 1998. Elevating the vitamin E content of plants through metabolic engineering. *Sci.*, 282(5396): 2098–2100.

Shutu, X., Z. Dalong, C. Ye, Z. Yi, T. Shah, F. Ali, L. Qing, L. Zhigang, W. Weidong, L. Jiansheng, and Y. Xiaohong. 2012. Dissecting tocopherols content in maize (*Zea mays* L.), using two segregating populations and high-density single nucleotide polymorphism markers. *BMC Plant Bio.*, 12(1): 1–14.

Srinivasan, A., V. Sundaram, Vidya M. Muthulakshmi, and S. Srivastava. 2022. Multi-fold enhancement in vitamin E (alpha-tocopherol) production via integration of bioprocess optimization and metabolic engineering in cell suspension of sunflower. *J. Plant Biochem. Biotechnol.*, 1–14.

Stacey, M.G., R.E. Cahoon, H.T. Nguyen, Y. Cui, S. Sato, C.T. Nguyen, N. Phoka, K.M. Clark, Y. Liang, J. Forrester, and J. Batek. 2016. Identification of homogentisate dioxygenase as a target for vitamin E biofortification in oilseeds. *Plant Physiol.*, 172(3): 1506–1518.

Stahl, W., H. Van Den Berg, J. Arthur, A. Bast, J. Dainty, R.M. Faulks, C. Gärtner, G. Haenen, P. Hollman, B. Holst, and F.J. Kelly. 2002. Bioavailability and metabolism. *Mol. Aspects Med.*, 23(1–3): 39–100.

Sundararajan, S., V. Rajendran, H.P. Sivakumar, S. Nayeem, H.M. Chandra, A. Sharma, and S. Ramalingam. 2021. Enhanced vitamin E content in an Indica rice cultivar harbouring two transgenes from Arabidopsis thaliana involved in tocopherol biosynthesis pathway. *Plant Biotechnol. J.*, 19(6): 1083.

Syväoja, E.L., V. Piironen, P. Varo, P. Koivistoinen, and K. Salminen. 1986. Tocopherols and tocotrienols in Finnish foods: Oils and fats. *JAOCS*, 63(3): 328–329.

Tanaka, H., Y. Yabuta, M. Tamoi, N. Tanabe, and S. Shigeoka. 2015. Generation of transgenic tobacco plants with enhanced tocotrienol levels through the ectopic expression of rice homogentisate geranylgeranyl transferase. *Plant Biotechnol.*, 32(3): 233–238.

Tavva, V.S., Y.H. Kim, I.A. Kagan, R.D. Dinkins, K.H. Kim, and G.B. Collins. 2007. Increased α-tocopherol content in soybean seed overexpressing the Perilla frutescens γ-tocopherol methyltransferase gene. *Plant Cell Rep.*, 26: 61–70.

Tekaya, M., B. Mechri, H. Cheheb, F. Attia, I. Chraief, M. Ayachi, D. Boujneh, and M. Hammami. 2014. Changes in the profiles of mineral elements, phenols, tocopherols and soluble carbohydrates of olive fruit following foliar nutrient fertilization. *LWT - Food Sci. Technol.*, 59(2): 1047–1053.

Ujiie, A., T. Yamada, K. Fujimoto, Y. Endo, and K. Kitamura. 2005. Identification of soybean varieties with high α-tocopherol content. *Breeding Sci.*, 55(2): 123–125.

Upadhyaya, D.C., D.S. Bagri, C.P. Upadhyaya, A. Kumar, M. Thiruvengadam, and S.K. Jain. 2021. Genetic engineering of potato (*Solanum tuberosum* L.) for enhanced α-tocopherols and abiotic stress tolerance. *Physiol. Plant*, 173(1): 116–128.

Van Eenennaam, A.L., K. Lincoln, T.P. Durrett, H.E. Valentin, C.K. Shewmaker, G.M. Thorne, J. Jiang, S.R. Baszis, C.K. Levering, E.D. Aasen, and M. Hao. 2003. Engineering vitamin E content: From Arabidopsis mutant to soy oil. *Plant Cell*, 15(12): 3007–3019.

Vom Dorp, K., G. Hölzl, C. Plohmann, M. Eisenhut, M. Abraham, A.P. Weber, A.D. Hanson, and P. Dörmann. 2015. Remobilization of phytol from chlorophyll degradation is essential for tocopherol synthesis and growth of Arabidopsis. *Plant Cell*, 27(10): 2846–2859.

Wang, W., H.L. Wang, X.Z. Xiao, and X.Q. 2019. Chemical composition analysis of seed oil from five wild almond species in China as potential edible oil resource for the future. *S. Afr. J. Bot.*, 121: 274–281.

Wang, X., Y.E., Song, and J.Y. Li. 2013. High expression of tocochromanol biosynthesis genes increases the vitamin E level in a new line of giant embryo rice. *J. Agric Food Chem.*, 61(24): 5860–5869.

Warner, K., and T.L. Mounts. 1990 Analysis of tocopherols and phytosterols in vegetable oils by HPLC with evaporative light-scattering detection. *J. Am. Oil Chem. Soc.*, 67(11): 827–831.

Wu, D., X. Li, R. Tanaka, J.C. Wood, L.E. Tibbs-Cortes, M. Magallanes-Lundback, N. Bornowski, J.P. Hamilton, B. Vaillancourt, C.H. Diepenbrock, and X. Li. 2022. Combining GWAS and TWAS to identify candidate causal genes for tocochromanol levels in maize grain. *Genet.*, 221(4): iyac091.

Wu, Y., R. Zhou, Z. Wang, B. Wang, Y. Yang, X. Ju, and R. He. 2019. The effect of refining process on the physicochemical properties and micronutrients of rapeseed oils. *PLoS One*, 14(3): e0212879.

Xiao, Y., Y. Yu, G. Li, L. Xie, X. Guo, J. Li, Y. Li, and J. Hui. 2020. Genome-wide association study of vitamin E in sweet corn kernels. *Crop J.*, 8(2): 341–350.

Yabuta, Y., H. Tanaka, S. Yoshimura, A. Suzuki, M. Tamoi, T. Maruta, and S. Shigeoka. 2013. Improvement of vitamin E quality and quantity in tobacco and lettuce by chloroplast genetic engineering. *Transgenic Res.*, 22: 391–402.

Zaaboul, F.Y. Liu. 2022. Vitamin E in foodstuff: Nutritional, analytical, and food technology aspects. *Comp. Rev. Food Sci. Food Saf.*, 21(2): 964–998.

Zeng, Z., N. Han, C. Liu, B. Buerte, C. Zhou, J. Chen, M. Wang, Y. Zhang, Y. Tang, M. Zhu, and J. Wang. 2020. Functional dissection of HGGT and HPT in barley vitamin E biosynthesis via CRISPR/Cas9-enabled genome editing. *Ann. Bot.*, 126(5): 929–942.

Zhang, G.Y., R.R. Liu, G. Xu, P. Zhang, Y. Li, K.X. Tang, G.H. Liang, and Q.Q. Liu. 2013b. Increased α-tocotrienol content in seeds of transgenic rice overexpressing arabidopsis γ-tocopherol methyltransferase. *Transgenic Res.*, 22: 89–99.

Zhang, L., Y. Luo, B. Liu, L. Zhang, W. Zhang, R. Chen, and L. Wang. 2020. Overexpression of the maize γ-tocopherol methyltransferase gene (ZmTMT) increases α-tocopherol content in transgenic arabidopsis and maize seeds. *Transgenic Res.*, 29: 95–104.

Zhang, L., Y. Luo, Y. Zhu, L. Zhang, W. Zhang, R. Chen, M. Xu, Y. Fan, and L. Wang. 2013a. GmTMT2a from soybean elevates the α-tocopherol content in corn and Arabidopsis. *Transgenic Res.*, 22: 1021–1028.

Zhang, H., Y. Shi, M. Sun, X. Hu, M. Hao, Y. Shu, X.R. Zhou, Q. Hu, C. Li, and D. Mei. 2022. Functional differentiation of BnVTE4 gene homologous copies in α-tocopherol biosynthesis revealed by CRISPR/Cas9 editing. *Front. Plant Sci.*, 13: 850924.

CHAPTER

11

Dietary Fiber: Balancing Taste and Health – Can We Have Both?

Vijeta Sagwal[a], Pooja Sihag[a], Sourav Panigrahi[a], Yogita Singh[a] and Upendra Kumar[a,b*][0000-0002-1137-9103]

[a] Department of Molecular Biology and Biotechnology, College of Biotechnology, CCS Haryana Agricultural University, Hisar - 125004, India

[b] Department of Plant Science, Mahatma Jyotiba Phule Rohilkhand University, Bareilly - 234001, India

1. Introduction

Globally, consumers are displaying a heightened concern for maintaining a healthful diet that boasts a substantial content of dietary fibers (DFs). DFs, encompassing polysaccharides such as glucan, cellulose, hemicellulose, pectin, and lignin, demonstrate resistance to the digestive enzymes present in the gut. As a result of this resistance, they remain unhydrolyzed and unabsorbed within the stomach and small intestine, eventually finding their way directly into the large intestine (Kaoutari et al., 2013).

These polysaccharides, forming part of the carbohydrate makeup, are divided into two classifications based on their solubility in buffers at specific pH levels: soluble dietary fiber (SDF) and insoluble dietary fiber (IDF) (He et al., 2022). Soluble fibers, encompassing substances like guar gums, mucilage, and pectin, contribute to the reduction of blood cholesterol levels. Conversely, IDF includes cellulose, hemicellulose, and lignin. The presence of SDF aids in improved digestion and lowers the risk of diabetes, while IDF, through water absorption in the gut, facilitates smoother bowel movements, preventing constipation.

Natural sources of DFs encompass fruits, vegetables, cereals, and pulses. Among these, cereals alone account for 50% of DF intake, with vegetables and

*Corresponding author: baliyan.upendra@gmail.com

fruits contributing 35–40% and 15–16%, respectively (Lambo et al., 2005). Polysaccharides found in wheat primarily consist of soluble and insoluble pentosans or arabinoxylans. Meanwhile, barley and oats are rich in soluble β-glucans. The recommended daily DF intake stands at 25g for women and 35g for men (Dhingra et al., 2012).

To bolster fiber intake and enhance DF quality, the food industry employs various processing methods, including heating, cooking, and grinding, which manipulate the composition and texture of DF (Liu et al., 2022). A diet enriched with DF yields manifold health benefits, subsequently leading to a reduction in the incidence of numerous diseases. Therefore, these DFs are introduced to foods during processing or incorporated into food crops via biofortification, augmenting the diet's nutritional value. Distinct DFs exhibit varying physicochemical properties and degrees of polymerization contingent upon their structural origins. Thus, there arises a necessity to delve into the structural attributes and polymerization degrees of DFs, along with their roles in disease prevention (He et al., 2022).

In their research, Shafi et al. (2022) investigated the effects of incorporating banana peel flour into wheat flour. Additionally, Burton et al. (2011) overexpressed the cellulose synthase-like gene (Cslf6) in barley, yielding an approximate 80% increase in barley grain β-glucan content. Within this chapter, we comprehensively cover the nutritional advantages of DFs for human health, the methodologies employed for their biofortification, the targeted crops, the concept of bioavailability, and the accompanying challenges.

2. Contribution to Human Health

Dietary fiber (DF) constitutes a non-digestible form of carbohydrates, with its quantity and composition varying across different foods. DF serves diverse functions within the human body, displaying a range of activities. The 30%–50% ratio of SDF and TDF is considered suitable for various physiological and cholesterol-reducing effects (Ajila et al., 2007). Its consumption fosters the growth of microbial biomass in the human gut, diminishes cholesterol levels and weight, enhances intestinal function by augmenting water absorption and microbial biomass, and governs glucose and insulin levels within the body (Rao and Fedewa, 2015).

Soluble fibers augment the viscosity of the food bolus, thereby accelerating intestinal absorption. Furthermore, DFs aid in the prevention of colon cancer, reduction of heavy metal toxicity, and mitigation of cardiovascular and female-related ailments, as suggested by multiple studies (Barber, 2020). A study by He et al. (2004) investigated the impact of DF intake on blood pressure across 110 participants and found that a diet rich in DF moderated blood pressure. In a cohort of 345 participants, DF intake contributed to weight loss independently of macronutrient and caloric intake (Miketinas et al., 2019). Solah et al. (2017) examined the effects of supplementing a proprietary soluble DF product called

PolyGlycopleX® (PGX), composed of konjac powder, sodium alginate, and xanthan gum, on body weight. The study, involving 118 overweight participants across three dosage groups (4.5 g PGX soft gels, 5 g PGX granules, or 5 g rice flour control), revealed that PGXG led to reduced body weight and body mass index at the recommended dosage.

Among the many beneficial impacts of DFs on human health, the enhancement of gut motility stands out, playing a pivotal role in preventing constipation and irritable bowel syndrome (IBS) (Rao and Fedewa, 2015). The supplementation of FODMAPs (fermentable oligosaccharide, disaccharide, monosaccharide, and polyol) in the diet has demonstrated effectiveness in managing chronic constipation and IBS.

Type 2 diabetes, a consequence of insulin resistance, is associated with a growing array of diseases and has witnessed an escalating global prevalence. The network structure of DF functions as a physical barrier, entrapping glucose and thereby reducing glucose diffusion. A lower carbohydrate-to-DF intake ratio has been linked to a reduced risk of type 2 diabetes (Barber et al., 2020). In a study involving adults with type I and II diabetes, an average DF intake of 19g resulted in reductions in glycated hemoglobin, insulin, total cholesterol, and low-density lipoprotein (Reynolds et al., 2020).

Thus, a prudent approach involves favoring whole grain foods with high fiber content over refined foods, to mitigate the aforementioned health issues.

3. Target Crops

The growing recognition of the significant impact of dietary fiber (DF) on human health has spurred intensive research in this area. Therefore, work is done to increase the content of DF in different crops. The plant-based DFs are mainly obtained from cereals, fruits, vegetables, and legumes. Efforts are underway to increase DF content in various crops, particularly those sourced from plants such as cereals, fruits, vegetables, and legumes. It's important to note that the content and composition of DFs can vary based on the source and the methods used for processing (Tejada-Ortigoza et al., 2020).

Cereals hold a prominent position as a vital source of both micronutrients and macronutrients, including dietary fibers. These have been targeted to improve the DF content by various approaches. β-glucans, a DF composed of D-glucose, is found in the aleurone layer or cell walls of the endosperm in major cereals, with oats containing notably higher levels. Another class of DF, arabinoxylan, characterized by a linkage between arabinose and xylose, is more abundant in rye (Joye et al., 2020). Arabinoxylan content can range from 11% to 16.4% (w/w) in bran and 1.5% –1.8% in the starchy endosperm. In tetraploid wheat, β-glucans content varies from 0.18% to 0.89% (w/w) dry weight (Marcotuli et al., 2016) while oat and barley are well known for their high β-glucans (4.5%) therefore wheat is a target crop for β-glucans enhancement. Factors like cropping systems and weather conditions have an impact on the DF content of crops (Korge et al., 2023). Table 1 provides an overview of the DF content in various cereals.

Table 1: Dietary fiber content (g/100gm)of cereals, fruits pulp, and raw vegetables

Crop	TDF	IDF	SDF	Reference
Cereals				
Barley	14.6-27.1	12.0-22.1	2.6-5.0	Vitaglione et al., 2008
Maize	13.1-19.6	11.6-16.0	1.5-3.6	Vitaglione et al., 2008
Oats	11.5-37.7	8.6-33.9	2.9-3.8	Vitaglione et al., 2008
	10.3	6.5	3.8	Dhingra et al., 2012
Rice (miled)	2.7-4.9	1.9-4.2	0.6-1.1	Prasad et al., 2018
Wheat	11.6-17.0	10.2-14.7	1.4-2.3	Vitaglione et al., 2008
Fruits				
Apple	2	0.2	1.8	Dhingra et al., 2012
Banana	10.5	8.64	1.86	Garcia-Amezquita et al., 2018
	1.7	0.5	1.2	Dhingra et al., 2012
Date palm	18.4	6.7	11.7	Mrabet et al., 2019
Grapefruit	1.2	0.5	0.7	Dhingra et al., 2012
Mango	6.71	4.01	2.7	Garcia-Amezquita et al., 2018
	1.8	0.74	1.06	Dhingra et al., 2012
Orange	1.8	1.1	0.7	Dhingra et al., 2012
Pineapple	1.2	0.1	1.1	Dhingra et al., 2012
Watermelon	7.2	2.36	4.84	Garcia-Amezquita et al., 2018
Vegetables				
Broccoli	36	4.8	31.2	Kalala et al., 2018
Carrot	25.3	5.3	20	Kalala et al., 2018
Cauliflower	29.7	2.7	27	Kalala et al., 2018
Onion	2.2	0	2.2	Dhingra et al., 2012
Pumpkin	27.7	8.3	19.4	Kalala et al., 2018
Radish	33.5	2.4	31.1	Kalala et al., 2018
Tomato	1.2	0.4	0.8	Dhingra et al., 2012
Turnips	28.9	2.5	26.4	Kalala et al., 2018

TDF, total dietary fiber; SDF, soluble dietary fiber; IDF, insoluble dietary fiber

Fruits and vegetables also serve as valuable sources of dietary fiber (Abraão et al., 2023; Galvão et al., 2023). The inclination to incorporate raw fruits and vegetables into daily diets is rising due to their high DF content. It's worth noting that although fruits and vegetables have higher moisture content, making their total DF content appear lower compared to cereal grains on an equal weight basis, they remain popular choices. Abraão et al. (2023) recently reported the nutritional content of wild Prunus fruit, revealing an estimated DF content of 10g/100g which is similar to the cultivated Prunus. Refer to Table 1 for a summary of dietary fiber content in selected fruits and vegetables. Lambo et al. (2005) found that grains are one of the main sources of dietary fiber, making up about 50% of fiber intake in western countries. 30–40% of dietary fiber may come from veggies, 16% from fruits, and the remaining 3% from other minor sources.

4. Strategies for Biofortification

Biofortification is an approach to enhance the nutrient amount in the edible parts of plants to increase their availability to humans (Kapoor et al., 2022). As starch is a major component of DF in the human diet, more focus has been given to biofortifying the crops with DF by modifying the amylose to amylopectin ratio using breeding and transgenic approaches. These approaches mainly target the starch biosynthesizing enzymes such as starch branching enzyme (*SBEI, SBEIIa*), granule-bound starch synthase (*GBSSI*), starch synthase (*SSIIa, SSIIIa*), and cellulose synthase-like (*Csl*) genes, that ultimately enhance the amylose content (AC), resistant starch (RS), and β-glucan content of the food (Botticella et al., 2016).

4.1. Breeding Approach

Remarkable successes have been achieved using TILLING (targeting induced local lesions in genomes) to biofortify crops with enhanced DF content. Borticella et al. (2011) employed TILLING to target the *SBEIIa* gene, leading to a substantial increase of up to 21% in amylose content (AC) in double null mutants compared to control varieties. In a subsequent study, Borticella et al. (2016) developed null wheat lines for the *SSIIa* gene, resulting in an impressive 37–46.2% increase in AC compared to the control variety's 30.09%. The null wheat lines also exhibited significant alterations in resistant starch (RS) and β-glucan content. Borticella et al. (2018) further utilized TILLING, coupled with marker-assisted breeding, to generate mutant wheat lines harboring mutations in *GBSSI*, *SSIIa*, and *SBEIIa* genes. This approach yielded noteworthy enhancements in amylose, β-glucans, fructans, and arabinoxylans content across all the mutant lines (Figure 1).

In a parallel effort, Gurunathan et al. (2019) employed gamma ray-induced mutagenesis to develop mutant rice lines with deleterious mutations in starch-synthesizing genes GBSSI, SSIIA, and SSIIIa. These mutants were then crossbred with wild lines to create an F2 population, resulting in high AC (26.18%) and RS (8.68%) content in lines carrying mutations in *GBSSI*, *SSIIa*, and *SSIIIa* genes.

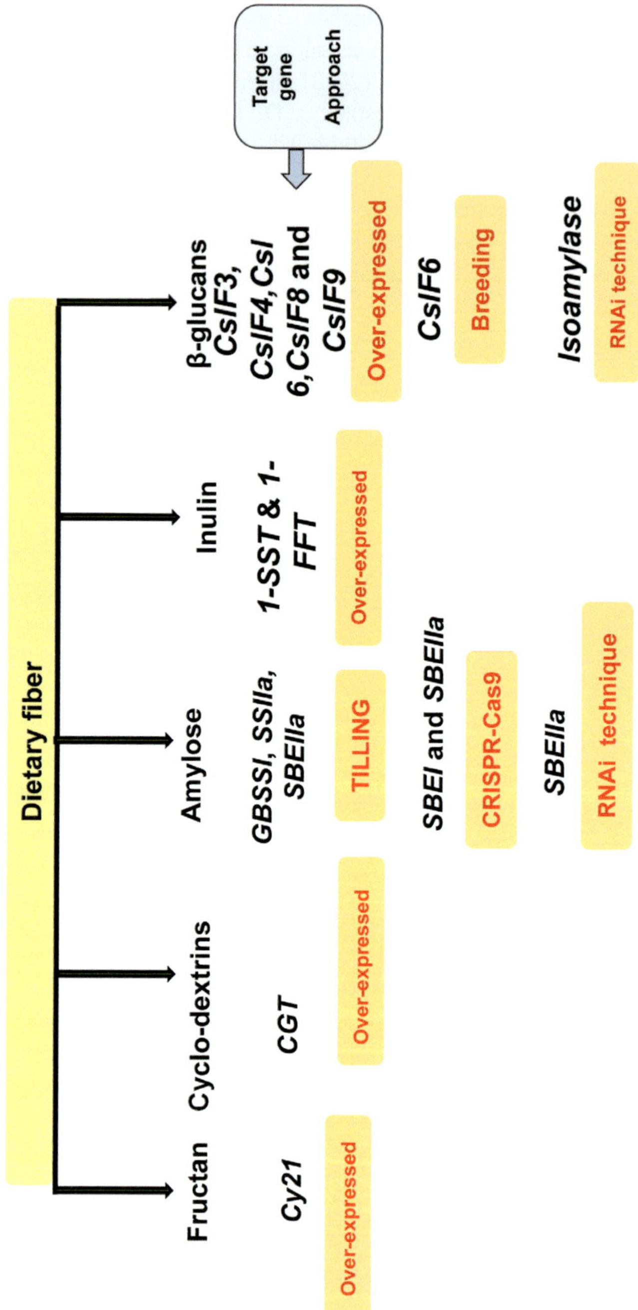

Figure 1: Enhancing soluble dietary fibers through biofortification strategies.

Furthermore, the integration of genes from the Csl family, particularly cslF6, known to play a role in β-glucan biosynthesis in barley, has proven effective in elevating grain β-glucan content when introgressed into wheat genomes. Similarly, the incorporation of Aegilops U and M chromosomes into wheat backgrounds has been successful in significantly increasing the production of dietary fibers in these addition lines (Rakszegi et al., 2017). Marker-assisted breeding could be used for the selection of crop varieties with a high content of DF. Recently, 562 hexaploid wheat varieties were collected from 46 countries for genome-wide association studies to identify genomic regions and candidate genes for arabinoxylan (Lee et al., 2023).

In sum, TILLING and mutational breeding strategies hold substantial promise in augmenting the DF content of crops, paving the way for more nutritionally enriched food sources.

4.2. Transgenic Approach

RNA interference (RNAi) is another reverse genetics approach used to knock down the expression of target gene using dsRNA (double-stranded RNA) (Fu et al., 2007). The content of amylose has been improved in barley and wheat using the RNAi approach in which the silencing of *SBE* was carried out. Sestili et al. (2015) silenced the expression of *SBEIIa* in durum wheat cultivars using the RNAi technique. An increment of 30.8%– 75% in the AC was observed in transgenic wheat lines and it was highly concentrated (>70%) in the grain endosperm. Carciofi et al. (2012) increased the RS by suppressing the expression of amylopectin synthesis genes including *SBEI*, *SBEIIa*, and *SBEIIb*. An increment of 2.2-fold increase in RS content of transgenic lines (65%) than control (29%) was reported in barley. Sestili et al. (2016) suppressed the Isoamylase 1 using RNAi technique in durum wheat and observed a 7% reduction in starch, whereas enhancement in β-glucan and phytogylcogen content by 1.3–1.4 and 2.1–2.4-fold, respectively, in transgenic wheat lines. The downregulation of *SBEIIa* and *SBEIIb* in wheat endosperm resulted in starch having higher amylose content (>70%) (Regina et al., 2010, 2015). Recently Sun et al. (2017) mutated *SBEI* and *SBEIIa* genes using CRISPR-Cas9 technology in rice and reported an increment of 25.0 and 9.8% in AC and RS content, respectively. In *cynara scolymus*, fructan content was also improved by using transgenic potato tubers expressing *Cy21* which accumulate high levels of 1-kestose and nystose (Hellwege et al., 1997). Besides this, inulin content was improved in *Cynara scolymus* (artichoke) by using constitutive expression of the *1-SST* (sucrose: sucrose 1-fructosyltransferase) and the *1-FFT* (fructan: fructan 1-fructosyltransferase) genes (Hellwege et al., 2000).

Burton et al. (2011) biofortified the barley crop with DF including β-glucan, and arabinoxylan. The transgenic barley lines were developed to express *CslF3*, *CslF4*, *CslF6*, *CslF8*, and *CslF9* genes under the regulation of ProASGLO promoter. An increment of more than 80% in the grain (1,3; 1,4)-β-D-glucan content was observed in transgenic barley lines due to high expression of *CslF6*, and *CslF4* genes. The total DF content also increased from 13.4 % to 21% in *CslF6* containing transgenic lines as compared to the control. High-value carbohydrate-rich potato tubers have been synthesized by expressing the cyclodextrin glycosyltransferases (*CGT*) gene, which results in the production of multipurpose dietary fiber cyclodextrins from starch (Oakes et al., 1991).

5. Bioavailability

Bioavailability refers to that portion of the food that is absorbed in the body. DF are resistant to digestion in the small intestine due to the absence of enzymes which can degrade their glycosidic bonds, therefore these are not absorbed and passed directly to the large intestine (Englyst and Englyst, 2005). Recently, Pereira-Caro et al. (2023) studied the effect of oat bran enriched with β-glucan on the bioavailability of flavanones in orange juice. Volunteers consumed 500 ml of orange juice with and without 22g of oat bran containing 6g of β-glucan (OB-6). With the consumption of OB-6 bioavailability of flavanones was found to be increased (Figure1).

6. Challenges

DF has a lot of benefits for human health. It helps to reduce the health-associated risk of diseases like cardiovascular, diabetes, obesity, constipation, colorectal cancer, etc. The recommended daily intake of DF ranged from 25–30 grams or more for adults (Stephen et al., 2017). Despite health benefits, people around the world are less aware of the intake of fiber-rich food. Therefore, campaigns should be started to educate people to consume fiber-rich food and to use the best food processing methods. They should encourage people to add whole grains and fruits to their diet instead of refined food (Buttriss and Stokes, 2008). The fiber richness of food also depends on the food processing approach. The physiochemical properties, structure, and nutritional value of DF are directly affected by the methods used for food processing. Thermal treatment such as boiling and cooking wheat bran was found to enhance the fiber content of wheat (Caprez et al., 1986). Different DFs have different physiochemical properties and degree of polymerization which alters their structure and helps in the prevention of disease. Therefore, the polymerization degree of DF from different sources is necessary to explore disease prevention (He et al., 2022).

Although the above studies suggested the beneficial effects of DF on human health, DF also negatively affects human health by interfering with the absorption of minerals and micronutrients (Baye et al., 2015). These were found to reduce the nutrients and vitamins' bioavailability and bioaccessibility by forming a complex that is not digested in the small intestine (Khokhar and Kapoor, 1990). Kurek et al. (2017) reported that the reduced particle size negatively affects the bioaccessibility of thiamine and riboflavin by 69.1–91.2% and 40.9–50.2%, respectively. Therefore, it's challenging to use strategies that minimize the adverse effect of DF on nutrient bioavailability.

The approval and release of crops is the major challenge associated with crops biofortified using a transgenic approach. According to Paarlberg (2002), almost all developing countries have established biosafety guidelines related to genetically modified organisms (GMO), but very few legally approved the commercial cultivation of GM crops by farmers (Paarlberg, 2002).

7. Conclusion

Dietary fiber biofortification holds great potential to improve human health by addressing various health issues associated with fiber deficiency. This chapter highlighted the significance of dietary fiber, its health benefits, strategies for biofortification using breeding and transgenic approaches, and the challenges associated with implementation and acceptance. As research and technological advancements continue, collaboration among scientists, policymakers, and the food industry is essential to develop and promote biofortified crops that can contribute to a healthier population and a better future.

References

Abraão, A., M. Yu, I. Gouvinhas, L. Ferreira, A.M. Silva, R. Domínguez-Perles, and A. Barros. 2023. L. *Prunus lusitanica*. Fruits: A promising underexploited source of nutrients with potential economic value. *Foods*, 12(5): 973.

Ajila, C.M., S.G. Bhat, and, U.P. Rao. 2007. Valuable components of raw and ripe peels from two Indian mango varieties. *Food Chemistry*, 102(4): 1006–1011.

Barber, T.M., S. Kabisch, A.F.H. Pfeiffer, and M.O.Weickert. 2020. The health benefits of dietary fiber. *Nutrients*, 12(10): 3209.

Baye, K., J.P. Guyot, C. Icard-Vernière, I. Rochette, and C. Mouquet-Rivier. 2015. Enzymatic degradation of phytate, polyphenols, and dietary fibers in Ethiopian injera flours: Effect on iron bioaccessibility. *Food Chemistry*, 174: 60–67.

Botticella, E., F. Sestili, F. Sparla, S. Moscatello, L. Marri, J.A. Cuesta-Seijo, ... and D. Lafiandra. 2018. Combining mutations at genes encoding key enzymes involved in starch synthesis affects the amylose content, carbohydrate allocation, and hardness in the wheat grain. *Plant Biotechnol. J.*, 16(10): 1723–1734.

Botticella, E., F. Sestili, G. Ferrazzano, P. Mantovani, A. Cammerata, M.G. D'Egidio, and D. Lafiandra. 2016. The impact of the SSIIa null mutations on grain traits and composition in durum wheat. *Breed. Sci.*, 66(4): 572–579.

Botticella, E., F. Sestili, A. Hernandez-Lopez, A. Phillips, and D. Lafiandra. 2011. High resolution melting analysis for the detection of EMS induced mutations in wheat Sbella genes. *BMC Plant Biol.*, 11(1): 1–14.

Burton, R.A., H.M. Collins, N.A. Kibble, J.A. Smith, N.J. Shirley, S.A. Jobling, and G.B. Fincher. 2011. Over expression of specific HvCslF cellulose synthase like genes in transgenic barley increases the levels of cell wall (1, 3; 1, 4) β d glucans and alters their fine structure. *Plant Biotechnol J.*, 9(2): 117–135.

Buttriss, J.L., and C.S. Stokes. 2008. Dietary fiber and health: An overview. *Nutr. Bull.*, 33(3): 186–200.

Caprez, A., E.V.A. Arrigoni, R. Amadò, and H. Neukom. 1986. Influence of different types of thermal treatment on the chemical composition and physical properties of wheat bran. *J. Cereal Sci.*, 4(3): 233–239.

Carciofi, M., A. Blennow, S.L. Jensen, S.S. Shaik, A. Henriksen, A. Buléon, P.B. Holm, and K.H. Hebelstrup. 2012. Concerted suppression of all starch-branching enzyme genes in barley produces amylose-only starch granules. *BMC Plant Biol.*, 12: 223.

Dai, F.J., and C.F. Chau. 2017. Classification and regulatory perspectives of dietary fiber. *J. Food & Drug Anal.*, 25(1): 37–42.

Dhingra, D., M. Michael, H. Rajput, and R.T. Patil. 2012. Dietary fiber in foods: A review. *J. Food Sci. Technol.*, 49: 255–266.

Englyst, K.N., and H.N. Englyst. 2005. Carbohydrate bioavailability. *Br. J. Nutr.*, 94(1): 1–11.

Fu, D., C. Uauy, A. Blechl, and J. Dubcovsky. 2007. RNA interference for wheat functional gene analysis. *Transgenic Res.*, 16: 689–701.

Galvão, F., E. Pinto, Z.E. Martins, A. Almeida, I. Ferreira, V.A. Lima, and M.L. Felsner. 2023. Nutritional composition and minerals bioaccessibility of commercial fruit flours. *J. Food Meas. Charact.*, 17(3): 1–8.

Gurunathan, S., B.R. Ramadoss, V. Mudili, C. Siddaiah, N.K. Kalagatur, J.R.K Bapu, C.D. Mohan, and E.F. Abduallah. 2019. Single nucleotide polymorphisms in starch biosynthetic genes associated with increased resistant starch concentration in rice mutant. *Front Genet.*, 10: 946.

He, J., R.H. Streiffer, P. Muntner, M.A. Krousel-Wood, and P.K. Whelton. 2004. Effect of dietary fiber intake on blood pressure: A randomized, double-blind, placebo-controlled trial. *J. Hypertens.*, 22(1): 73–80.

He, Y., B. Wang, L. Wen, F. Wang, H. Yu, D. Chen., ... and C. Zhang. 2022. Effects of dietary fiber on human health. *Food Sci. Hum. Wellness*, 11(1): 1–10.

Hellwege, E.M., S. Czapla., Jahnke, L. Willmitzer, and A.G. Heyer. 2000. Transgenic potato (*Solanum tuberosum*) tubers synthesize the full spectrum of inulin molecules naturally occurring in globe artichoke (*Cynara scolymus*) roots. *In: Proc. Natl. Acad. Sci. U.S.A.*, 97(15): 8699–8704.

Hellwege, E.M., D. Gritscher, L. Willmitzer, and A.G. Heyer. 1997. Transgenic potato tubers accumulate high levels of 1-kestose and nystose: Functional identification of a sucrose 1-fructosyltransferase of artichoke (*Cynara scolymus*) blossom discs. *Plant J.*, 12(5): 1057–1065.

Joye, I.J. 2020. Dietary fiber from whole grains and their benefits on metabolic health. *Nutrients*, 12(10): 3045.

Kaoutari, A.E., F. Armougom, J.I. Gordon, D. Raoult, and B. Henrissat. 2013. The abundance and variety of carbohydrate-active enzymes in the human gut microbiota. *Nat. Rev. Microbiol.*, 11(7): 497–504.

Kapoor, P., R.K. Dhaka, P. Sihag, S. Mehla, V. Sagwal, Y. Singh, ... and U. Kumar. 2022. Nanotechnology-enabled biofortification strategies for micronutrients enrichment of food crops: Current understanding and future scope. *NanoImpact*, 26: 100407.

Khokhar, S., and A.C. Kapoor, 1990. Effect of dietary fibers on bioavailability of vitamin A and thiamine. *Plant Foods Hum. Nutr.*, 40: 259–265.

Korge, M., M. Alaru, I. Keres, B. Khaleghdoust, K. Möll, I. Altosaar, and E. Loit. 2023. The influence of cropping system, weather conditions and genotype on arabinoxylan content in wheat and barley grains. *J. Cereal Sci.*, 110: 103650.

Kurek, M.A., J. Wyrwisz., S. Karp, and A. Wierzbicka. 2017. Particle size of dietary fiber preparation affects the bioaccessibility of selected vitamin B in fortified wheat bread. *J. Cereal Sci.*, 77: 166–171.

Lambo, A.M., R. Öste, and M.E.L. Nyman. 2005. Dietary fiber in fermented oat and barley β-glucan rich concentrates. *Food Chem.*, 89(2): 283–293.

Lee, M.H., J. Park, K.H. Kim, K.M. Kim, C.S. Kang, G.E. Lee, ... and C. Choi. 2023. Genome-wide association study of arabinoxylan content from a 562 hexaploid wheat collection. *Plants*, 12(1): 184.

Liu, H., J. Liang, C. Liang, G. Liang, J. Lai, R. Zhang, ... and, G. Xiao. 2022. Physicochemical properties of dietary fiber of bergamot and its effect on diabetic mice. *Front Nutr.*, 9.

Marcotuli, I., K. Houston, J.G. Schwerdt, R. Waugh, G.B. Fincher, R.A. Burton, ... and Gadaleta, A. 2016. Genetic diversity and genome-wide association study of β-glucan content in tetraploid wheat grains. *PLoS One*, 11(4): e0152590.

Miketinas, D.C., G.A. Bray, R.A. Beyl, D.H. Ryan, F.M. Sacks, and C.M. Champagne. 2019. Fiber intake predicts weight loss and dietary adherence in adults consuming calorie-restricted diets: The POUNDS lost (preventing overweight using novel dietary strategies) study. *J. Nutri.*, 149(10): 1742–1748.

Oakes, J.V., C.K. Shewmaker, and D.M. Stalker. 1991. Production of cyclodextrins, a novel carbohydrate, in the tubers of transgenic potato plants. *Bio/technology*, 9(10): 982–986.

Paarlberg, R.L. 2002. The real threat to GM crops in poor countries: Consumer and policy resistance to GM foods in rich countries. *Food Policy*, 27(3): 247–250.

Pereira-Caro, G., T.M. Almutairi, S. Cáceres-Jiménez, J.M. Moreno-Rojas, D. Malkova, A.L. García, and A. Crozier. 2023. Bioavailability of orange juice (poly) phenols: β-glucan-rich oat bran decreases urinary excretion of flavanone phase II metabolites and enhances excretion of microbiota-derived phenolic catabolites. *J. Freeradbiomed*, 199: 34–43.

Rakszegi, M., I. Molnár, A. Lovegrove, É. Darkó, A. Farkas, L. Láng, ... and P. Shewry. 2017. Addition of Aegilops U and M chromosomes affects protein and dietary fiber content of wholemeal wheat flour. *Front in Plant Sci.*, 8: 1529.

Rao, S.S., S. Yu, and A. Fedewa. 2015. Systematic review: Dietary fiber and FODMAP-restricted diet in the management of constipation and irritable bowel syndrome. *Aliment Pharmacol and Ther.*, 41(12): 1256–1270.

Regina, A., P. Berbezy, B. Kosar-Hashemi, S. Li, M. Cmiel, O. Larroque, ... and M. Morell. 2015. A genetic strategy generating wheat with very high amylose content. *Plant Biotechnology J.*, 13(9): 1276–1286.

Regina, A., B. Kosar-Hashemi, S. Ling, Z. Li, S. Rahman, and M. Morell. 2010. Control of starch branching in barley defined through differential RNAi suppression of starch branching enzyme IIa and IIb. *J. Exp. Bot.*, 61(5): 1469–1482.

Reynolds, A.N., A.P. Akerman, and J. Mann. 2020. Dietary fiber and whole grains in diabetes management: Systematic review and meta-analyses. *PLoS Medicine*, 17(3): e1003053.

Sestili, F., F. Sparla, E. Botticella, M. Janni, R. D'Ovidio, G. Falini, ... and D. Lafiandra. 2016. The down-regulation of the genes encoding Isoamylase 1 alters the starch composition of the durum wheat grain. *Plant Sci.*, 252: 230–238.

Sestili, F., S. Palombieri, E. Botticella, P. Mantovani, R. Bovina, and D. Lafiandra. 2015. TILLING mutants of durum wheat result in a high amylose phenotype and provide information on alternative splicing mechanisms. *Plant Sc.*, 233: 127–133.

Shafi, A., F. Ahmad, and Z.H. Mohammad. 2022. Effect of the addition of banana peel flour on the shelf life and antioxidant properties of cookies. *ACS Food Sci. Technol.*, 2(8): 1355–1363.

Solah, V.A., D.A. Kerr, W.J. Hunt, S.K. Johnson, C.J. Boushey, E.J. Delp, ... and S. Wood. 2017. Effect of fiber supplementation on body weight and composition, frequency of eating, and dietary choice in overweight individuals. *Nutrients*, 9(2): 149.

Stephen, A.M., M.M. Champ, S.J. Cloran, M. Fleith, L. Van Lieshout, H. Mejborn, and V.J. Burley. 2017. Dietary fiber in Europe: Current state of knowledge on definitions, sources, recommendations, intakes, and relationships to health. *Nut. Res. Rev.*, 30(2): 149–190.

Sun, Y., G. Jiao, Z. Liu, X. Zhang, J. Li, X. Guo, ... and L. Xia. 2017. Generation of high-amylose rice through CRISPR/Cas9-mediated targeted mutagenesis of starch-branching enzymes. *Front. Plant Sci.*, 8: 298.

Tejada-Ortigoza, V., L.E. Garcia-Amezquita, J.A.Torres, and J. Welti-Chanes. 2020. Chemical processes for the extraction and modification of dietary fiber. *Science and Technology of Fibers in Food Systems*, 343-361.

CHAPTER
12

Starch Quality: High Energy or Low Calories – Both Matter

Joy Roy[a*], Akansha Madhawan[a,b], Mohammed Saba Rahim[c] and Prashant Kumar[a,b]

[a] Agri-Food Biotechnology Division, National Agri-Food Biotechnology Institute, Mohali - 140306, Punjab, India

[b] Regional Centre for Biotechnology (RCB), NCR Biotech Science Cluster, 3rd Milestone, Faridabad-Gurgaon Expressway, Faridabad, Haryana - 121001, India

[c] Molecular Genetic and Genomics Laboratory, Biotechnology Division, CSIR-Institute of Himalayan Bioresource Technology, Palampur - 176061, India

1. Introduction

Modern sedentary lifestyles and unhealthy eating habits have led to a significant rise in global diseases like obesity, diabetes and obesity rates doubling since 1980, as per the World Health Organization (WHO). Carbohydrates, mainly starch, found in cereals, legumes, and starchy vegetables are a major part of our diet. Starch is composed of two components: amylose and amylopectin. Amylose consists of linear chains connected by α-1,4-linkages, resulting in an amorphous structure. In contrast, amylopectin has α-1,4-linkages and α-1,6-linkages occurring after about 20 units, making it a highly branched and crystalline structure (Pfister and Zeeman, 2016). The ratio of amylose to amylopectin, ratio of long and short chains of amylopectin, the starch granule morphology, crystallinity, processing methods, and the plant source, all impact starch digestibility (Zhao et al., 2018). High amylose starch, known as resistant starch type-II, is slowly digested in the intestines and promotes gut health (Srikaeo and Sangkhiaw, 2014). It also helps maintain a lower glycemic index, reducing diabetes risk (Maki et al., 2012). Several studies indicate that high amylose wheat starch can enhance insulin sensitivity and contribute to a lower

*Corresponding author: joykroy@nabi.res.in

glycemic index (Li et al., 2023). Additionally, high amylose wheat lines also show potential in managing metabolic disorders (Rahim et al., 2022).

Despite its health benefits, resistant starch content in cereal grains is typically low. Hence, there is a pressing need to biofortify staple foods with resistant starch to enhance nutritional security. This chapter explores recent advances and approaches in starch biofortification, highlighting the documented health impacts. Efforts, including molecular genetics and targeted genome editing, are emphasized. Biofortification offers a realistic, cost-effective means to enhance starch quality and quantity in cereal crops, thereby improving their nutritional value.

2. Classification of Starch

There are three main types of starch-based on their rate of digestion.

1. **Rapidly Digestible Starch (RDS):** This starch rapidly breaks down within 20 minutes when exposed to enzymes, causing a sudden spike in blood glucose levels. This can trigger an insulin response and may lead to diet-related diseases like hyperglycemia (Raigond et al., 2015).
2. **Slowly Digestible Starch (SDS):** SDS is not fully accessible to enzymes, leading to a slower digestion process that can take up to 120 minutes. This gradual digestion results in a controlled increase in blood glucose levels, promoting a low glycemic index and reducing the risk of nutrition associated diseases (Raigond et al., 2015).
3. **Resistant Starch (RS):** A term coined by Englyst et al. (1982) describe starch that remains undigested for at least 120 minutes after enzymatic treatment. RS resists digestion and absorption in the small intestine and undergoes fermentation by gut microflora in the large intestine. In the colon, RS is transformed into short-chain fatty acids (SCFAs), which support immunity and gut health (Englyst et al., 1982). While there is no official daily RS recommendation, research suggests that consuming 6–20 grams per day can yield significant health advantages. This growing importance of RS has spurred research focused on increasing its presence in various crop plants.

3. Classification of Resistant Starch

The classification RS was initially proposed by Englyst et al. (1992), and later extended to five categories; each type offers unique properties.

3.1. RS-I (Physically Embedded Starch)

RS-I is physically inaccessible starch found in coarsely ground cereals and pulses. The starch within endosperm cells is shielded by cell walls, inhibiting the action of digestive α-amylase enzymes. Additionally, cell wall components like cellulose provide a non-catalytic surface for α-amylase adsorption. Proteins capable of forming disulfide bonds slow down starch digestion in cooked foods.

These factors collectively limit moisture content, hindering starch granule swelling and gelatinization.

3.2. RS-II (Natural Starch Granules)

RS-II exhibit a B-type crystalline polymorph, differentiating it from the A-type found in cereal starch. B-type polymorph includes 'blocklets' varying in size by species, reducing granule porosity, enzyme diffusion, and digestion. Cereals, with their A-type polymorph, enable enzyme diffusion through their crystalline structure. RS-II is present in raw potatoes, starch-rich fruits like green bananas, and high-amylose cereal mutants.

3.3. RS-III (Retrograde Starch)

RS-III forms when gelatinized starch reassociates at lower temperatures (4–5 °C), creating closely packed H-bonded double helical structures resistant to digestion. Linear amylose chains and amylopectin branches undergo this process, altering enzyme binding site conformations. Retrogradation of amylopectin requires longer time and a lower temperature (55 °C) compared to amylose (Birt et al., 2013; Roman and Martinez, 2019).

3.4. RS-IV (Chemically Modified Starch)

Chemical modifications like substitutions, oxidations, and cross-linkages alter starch confirmation, change enzyme attack sites and prevent digestion. Chemicals with substituting groups (e.g., acetyl, hydroxypropyl) or those leading to oxidation and cross-linking (mostly phosphate groups) prevent starch swelling, maintaining granular integrity during cooking. This renders starch unavailable for hydrolysis by digestive enzymes and bacterial fermentation, converting it into resistant starch (Birt et al., 2013; Roman and Martinez, 2019).

3.5. RS-V (Self-assembled-starch Complexes)

Lipids, when bound with starch, in the amylose chains, forms single-helical starch-lipid complexes (Rahim et al., 2020; Madhawan et al., 2023). These complexes, either naturally occurring (e.g., in high amylose starch) or added exogenously, can form either less ordered complexes melting at 95–105 °C or V-type crystals with higher melting points (110–120 °C). These complex fatty acids fill the helical chain cavities, preventing amylases from accessing glycosidic bonds. These complexes can quickly reform after cooking, regaining their resistant starch properties and thermostability (Birt et al., 2013; Panyoo and Emmambux, 2017; Roman and Martinez, 2019).

4. Biotechnology Intervention for High Amylose/ Resistant Starch Cereal Crops

Biotechnology interventions have been harnessed to boost high amylose and resistant RS content in cereal crops. The synthesis of starch involves multiple genes and enzymes, including ADP-glucose pyrophosphorylase (AGPase), granule-bound starch synthase (GBSS), starch-branching enzyme (SBE), starch synthase (SS), and debranching enzyme (DBE) (Figure 1). Various strategies, such as marker-assisted breeding, TILLING, chemical mutation, RNAi, and genome editing, has been explored to enhance amylose and RS in multiple crops (Figure 2, Table 1).

Cereals, particularly rice, are crucial for their varying amylose content, distinguishing between waxy and non-waxy varieties. Waxy rice primarily contains amylopectin, while non-waxy rice combines 7–33% amylose with 67–93% amylopectin. Significant progress has been made in enhancing rice's nutritional value. RS-rich rice mutants like RS111, Goami2, and Jiangtangdao1 were developed through mutagenesis (Yang et al., 2012). *OsSBE3* was identified as a key player in elevating RS content, particularly in Jiangtangdao1 (Yang et al., 2012). Using CRISPR/Cas9, researchers targeted two genes, *SBEI* and *SBEIIb*, to increase RS content in rice (Sun et al., 2017). Surprisingly, *SBEI* knockout had minimal impact on starch biosynthesis compared to non-transgenic controls (Sun et al., 2017). However, *SBEIIb* mutants led to higher amylose and RS content. Thus, *SBEIIb* emerges as a more effective gene for biofortified rice development with an elevated RS content.

Wheat plays a pivotal role in enhancing resistant starch (RS). Marker-assisted breeding has been employed to boost amylose content. Induced mutations using radiation or chemicals have introduced variations in amylose/RS-II (Mishra et al., 2016) and amylose-lipid complex/RS-IV (Madhawan et al., 2020). In a recent discovery, a novel mutant allele, GBSSI.L539P, was identified in a high amylose mutant line, underlining the impact of GBSSI mutations on starch composition (Sharma et al., 2022). Moreover, the use of TILLING has contributed to increased amylose content. Similarly, CRISPR/Cas9 targeting has focused on *SBEIIa* for amylose reduction (Li et al., 2021).

Maize, the world's leading cereal, has been a prime target for enhancing amylose content. Maize starches fall into three categories: waxy, normal, and high-amylose, determined by genetic mutations (Obadi and Xu, 2023). Four key mutant genes—amylose-extender (*ae*), dull (*du*), sugary-2 (*su2*), and waxy (*wx*)—significantly affect maize grain amylose content. The *ae* gene, in particular, has been widely utilized by researchers to elevate maize amylose content (Obadi and Xu, 2023). A ground-breaking achievement saw an 80% increase in amylose through the recurrent selection approach employing this gene (García-Lara and Serna-Saldivar, 2019).

Starch biofortification extends beyond major crops to include millets. Amylose synthesis in millets is controlled by the key enzyme GBSS1, which catalyzes amylose formation. High-throughput gene and transcript analysis has revealed that the waxy nature of GBSS1 is a result of transposable element insertion

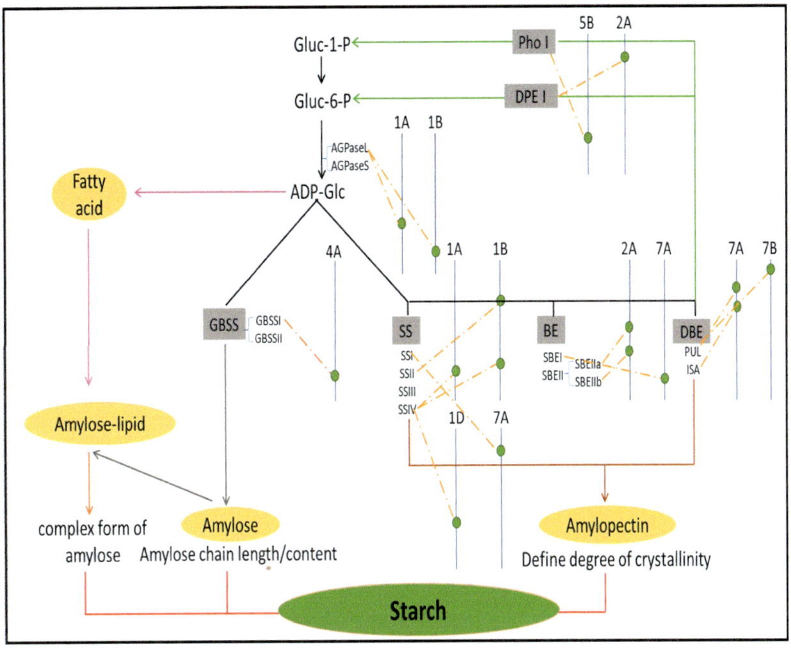

Figure 1: Chromosomal locations of starch biosynthesis genes in wheat that have been targeted for starch biofortification (Rahim et al., 2020; Wang et al., 2014).

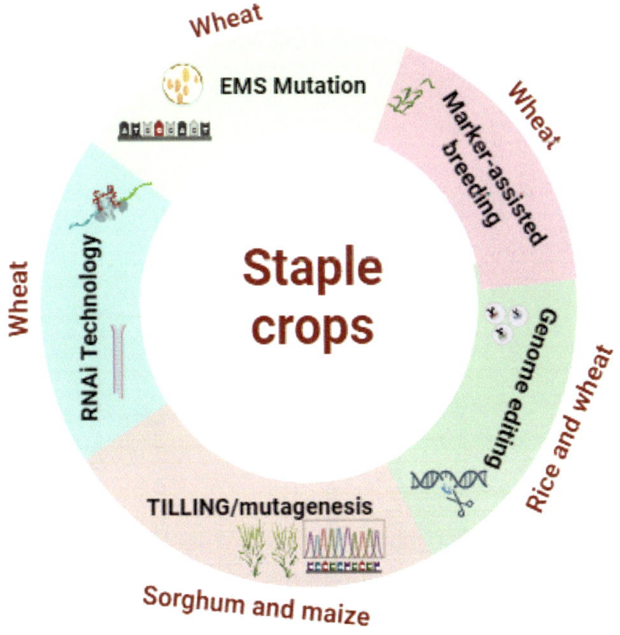

Figure 2: Strategies for increasing amylose content and resistant starch in different crops.

Table 1: Different reported biofortified cereal crops and the methods utilized to generate the high amylose lines

Technique	Crop (Cultivar)	Target Gene	Outcome	References
RNA interference technology	Wheat - durum (Svevo and Ofanto	*SBEIIa*	Amylose 30.8% to 75%	Sestili et al., 2010
TILLING	Wheat - Bread and durum	*SBEIIa*	Amylose 47 to 55%	Slade et al., 2012
Mutagenesis	Rice (RS111, Goami2 and Jiangtangdao1)	*SBE3*	Increased RS content	Yang et al., 2012
Ethyl methane sulphonate (EMS) mutants	Wheat - durum	*SSIIa*	Amylose 38% to 50%	Hogg et al., 2013
High throughput analysis of genes and transcripts	Millets	*Wx*	Major gene identification	-
Mutation and breeding	Wheat durum	*SSIIa or SGP-A1*	Amylose increased 3%	Martin et al., 2014
Marker-assisted breeding	Wheat	*SGP*	Amylose (40%)	Rakszegi et al., 2015
Plant breeding	Sorghum	*Wx, kafirin*	High amylopectin and protein digestibility	Elhassan et al., 2015
Ethyl methanesulfonate	Wheat	*GBSSI, BMY, SSIII, SBEI, SBEIII, ISA3*	Amylose ~3 to 76 %	Mishra et al., 2016
Activate hydrolyzing enzymes during brown rice germination	Rice	Hydrolytic enzymes	Increased starch content	Wunthunyarat et al., 2020
Mutation and hybridization	Wheat	*SBEII*	Amylose (~ 60%)	Schönhofen et al., 2017
CRISPR/Cas9	Rice	*SBEI, SBEIIb*	Increased RS content	Sun et al. 2017

TILLING followed by marker-assisted breeding	Wheat (Cadenza)	*GBSSI*, *SSIIa* and *SBEIIa*	Amylose 0%, 46% , 79%	Botticella et al., 2018
Recurrent selection	Maize	*ae*	Amylose up to 80%	García-Lara and Serna-Saldivar, 2019
CRISPR/Cas9	Wheat (Zhengmai 7698, Bobwhite)	*SBEIIa*	Amylose ZM line (45.4%) Amylose Bobwhite line (47.1%)	Li et al., 2021
Ethyl methanesulfonate (EMS)	Wheat (J411)	*GBSSI*, *AMY* and *ISA1*	Amylose increased 17%	Irshad et al., 2022

Abbreviations: Granule-bound starch synthases (GBSSs), Starch synthases (SS), Starch-branching enzymes (SBEs), α-amylase (AMY), Isoamylases (ISA), β-amylase (BMA), Starch granule proteins (SGP), Waxy (Wx), Amylose extender (*ae*).

(Bai et al., 2013). Elhassan et al. (2015) successfully developed biofortified sorghum by combining waxy (high amylopectin) starch with high-protein digestibility traits.

5. Potential Effects of RS on Prevention or Therapy of Some Chronic Diseases

Resistant starch especially RS-II has been associated with several health benefits and thus is recommended for renaming as dietary fibers.

Type 2 diabetes, often lifestyle-related, involves insulin resistance and high blood glucose levels. Studies have verified that consuming resistant starch in dietary supplements can lead to lower plasma blood glucose levels, improve insulin sensitivity. FDA has approved its consumption and stated that "high-amylose maize resistant starch may reduce the risk of type 2 diabetes" (https://www.regulations.gov/docket?D=FDA-2015-Q-2352). RS improves glucose uptake in muscles and enhances GLP-1 production, aiding insulin sensitivity. Meta-analyses have found that RS reduces the risk of metabolic syndrome and type 2 diabetes at a population level by reducing postprandial insulin level, cholesterol, and inflammatory markers (Bodinham et al., 2014). Resistant starch acts as a prebiotic, resisting digestion, fermenting within the gut, and fostering the growth of beneficial gut bacteria. Currently, the health benefit assessment of chapatti made from high resistant starch wheat flour was studied and found increased good bacteria such as *Lactobacillus* sp., *Bifidobacteria*, *Akkermansia muciniphila* and *Segmental filamentous bacterium* and reduced the abundance of harmful bacteria like *Firmicutes* (Rahim et al., 2022). Colorectal cancer, a major global cause of death, has links to unhealthy diet. RS is beneficial for colorectal health. By fermenting in the large intestine and producing SCFAs, RS reduces cell proliferation in the colon. SCFAs, especially butyrate, regulate genes, decreasing cancer risk (Tian et al., 2018; Bishop et al., 2017; Chapkin et al., 2020). Obesity results from excessive energy consumption, and RS helps manage weight. An Animal study on sorghum RS indicated inhibition of weight gain and improvement in lipid metabolism (Shen et al., 2015). Human and meta-analysis studies also show increased fat oxidation and reduced appetite with RS intake (Amini et al., 2021). The mechanisms behind RS-induced weight reduction remain unclear but may involve increased satiety hormones and reduced metabolic endotoxemia.

It also reduces oxidative stress and benefits chronic kidney disease (DKD) patients. Consumption of RS from high amylose maize and other sources has been linked with reduction in serum creatinine and p-cresol levels, uric acid, and urine β2-MG in clinical and meta-analysis studies. Studies have evaluated RS-enriched foods, considering sensory, textural, digestive, and glycemic aspects. A daily intake of 6–20 g of RS appears effective for health benefits. However, the impact may vary depending on RS type and composition (Du et al., 2022).

While several studies confirm RS's health benefits in humans, some animal studies suggest potential concerns related to kidney health and anxiety (Barouei et al., 2017). Therefore, to meet global nutritional security goals, it is imperative to investigate and validate the widespread health benefits and concerns of biofortified high amylose crops through clinical trials. Development and commercialization of high amylose starch sound possible for future food in managing health issues.

References

Amini, S., M. Anahita, and M.N. Leila. 2021. The effect of acute consumption of resistant starch on appetite in healthy adults: A systematic review and meta-analysis of the controlled clinical trials. *Clinical Nutrition ESPEN*, 41: 42-48,

Barouei, J., Z. Bendiks, A. Martinic, D. Mishchuk, D. Heeney, Y.H. Hsieh, D. Kieffer, J. Zaragoza, R. Martin, C. Slupsky, and M.L. Marco. 2017. Microbiota, metabolome, and immune alterations in obese mice fed a high-fat diet containing type 2 resistant starch. *Mol. Nutr. Food Res.*, 61(11).

Birt, D.F., T. Boylston, S. Hendrich, J.L. Jane, J. Hollis, L. Li, J. McClelland, S. Moore, G.J. Phillips, M. Rowling, K. Schalinske, M. Paul Scott, and E.W. Whitley. 2013. Resistant starch: Promise for improving human health. *Adv. Nutr.*, 4(6).

Bishop, K.S., H. Xu, and G. Marlow. 2017. Epigenetic regulation of gene expression induced by butyrate in colorectal cancer: Involvement of microRNA. *Genetics Epigenetics*, 9: 1179237X1772990.

Bodinham, C., L. Smith, E.L. Thomas, J.D. Bell, J.R. Swann, A. Costabile, D. Russell-Jones, A.M. Umpleby, and M.D. Robertson. 2014. Efficacy of increased resistant starch consumption in human type 2 diabetes. *Endocr. Connect.*, 3(2): 75–84.

Botticella, E., F. Sestili, F. Sparla, S. Moscatello, L. Marri, J.A. Cuesta-Seijo, G. Falini, A. Battistelli, P. Trost, and D. Lafiandra. 2018. Combining mutations at genes encoding key enzymes involved in starch synthesis affects the amylose content, carbohydrate allocation, and hardness in the wheat grain. *Plant Biotechnol. J.*, 16(10): 1723–1734.

Chapkin, R.S., S.L. Navarro, M.A. Hullar, and J.W. Lampe. 2020. Diet and gut microbes act coordinately to enhance programmed cell death and reduce colorectal cancer risk. *Dig. Diseases Sci.*, 65: 840–851.

Chung, H.J., D.H. Shin, and D.T. Lim. 2008. In vitro starch digestibility and estimated glycemic index of chemically modified corn starches. *Food Res. Int.*, 41(6): 579–585.

Elhassan, M.S.M., M.N. Emmambux, B.H. Dirk, P.C. Gary, and J.R.N. Taylor. 2015. Novel biofortified sorghum lines with combined waxy (high amylopectin) starch and high protein digestibility traits: Effects on endosperm and flour properties. *J. Cereal Sci.*, 65: 132–139.

Englyst, H.N., S. Kingman, and J. Cummings. 1992. Classification and measurement of nutritionally important starch fractions. *European J. Clinical Nutr.*, 46: S33–S50.

Englyst, H., H.S. Wiggins, and J. Cummings. 1982. Determination of the non-starch polysaccharides in plant foods by gas-liquid chromatography of constituent sugars as alditol acetates. *Analyst*, 107(1272): 307–318.

García-Lara, S., O. Sergio, and Serna-Saldivar. 2019. Chapter 1: *In:* García-Lara, Serna-Saldivar (Eds.), *Corn History and Culture*. 1–18. AACC International Press.

Hogg, A.C., K. Gause, P. Hofer, J.M. Martin, Graybosch, R.A., Hansen, L.E., and M.J. Giroux. 2013. Creation of a high-amylose durum wheat through mutagenesis of starch synthase II (SSIIa). *J. Cereal Sci.*, 57(3): 377–383.

Irshad, A., H. Guo, S. Ur Rehman, J. Gu, C. Wang, H. Xiong, Y. Xie, S. Zhao and L. Liu. 2022. Screening of induced mutants led to the identification of starch biosynthetic genes associated with improved resistant starch in wheat. *Int. J. Mol. Sci.*, 23(18): 10741.

Kumar, P., A. Mishra, M.S. Rahim, V.M.S. Sharma, A. Madhawan, A. Parveen, V. Fandade, H. Sharma, and J. Roy. 2022. Comparative transcriptome analyses revealed key genes involved in high amylopectin biosynthesis in wheat. *3 Biotech.*, 12(11): 295.

Li, C., S. Dhital, and M.J. Gidley. 2023. High-amylose wheat tortillas with reduced *in vitro* digestion and enhanced resistant starch content. *Food Hydrocoll.*, 137: 108321.

Li, J., G. Jiao, Y. Sun, J. Chen, Y. Zhong, L. Yan, D. Jiang, Y. Ma and L. Xia. 2021. Modification of starch composition, structure and properties through editing of TaSBEIIa in both winter and spring wheat varieties by CRISPR/Cas9. *Plant Biotechnol. J.*, 19(5): 937–951.

Madhawan, A., A. Sharma, A. Bhandawat, M.S. Rahim, P.M.S., Kumar, A. Mishra, A. Parveen, H. Sharma, S.K. Verma, and J. Roy. 2020. Identification and characterization of long non-coding RNAs regulating resistant starch biosynthesis in bread wheat (*Triticum aestivum* L.) *Genomics*, 112(5): 3065–3074.

Maki, K.C., C.L. Pelkman, E.T.C.L. Finocchiaro, K.M.E.T. Kelley, A.L.K.M. Lawless, A.L.A.L. Schild, and T.M. Rains. 2012. Resistant starch from high-amylose maize increases insulin sensitivity in overweight and obese men. *J. Nutr.*, 142(4): 717–723.

Martin, J., M.A.C. Hogg, P.A.C. Hofer, F.A. Manthey, and M.J. Giroux. 2014. Impacts of SSIIa-A null allele on durum wheat noodle quality. *Cereal Chem. J.*, 91(2): 176–182.

Mishra, A., A. Singh, M. Sharma, P. Kumar, and J. Roy. 2016. Development of EMS-induced mutation population for amylose and resistant starch variation in bread wheat (*Triticum aestivum*) and identification of candidate genes responsible for amylose variation. *BMC Plant Bio.*, 16(1): 217.

Obadi, M., Q. Yajing, and X. Bin. 2023. High-amylose maize starch: Structure, properties, modifications, and industrial applications. *Carbohydrate Polymers*, 299: 120185.

Panyoo, A.E., and M.N. Emmambux. 2017. Amylose–lipid complex production and potential health benefits: A mini-review. *Starch-Stärke*, 69(7–8): 1600203.

Pfister, B., and S.C. Zeeman. 2016. Formation of starch in plant cells. *Cellular and Mol. Life Sci.*, 73(14): 2781–2807.

Rahim, M.S., V. Sharma, P. Yadav, A. Parveen, A. Kumar, J. Roy, and V. Kumar. 2023. Rethinking underutilized cereal crops: Pan-omics integration and green system biology. *Planta*, 258: 91.

Rahim, M.S., V. Kumar, A. Mishra, V. Fandade, V. Kumar, K. Kiran Kondepudi, M. Bishnoi, and J. Roy. 2022. High resistant starch mutant wheat 'TAC 35' reduced glycemia and ameliorated high-fat diet induced metabolic dysregulation in mice. *J. Cereal Sci.*, 105.

Rahim, M.S., A. Chauhan, A. Madhawan, A. Mishra, A. Parveen, M. Garg, M. Bishnoi, V. Kumar, M.D. Oak, N.K.M.D. Singhal, and J. Roy. 2020. Development and evaluation of chapatti quality of high amylose wheat mutants on the basis of physicochemical, textural, and sensory characteristics. *LWT*, 133: 110051.

Raigond, P., R. Ezekiel, and B. Raigond. 2015. Resistant starch in food: A review: *J. Sci. Food Agri.*, 95(10): 1968–1978.

Rakszegi, M., B.N. Kisgyörgy, T. Kiss, F. Sestili, L. Láng, D. Lafiandra, and Z. Bedő. 2015. Development and characterization of high-amylose wheat lines. *Starch – Stärke*, 67(3–4): 247–254.

Roman, L., and M.M. Martinez. 2019. Structural basis of resistant starch (RS) in bread: Natural and commercial alternatives. *Foods*, 8(7): 267.

Schönhofen, A., X. Zhang and J. Dubcovsky. 2017. Combined mutations in five wheat Starch Branching Enzyme II genes improve resistant starch but affect grain yield and bread-making quality. *J. Cereal Sci.*, 75: 165–174.

Sestili, F., M. Janni, A. Doherty, E. Botticella, R. D'Ovidio, S. Masci, H.D. Jones, and D. Lafiandra. 2010. Increasing the amylose content of durum wheat through silencing of the SBEIIa genes. *BMC Plant Bio.*, 10(1): 144.

Sharma, V., V. Fandade, P. Kumar, A. Parveen, A. Madhawan, M. Bathla, A. Mishra, H. Sharma, V. Rishi, S.B. Satbhai, and J. Roy. 2022. Protein targeting to starch 1, a functional protein of starch biosynthesis in wheat (*Triticum aestivum* L.). *Plant Mol. Bio.*, 109(1–2): 101–113.

Shen, R.-L., W.L. Zhang, J.L. Dong, G.X. Ren, and M. Chen. 2015. Sorghum resistant starch reduces adiposity in high-fat diet-induced overweight and obese rats via mechanisms involving adipokines and intestinal flora. *Food Agric. Immunology*, 26(1): 120–130.

Shevkani, K., N. Singh, R. Bajaj, and A. Kaur. 2017. Wheat starch production, structure, functionality and applications: A review. *International J. Food Sci. Technol.*, 52(1): 38–58.

Shewry, P.R., N.G. Halford, P.S. Belton, and A.S. Tatham. 2002. The structure and properties of gluten: An elastic protein from wheat grain. *Philosophical Transactions of the Royal Society of London. Series B: Bio. Sci.*, 357(1418): 133–142.

Slade, A.J., C. McGuire, D. Loeffler, J. Mullenberg, W. Skinner, G. Fazio, A. Holm, K.M. Brandt, M.N. Steine, J.F. Goodstal, and V.C. Knauf. 2012. Development of high amylose wheat through TILLING. *BMC Plant Bio.*, 12(1): 69.

Srikaeo, K., and J. Sangkhiaw. 2014. Effects of amylose and resistant starch on glycaemic index of rice noodles. *LWT - Food Sci. Technol.*, 59(2): 1129–1135.

Sun, Y., G. Jiao, Z. Liu, X. Zhang, J. Li, X. Guo, W. Du, J. Du, F. Francis, Y. Zhao, and L. Xia. 2017. Generation of high amylose rice through CRISPR/Cas9-mediated targeted mutagenesis of starch branching enzymes. *Front Plant Sci.*, 8: 298.

Tian, Y., Q. Xu, L. Sun, Y. Ye, and G. Ji. 2018. Short-chain fatty acids administration is protective in colitis-associated colorectal cancer development. *J. Nutr. Biochem.*, 57: 103–109.

Wunthunyarat, W., H.S. Seo, and Y.J. Wang. 2020. Effects of germination conditions on enzyme activities and starch hydrolysis of long-grain brown rice in relation to flour properties and bread qualities. *J. Food Sci.*, 85(2): 349–357.

Zhao, K., A.S.M. Saleh, B. Li, H. Wu, Y. Liu, G. Zhang, and W. Li. 2018. Effects of conventional and microwave pre-treatment acetylation on structural and physicochemical properties of wheat starch. *International J. Food Sci. Technol.*, 53(11): 2515–2524.

Zhong, Y., J.Z. Qu, X. Liu, L. Ding, Y. Liu, E. Bertoft, B.L. Petersen, B.R. Hamaker, K.H. Hebelstrup, and A. Blennow. 2022. Different genetic strategies to generate high amylose starch mutants by engineering the starch biosynthetic pathways. *Carbohydr. Polym.*, 287: 119327.

CHAPTER

13

Antinutrients and Underutilized Food Crops: Unearthing Hidden Potential

Dr. Moshe Goldsmith[0000-0002-4257-0509]

Department of Biomolecular Sciences, Weizmann Institute of Science, Rehovot, 7610001, Israel

1. Introduction

Many plant foods present a nutritional paradox as they contain both nutrients that are essential for sustenance and health, as well as antinutrients that compromise it. Antinutrients are secondary plant metabolites that inhibit the absorption of nutrients during food digestion and can interfere with normal physiological functions (Lopez-Moreno et al., 2022; Petroski and Minich, 2020). They are usually classified as lectins, phytates, tannins, phytoestrogens, goitrogens, trypsin inhibitors, oxalates, saponins, cyanogenic glycosides, and lathyrogens; however, additional classifications are sometimes applied. Antinutrients adversely affect health through different mechanisms when consumed and accumulated to high levels. This usually occurs when specific kinds of food crops are ingested in large amounts, without processing, as primary components of an unbalanced diet, and over prolonged periods of time. Different genetic backgrounds or physiological conditions, such as malnutrition, may increase susceptibility to the harmful effects of antinutrients. However, in some cases, these can be alleviated upon resumption of a balanced diet, reduced consumption of the antinutrients, and the use of dietary supplements.

Some antinutrients (e.g., cyanogenic glucosides) elicit their adverse effects since they have evolved as plant defense mechanisms against herbivores (Gleadow and Møller, 2014). In other cases, e.g., β-ODAP they are harmful simply because they constitute plant metabolites that are toxic to humans and animals (Tan et al., 2017). Initially, antinutrients had gained interest in light of their role in the

*Corresponding author: moshe.goldsmith@weizmann.ac.il

etiology of epidemics of spastic paralysis (e.g., Lathyrism and Konzo) or goiter (Sunil and Leena, 2021; Baguma et al., 2021). More recently, they have been the focus of research into the health benefits they can confer. Their anticarcinogenic, anti-inflammatory, antidiabetic, antioxidant, immunostimulatory, antimicrobial, and antifungal activities have been suggested to be helpful in preventing or curing a number of disease states. Among them are various forms of cancers, cardiovascular and autoimmune diseases, obesity, metabolic syndromes, type 2 diabetes, and cognitive disorders (Elekofehinti et al., 2021; Kaur et al., 2019; Cid-Gallegos et al., 2022; Petrine et al., 2021; Hamsalakshmi et al., 2022).

Here, we discuss the different classes of antinutrient compounds, their harmful and beneficial impacts on health, and the factors that determine them. We list different strategies used for their reduction or elimination from plant food and the challenges facing breeding and genetic engineering efforts aiming to do so. We also discuss several crops that are prime targets for such endeavors due to a combination of high levels of both nutrients and toxic antinutrients.

2. Influence on Human Health

Different classes of antinutrients can similarly affect our bodies through the same mechanism of action: Oxalates, phytates, tannins, and saponins can all bind to mineral ions, generating insoluble complexes that prevent their absorption from the gastrointestinal (GI) tract, thus depriving the body of essential minerals. Protease inhibitors, phytates, tannins, and saponins inhibit proteolytic enzymes in the stomach and gut, preventing protein digestion that is required for the absorption of essential amino acids from food. Amylase inhibitors act similarly towards ingested starch and glycogen and prevent their utilization as energy sources.

Some antinutrients can directly damage cells and organs: Lectins have a hemolytic effect as they can bind and agglutinate red blood cells; phytoestrogens and goitrogens disrupt the normal function of the thyroid gland; cyanogenic glycosides and β-ODAP act as neurotoxins that can cause spastic paralysis of the lower limbs; oxalate-calcium complexes can lead to the formation of kidney stones, and lectins or protease inhibitors can produce intestinal damage. Of all the above-mentioned antinutrients, the most dangerous are cyanogenic glycosides, as they produce hydrocyanic acid (HCN) that is highly toxic and can lead to instant death.

Despite the above list of physiological influences exerted by antinutrients, they can also be highly beneficial to human health if consumed at the appropriate concentrations or applied in the correct context. Thus, antinutrients such as lectins, tannins, phytoestrogens, goitrogens, protease inhibitors, and saponins have all been found to have anticarcinogenic activities. Lectins, saponins, and phytoestrogens have antidiabetic properties, while tannins and protease inhibitors can act as anti-inflammatory agents. Other beneficial properties, such as antimicrobial, antioxidant, and immunostimulatory properties have been ascribed to lectins, phytates, and saponins.

2.1. Lectins

Lectins are a diverse class of multivalent, non-enzymatic, carbohydrate-binding proteins that bind specific sugar structures (Figure 1) and play a role in defense from pathogens and cell recognition in organisms from all kingdoms of life (Sharon and Lis, 2007). In plants, lectins function in biological processes such as plant development, immunity, stress signalling, and regulation of gene expression (De Coninck and Van, 2021). They are present in plant foods such as nuts, cereals, legumes, fruits, and tubers, where they have been found to have both deleterious and beneficial effects (Muramoto, 2017). Lectins are relatively resistant to proteolytic degradation in the digestive tract, enabling them to interact with epithelial cells in the intestine. This interaction interferes with nutrient absorption by altering the integrity and permeability of the small intestine and its mucosa and by activating the immune response (Muramoto, 2017). Lectins can inhibit exocytosis and repair of the plasma membrane, unbalance the normal gut flora, and induce allergic responses (Barre et al., 2020). Their ability to bind cell-surface glycoproteins and glycolipids enables them to agglutinate red blood cells when in contact with the blood (Sharon, 2007). On the other hand, they have been shown to display antitumor activities by activating the immune system and inducing apoptosis (Fu et al., 2011), antiangiogenic, antimetastatic, and antiproliferative activities (Lopez-Moreno et al., 2022; Panda et al., 2018). They have antimicrobial, antifungal, and antiviral activities by binding glycans on the cell wall or viral coat proteins and potential antidiabetic activities (Lopez-Moreno et al., 2022; Mishra et al., 2019).

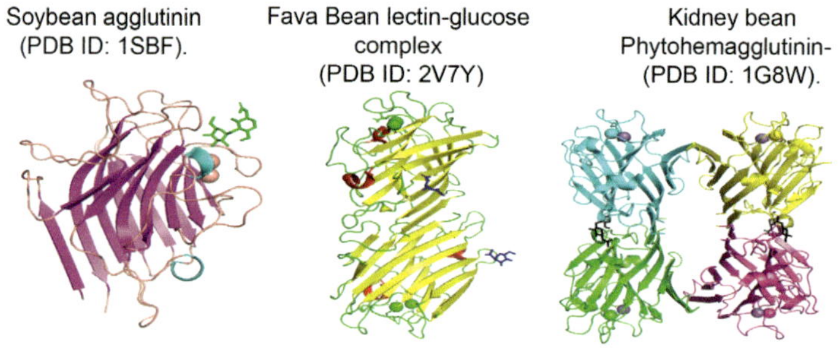

Figure 1: Structures of legume lectins. **a.** Soybean (*Glycin max*) agglutinin (PDB ID: 1SBF). **b.** Fava Bean (*Vicia fava*) lectin-glucose complex (PDB ID: 2V7Y). **c.** Phytohemagglutinin-L from kidney bean (*Phaseolus vulgaris*, PDB ID: 1G8W). (From left to right)

2.2. Protease and α-amylase Inhibitors

Similar to other antinutrients, protease inhibitors (PIs) can have both negative and positive effects on health (Figure 2). On the one hand, PIs diminish our ability to digest proteins into essential amino acids and absorb them in the digestive tract. On the other, they have been shown to have anticarcinogenic and anti-inflammatory properties that can be used to treat obesity, cardiovascular, autoimmune, and inflammatory diseases, as well as several types of cancer (Shan et al., 2022). The major families of protease inhibitors found in edible plants are the Kunitz, Bowman-Birk, serpin, potato type 1, and α-amylase/trypsin inhibiter families (Bendre et al., 2018). Although they are small proteins by themselves, they are resistant to enzymatic and acidic proteolysis in our stomach. Kunitz PIs and Bowman-Birk PIs inactivate serine, cysteine, and aspartic acid proteases such as trypsin, chymotrypsin, elastase, cathepsin G, and duodenase. As a result, they reduce the breakdown of proteins from food in the GI tract and reduce their nutritional value (Karlund et al., 2021). This inhibition can also induce a feedback loop in which the decrease in levels of digestive enzymes causes the pancreas to synthesize more enzymes resulting in pancreatic hyperplasia and hypertrophy (Karlund et al., 2021). Bowman-Birk and Kunits PIs are common in legumes such as soybeans, chickpeas, red kidney beans, mung beans, and black-eyed peas, but can also be found in wheat, barley, and rice. Serpins, which inhibit serine and some cysteine proteases, are found in raw cereals such as barley and wheat. Trypsin inhibitors of the alfa-amylase/trypsin family I (ATI) inhibit serine proteases and are found in cereals such as barley and rye. Potato tubers contain protease inhibitors I, II, and potato carboxypeptidase inhibitors (PCI), which are serine, chymotrypsin, and carboxypeptidase inhibitors. Alpha amylases enable the digestion of starch and glycogen by cleaving them into low molecular weight sugars such as glucose, maltose, dextrin, and maltotriose. It is secreted mainly in saliva and pancreatic juice and enables the utilization of starch as an energy source from food. Inhibitors of α-amylase are small proteins found in common beans (*Phaseolus vulgaris*), maize, barley, sorghum, millet, rye, peanuts, mangoes, and other food plants (Whitaker, 1997).

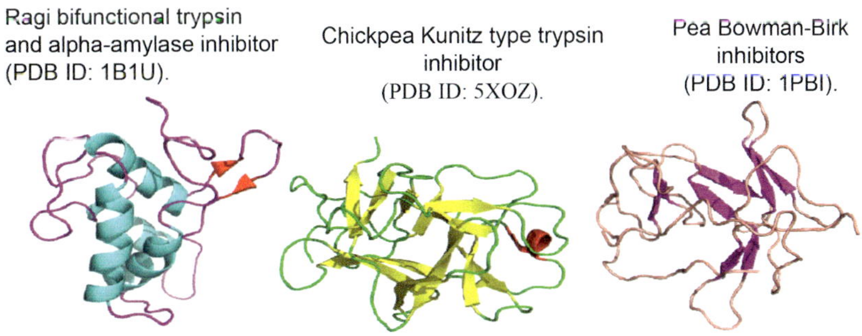

Figure 2: Protease and α-amylase inhibitors.

2.3. Phytates

Phytic acid, also called myoinositol hexaphosphate (IP_6), can be found in high concentrations, mainly in cereals, legumes, nuts, and seeds (1–3%) and at lower concentrations in several fruits and vegetables. It is a storage molecule for phosphate and inositol in plants, an antioxidant, and an energy source (Kumar et al., 2010; Fuster et al., 2017). Structured as a central 6-membered inositol ring bound to six phosphate groups, it has the ability to bind up to 12 protons. Phytic acid acts as a strong chelator of di- and trivalent metal ions such as Zn^{2+}, Cu^{2+}, Ca^{2+}, Fe^{2+}, Mg^{2+}, Mn^{2+}, and Fe^{3+}, and forms insoluble complexes in physiological pH values of 6-7. Humans lack the phytic acid degrading enzyme (phytase) that exists in plants and microorganisms. As a result, it prevents the absorption and uptake of essential minerals from foods (Lesjak and Srai, 2019). Phytic acid also forms complexes with proteins that prevent their enzymatic degradation and absorption in the digestive tract (Kaspchak et al., 2018). The ability of phytates to chelate iron can be beneficial as it prevents the generation of radical formation and acts as an antioxidant (Sanchis et al., 2018). For a recent comprehensive review of phytic acid, see Pramitha et al. (2021).

2.4. Tannins

Tannins are the fourth most common secondary plant metabolites and can amount to 20% of the dry weight of plants (Adamczyk et al., 2017). They are water-soluble polyphenols of high molecular weights (<500 Da) present in many food plants such as fruits, nuts, seeds, legumes, cereal grains, tea, wine, and cocoa beans. Tannins are commonly divided into two groups: hydrolysable tannins (e.g., gallotannins, ellagitannins and hexahydroxydiphenic acid esters) and condensed tannins (e.g., catechins, flavanols, or proanthocyanidins) (Bule et al., 2020). Condensed tannins have a more complex structure and are more prevalent in human diets than hydrolysable tannins. They can be found in apples, berries, red wine and chocolate, and more. The biological activities of tannins result from their unique ability to bind and aggregate proteins. This results in the formation of precipitates that inhibit enzymatic activity and proteolysis (Serrano et al., 2009). They can also form insoluble complexes with carbohydrates, polysaccharides, and metal ions that reduce nutrient availability (Adamczyk et al., 2017). When tannins bind to salivary proteins, they reduce oral lubrication and have an astringent taste that is common to many fruits and vegetables (De Jesus et al., 2012). However, there are many beneficial properties of tannins for human health, resulting from their anticarcinogenic, antimutagenic, antimicrobial, and antioxidant activities (Bule et al., 2020). They can promote healing of peptic ulcers by precipitating in sites of tissue damage, forming a protective layer, promoting tissue repair, and protecting them from *Helicobacter pylori* (De Jesus et al., 2012). In addition, they can alleviate neuroinflammation and balance serotonin metabolism through effects on the microbiome (Westfall and Pasinetti, 2019) (Figure 3).

2.5. Phytoestrogens

Phytoestrogens are polyphenolic compounds that are similar in structure to the female sex hormone 17-b-estradiol (E2) (Zaheer and Humayoun, 2017). They are

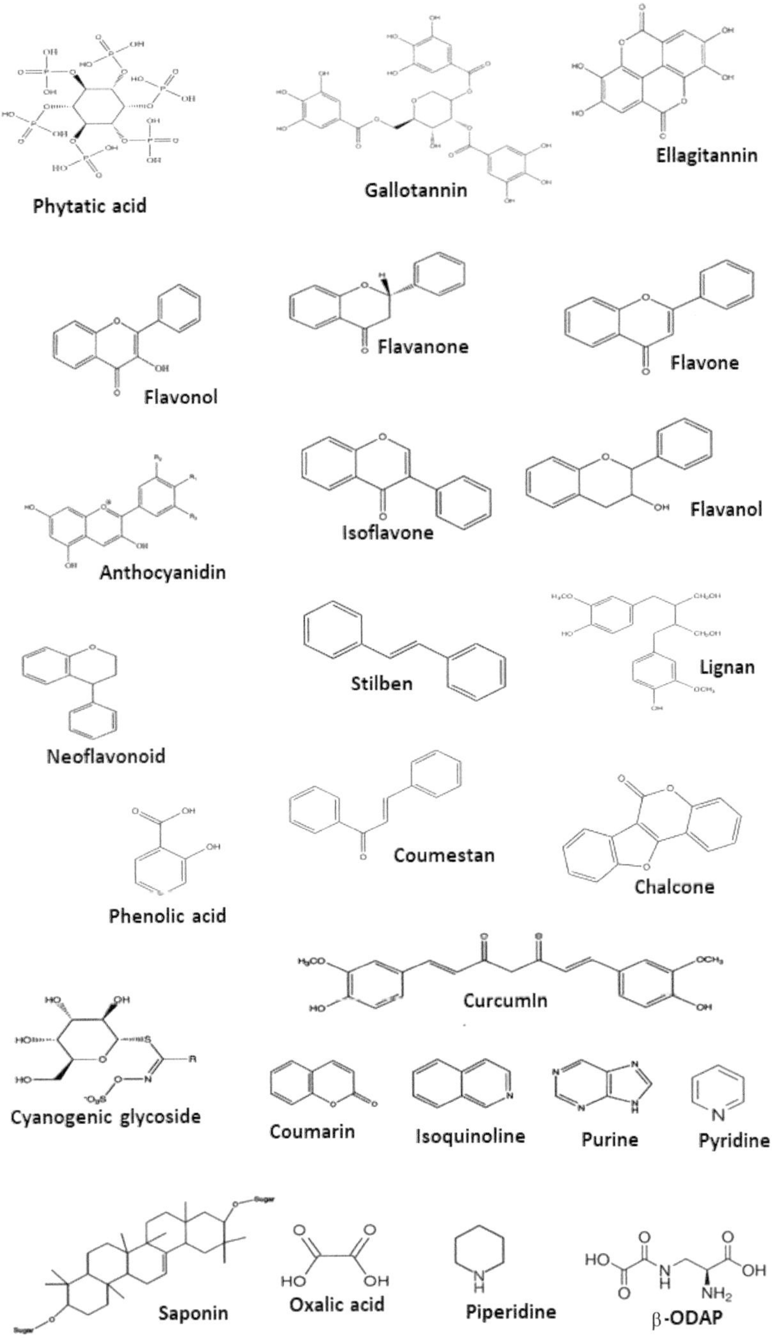

Figure 3: Chemical structures of phytochemical compounds.

able to elicit an endocrine response by binding and activating estrogen receptors ER-a and ER-b and can impact physiological processes such as reproduction, bone remodeling, and metabolism. They have been shown to affect the skin as well as the cardiovascular, nervous, and immune systems (Hamsalakshmi et al., 2022). Phytoestrogens can be divided into two main groups: flavonoids and non-flavonoids. The first group includes flavanols, flavones, isoflavonoids, neoflavonoids, and anthocyanidins. The second group contains stilbenes, lignans, phenolic acids, coumestans, and chalcones. Isoflavones, lignans, and coumestanes are the most relevant groups with respect to human diets. Isoflavones are present in soybean, chickpea, beans, peanut, coffee, tea, and red wine. Lignans are found in fiber-rich plants and grains such as wheat, barley, oats, beans, lentils, garlic, asparagus, carrots, broccoli, and flaxseeds. Coumestanes can be found in peas, pinto and lima beans, and alfalfa shoots. Phytoestrogens have been found to have either beneficial or deleterious effects on human health. On the one hand, they can reduce menopausal symptoms and the risks of cardiovascular diseases, obesity, metabolic syndromes, type 2 diabetes, cognitive disorders, and various forms of cancers (Hamsalakshmi et al., 2022; Messina, 2016). On the other hand, they can disrupt thyroid function, increase the risk of estrogen-sensitive and uterine cancers, increase the risk of uterine fibroids and central precocious puberty in girls, and have an endocrine-disrupting potential for babies and infants (Tagorti et al., 2023; Canivenc-Lavier and Bennetau-Pelissero, 2023). Most of the literature attempting to weigh the advantages and disadvantages of phytosteroids on human health has obtained, so far, mixed results due to the considerable heterogeneity between studies and the effect of hormesis, the phenomenon in which cells or whole organisms react in a biphasic manner to compounds depending on their concentration. While low concentrations can be stimulatory, high concentrations can be inhibitory. Thus, the effects of phytosteroids on human health may be context and concentration-dependent, beneficial in some cases and deleterious in others (Tagorti et al., 2023) (Figure 3).

2.6. Goitrogens

As their name implies, goitrogens are compounds that cause goiter, which is the swelling of the thyroid gland. Thyroid enlargement results from disruption of the normal function of the thyroid gland and decreased production of the thyroid hormones T_3 and T_4. As a result, levels of the pituitary hormone thyroid stimulating hormone (TSH) increase, leading to increased cell proliferation in the thyroid gland (hyperplasia) and its enlargement (Di Dalmazi and Giuliani, 2021). Thus, goiter forms as a compensatory response of the thyroid gland to a disruption in its hormonal secreting functions. Plant metabolites such as cyanogenic glucosides (e.g., glucosinolates), polyphenols (e.g., flavonols, flavones, isoflavones, flavanones, flavanols, stilbenoids, lignans and curcuminoids), coumarin and phenolic acids, and certain alkaloids (e.g., pyridines, isoquinoline purines, piperidines) have all been shown to act as goitrogens using *in vitro*, animal experiments, and epidemiological studies (Di Dalmazi and Giuliani, 2021). Cyanogenic glucosides are abundant in cruciferous plants such as broccoli, cabbage, Brussels sprouts, rape seed, mustard, turnip, and cauliflower. When consumed without processing, they are hydrolyzed by the enzyme myrosinase (thioglucoside glucohydrolase) to thiocyanates, isothiocyanates, nitriles, oxazolidine-2-thiones, and indole-3-carbinols (Prieto et al., 2019). Thiocyanate is also produced from cyanogenic glucosides in cassava, lima beans, and bamboo shoots. It competitively inhibits iodine uptake by the thyroid, which is required for the synthesis of iodine-containing thyroid hormones (Zimmermann and Boelaert, 2015). In addition, it inhibits a key enzyme in hormone synthesis, thyroid peroxidase (TPO). Flavones and flavonoids were shown to inhibit TPO activity and also interfere with the expression of thyroid-specific genes (e.g., TG, TPO, and TSHR). Additional mechanisms include inhibition of tyrosine iodination, inhibition of thyroid D1 enzyme activity, stimulation of D2 activity, induction of microfollicular changes in thyroid tissue, increase in TSH levels, induction of thyroid morphological changes, and decrease of liver D1 activity (Di Dalmazi and Giuliani, 2021). Stilbenoids, such as resveratrol, decrease iodine uptake and expression of thyroid-specific genes. Lignans, such as apigenin and pinoresinol, can act as antagonists of human thyroid hormone receptor β (TR_b), and coumarins can decrease the levels of serum T3 and T4 and liver D1 activity. Phenolic acids such as coumaric, ferulic, and caffeic acids may induce thyroid hypertrophy, and alkaloids such as nicotine, arecoline, harmine, and piperine have been shown to decrease thyroid iodide uptake, reduce serum T3 and T4 levels and inhibit liver D1 activity. Hypothyroidism due to iodine deficiency can lead to increased rates of spontaneous abortions, increased infant mortality, and impaired neurodevelopment in children (Zimmermann, 2011). As a result, children are at high risk of suffering from cretinism. Iodine deficiency occurs mainly due to the consumption of crops that grow on iodide-depleted soils, yet goitrogens may aggravate the situation and it is estimated that about 7% of hypothyroidism results from exposure to goitrogens (Sunil and Leena, 2021). The goitrogenic effects of all the above-mentioned compounds can be used to treat cases of thyroid cancer (Pasko et al., 2022) and thyroid autoimmunity (Hamsalakshmi et al., 2022). For example, sulforaphane, an isothiocyanate, has been shown to have apoptotic antiproliferative effects in thyroid cancer cells (Chatterjee et al., 2018) (Figure 3).

2.7. Oxalates

Oxalic acid is highly abundant in spinach, chard, beet, and rhubarb, as well as in raw legumes, whole grains, nuts, berries, beans, and cruciferous vegetables such as kale, cauliflower, broccoli, and radishes (Attalla and Monga, 2014). It is a highly acidic organic acid that can form soluble and insoluble salts depending on the cations it binds. Binding to sodium or potassium ions forms soluble oxalate salts, while binding to calcium, iron, or magnesium generates insoluble oxalates. Thus, similar to phytic acid, oxalates behave as metal ion chelators that prevent the absorption and uptake of essential minerals from foods. However, the primary health issue with oxalates results from the propensity of insoluble calcium-oxalate to precipitate into solid crystals in the urinary tract or the kidney, forming kidney stones (Mitchell et al., 2019). Thus, overconsumption of oxalate and its absorption in the GI tract may result in a condition termed hyperoxaluria, which is a risk factor for the formation of kidney stones. However, there is no simple correlation between oxalate consumption and kidney stone formation (Taylor and Curhan, 2007). Studies have found only a modest association between dietary oxalate and kidney stone formation, and high dietary calcium intake has actually been shown to reduce the risk of kidney stone formation. In addition, it has been found that soluble oxalates have a more significant impact on the risk of kidney stone formation than insoluble oxalates (Chai and Liebman, 2005). Thus, oxalate-rich foods that also contain magnesium and potassium are associated with a reduced risk of kidney stones. An important determinant in the fate of digested oxalate is the gut microbiome, as some anaerobic bacterial species, such as *Oxalobacter formigenes* can degrade oxalate (Chamberlain et al., 2020). Short-chain fatty acids produced by the gut microbiome can reduce the risk of kidney stones (Liu et al., 2020) (Figure 3).

2.8. Saponins

Saponins are steroid and triterpene glycosides that are named so due to their emulsifying properties that resembles those of soap. They are secondary plant metabolites that have been identified in a large number of species (e.g., herbs, grasses, shrubs, trees, legumes, etc.) and shown to have diverse activities (e.g., plant defense, development, root nodulation) (Faizal and Geelen, 2013). Their sweet and bitter tastes, combined with their surface-active properties, have made them important in the food industry. Saponins can have either beneficial or deleterious effects on health. As antinutrients, they bind and form insoluble complexes with iron, zinc, and calcium that reduce mineral uptake and can have strong hemolytic and membranolytic activities. The administration of a high dose of saponins has been shown to be toxic in mammals and can result in death (Diwan et al., 2000). Yet in most cases the oral toxicity of saponins is considered to be low (Lin et al., 2021). They can also inhibit protein digestion in the GI tract. However, they also have hypocholesterolemic, immunostimulatory, antidiabetic, and anticarcinogenic properties that are beneficial to health (Elekofehinti et al., 2021). They can reduce the risk of cardiovascular diseases, are used as adjuvants to boost the immune response in vaccines (Sun et al., 2009), reduce the infectivity of HIV (*in vitro*) (Nguyen et al., 2018), and have been shown to prevent obesity by lowering the levels of insulin in the serum. Saponins bind cholesterol and act as antioxidants that protect it from oxidation while preventing its absorption and increasing its extraction from the body (Shi et al., 2004). They can stimulate insulin secretion, decrease insulin resistance, and increase glucose uptake in skeletal muscles and the intestine. Finally, their cytotoxic activities can inhibit the growth and proliferation of cancer cells

(Elekofehinti et al., 2021). Dietary saponins are abundant in legumes such as soybean, chickpeas, peas, and a variety of beans (e.g., butter beans, broad beans, haricot beans, kidney beans, black-eyed beans, etc.) (Shi et al., 2004). They can also be found in cassava (yuca), ginseng, and quinoa (Figure 3).

2.9. Cyanogenic Glycosides

Cyanogenic glycosides were described above as goitrogens since they can be metabolized to thiocyanates that interfere with the normal function of the thyroid gland. However, the initial hydrolysis product of cyanogenic glycosides is α-hydroxynitriles, which are further hydrolyzed to aldehydes or ketones and hydrogen cyanide (HCN) by the enzyme hydroxynitrile lyase or spontaneously at high pH in the duodenum (Gleadow and Møller, 2014). In fact, thiocyanates are formed by the body as it attempts to detoxify HCN by converting it to the less toxic compound thiocyanate that is excreted in the urine (Logue et al., 2010). HCN is highly toxic as it inhibits respiratory chain enzymes such as cytochrome oxidases. Ingestion of unprocessed plant foods that contain high levels of cyanogenic glucosides can lead to death by HCN intoxication. Examples of cyanogenic glucosides in plant foods include linamarin and lotaustralin in cassava (*Manihot esculenta*) and lima beans, dhurrin in sorghum, taxiphyllin in bamboo shoots, sambunigrin in elderberries, neolinustatin and linustatin in flax seeds, and amgydalin in the kernels of almond, apricot, peach, plum, apple and cherry (Sharma, 2021).

2.10. β-ODAP

Seeds of the annual legume crop grass pea (*Lathyrus sativus* L.) (Kumar et al., 2013) contain the compound β-N-oxalyl-L-α,β-diaminopropionic acid (β-ODAP), which is a neurotoxin that can cause neurolathyrism (Yan et al., 2006). Neurolathyrism is a motor neuron disease characterized by spastic, lower-limb paralysis (Spencer and Schaumburg, 1983). It results from a chronic intoxication caused by the long-term ingestion of seeds or flour made from grass pea as a primary diet component (e.g., >30% of caloric intake for 3–4 months) (Lambein et al., 2019). The accumulation of ingested β-ODAP can result in the overexcitation of AMPA glutamate receptors and ultimately lead to neuronal cell death in the spinal cord and permanent lower limb paralysis (Tan et al., 2017). The neurotoxic effects of β-ODAP are exacerbated by a lack of dietary sulfur-containing amino acids and antioxidants that can decrease oxidative stress. Thus, β-ODAP is a crippling toxin but not a lethal one. As for health benefits, β-ODAP has some haemostatic properties that can be used to stop haemorrhages and induce wound healing (Sharma et al., 2018). Finally, β-ODAP has been found in the seeds of 17 species of *Acacia*, 13 species of *Crotalaria*, and 20 additional species of *Lathyrus*, as well as in several species of *Panax* (Quereshi et al., 1977; Long et al., 1996). Yet, the majority of them are not consumed as food, or are present in large amounts.

3. Target Crops

Similar to nutrients, food crops usually contain different kinds of antinutrients in varying concentrations. The methods used for their processing as foods, their level of consumption, and the nutritional context in which they are consumed also vary widely among cultures, geographic locations, and socio-economic conditions. As previously discussed, antinutrients may be beneficial when consumed at moderate doses and as part of balanced diets. Thus, the choice of crops targeted for biofortification by elimination or reduction of their antinutrient levels should be guided, primarily, by the health risks they present. These can be established based on the prevalence of case reports of individuals who suffered illnesses or health risks due to antinutrient consumption and the severity of their condition. Three main crops seem to have caused the most harm following the consumption of their antinutrients at high levels: cassava, grass pea, and millets. These also happen to be robust crops, able to withstand harsh environmental growth conditions, and are a rich source of nutrients. Eliminating the antinutrients they contain would increase the safety of their consumption and their global use as food crops.

3.1. Cassava (*Manihot esculenta*)

The roots of the plant cassava are a rich source of fiber and carbohydrates, making them a staple food in more than 100 countries in the tropical and subtropical regions of America, Africa, and Asia (Parmar et al., 2017). They contain more than 80% starch, vitamin C, carotenoids, and minerals (Montagnac et al., 2009; Li et al. 2017). However, cassava also contains tannins, nitrates, saponins, and especially cyanide glycosides that can have adverse effects on health when ingested in high quantities. The cyanogenic glucosides, linamarin and lotaustralin, are present in high amounts in cassava and are enzymatically hydrolyzed by β-glucosidase to produce sugars and cyanohydrins. The latter enzymatically and spontaneously decompose to hydrocyanic acid (HCN) and ketones (Panghal et al., 2021). As a result, the consumption of high amounts of unprocessed cassava roots has led to acute cyanide poisoning (Cheok et al., 1978), and long-term intake has been shown to result in tropical neuropathy (Otubogun et al., 2019) or epidemic spastic paralysis (Adamolekun et al., 2010). In fact, the association of cassava with cyanide poisoning has been described as the single most important constraint on the expansion of cassava utilization (Falade and Akingbala, 2010).

3.2. Grass Pea (*Lathyrus sativus* L.)

Grass pea is an annual legume crop grown for food and forage, mainly in South Asia and Sub-Saharan Africa as well as in South America. It has significant economic value, especially in India, Bangladesh, Pakistan, Nepal, and Ethiopia (Dixit et al., 2016). Its seeds have high nutritional values as they are enriched in protein (28%) and nutrients such as folic and linoleic acids and low in fat (<1%) and unsaturated fatty acids. The crop itself is highly robust to environmental stresses such as drought and flood and is cheap and easy to cultivate. However, grass pea seeds are not widely used for food and forage outside parts of Asia and Africa because they also contain a neurotoxic metabolite, β-ODAP (Xu et al., 2017), which can cause neurolathyrism. Neurolathyrism is a nutritional neurodegenerative disorder characterized by lower-limb paralysis that results from a chronic intoxication caused by the long-term ingestion of seeds or flour made from grass pea as a primary diet component (Fearon et al., 2016). Epidemics of neurolathyrism have occurred in various parts of India, Ethiopia, and Bangladesh (Bradbury and Lambein, 2011) and have caused the government of India to impose a ban on the sales of grass pea in 1961. The ban has since been lifted in many states in India with the introduction of low-β-ODAP cultivars, yet the safety issues concerning the consumption of grass pea are still subject to debate.

3.3. Millets

Millets (e.g., pearl millet, finger millet, foxtail millet) are nutritious cereal crops that sustain one-third of the world's population (Ratnavathi and Patil, 2017). Pearl millet, for example, is a staple food for more than 90 million people in Africa and Asia. They are rich in calcium, dietary fibers, polyphenols, and protein and have higher fat content than maize, rice, or sorghum (Taylor, 2017). Yet, millets also contain antinutrients such as phytates, phenols, tannins, trypsin inhibitors, and goitrogens (Dey et al., 2022). Phytic acid prevents protein digestion and mineral absorption. Tannins and phenolic compounds are also present in millets and can impede protein and carbohydrate digestion. However, the main concern in the consumption of millets comes from their high levels of glycosilflavones. Pearl millet, for example, contains six main glycosilflavones: glucosyl vitexin, vitexin, glucosyl orientin, orientin, rhamnosyl vitexin, and rhamnosyl orientin (Boncompagni et al., 2018). Another commonly consumed millet, fonio (*Digitaria exilis*), contains the glycosilflavones apigenin and luteolin (Sartelet et al., 1996). Glycosilflavones have anti-thyroid effects that, at high dietary intake, have been shown to cause goiter by inhibiting the enzyme thyroid peroxidase (TPO) (Gaitan et al., 1989, 1995). Consumption of millets as primary diet components has been directly implicated in goiter epidemics in different parts of the world (Elnour et al., 2000; Gaitan et al., 1991; Tripp 1992).

4. Contribution of Environmental Factors

Environmental factors such as humidity, soil composition, temperature, UV radiation, and the proximity and identity of neighboring plants and fungi, pathogens, herbivores, and all plant stresses influence the levels of antinutrients in food crops. For example, the concentrations of phenylpropanoids such as flavones and isoflavonoids increase in response to high light or UV radiation, pathogen attacks, and low nitrogen soil concentrations (Dixon and Paiva, 1995). Anthocyanins are induced by low phosphate soil concentrations or high light radiation and low temperatures, and stilbenes, 3-deoxyanthocyanidines, and coumarins are synthesized in response to pathogen attacks or plant wounding (Dixon and Paiva, 1995).

Dietary contexts also have their importance. Grass pea is a robust crop, resistant to drought, high salinity, and flooding. Sometimes, it is the only surviving crop in drought-prone (and often famine-ravaged) areas in Africa and Asia (Patto et al., 2006). Under conditions of environmental stress such as drought, the levels of β-ODAP in grass pea seeds have been shown to increase (Jiao et al., 2011). Thus, drought can create a situation in which grass pea seeds that are enriched in β-ODAP become the major diet component of resource-poor populations over a long period of time. Under such conditions, the availability of water for cooking and additional foods that could mitigate the effects of β-ODAP neurotoxicity (e.g., sulfur-containing amino acids) is also reduced. Consequently, drought can compound the risk of neurolathyrism epidemics in developing countries. The association between drought, famine, grass pea consumption, and an epidemic of neurolathyrism has been exemplified in the neurolathysrism epidemic reported in northeastern Ethiopia in 1999 (Getahun et al., 2002). Flood-related famines have also been found to result in excessive grass pea consumption and neurolathyrism epidemics, for example, in Bangladesh (Haque et al., 1996).

Cassava roots are also drought-tolerant, can remain underground for several months while still being edible, and may even survive the incidence of fires (Falade and Akingbala, 2010). As a result, they are regarded not only as food crops but also as security crops against unforeseen food shortages (Parmar et al., 2017). Chronic cyanide toxicity resulting from casava ingestion has been shown to be the causative agent of two motor-neuron diseases: tropical ataxic neuropathy (TAN) and Konzo (Panghal et al., 2021). Droughts and famines make cassava roots the primary source of nourishment under adverse conditions and increase the chances that their consumption will lead to Konzo and TAN epidemics (Ngudi et al., 2012). Similar to grass pea during droughts, the availability of water for cooking and additional foods that could mitigate the effects of cyanide neurotoxicity is also reduced. Specifically, the lack of foods containing sulphur amino acids (i.e., cysteine and methionine) are required for detoxifying cyanide. Therefore, drought can compound the risk of developing Konzo and TAN epidemics in developing countries (Bradbury and Lambein, 2011).

Finally, the drought resistance of millets, which are robust C4 plants that can grow under conditions of low rainfall, high temperatures, and sandy soils, can also increase the expression of their antinutrients (Saleem et al., 2021). The flavones luteolin and apigenin, which are the main constituents of the fonio millet, decrease the secretion of thyroid hormones (Sartelet et al., 1996). As a consequence, in places where much of the dietary energy comes from millets, such as the rural villages of the Darfur province in Sudan, the prevalence of goiter has also been reported to be high in the past (Boncompagni et al., 2018).

5. Strategies Used for Biofortification

5.1. Food Processing

The concentration of many antinutrients in foods can be reduced using various processing methods (Petroski and Minich, 2020; Sharma et al., 2021).

- **Cooking:** Including boiling, baking, roasting, steaming, autoclaving, etc., is the most effective way of significantly reducing the concentrations of all antinutrients, including phytates, tannins, phytoestrogens, goitrogens, oxalates, and cyanogenic glucosides. Lectins are especially sensitive to the denaturing effects of cooking and autoclaving.
- **Soaking:** Using acidic, alkaline, or salty water solutions is effective at removing oxalates which are highly water-soluble, as well as tannins, phytic acid, trypsin inhibitors, and saponins.
- **Fermentation:** Can eliminate phytates, phytoestrogens, goitrogens, protease inhibitors, tannins, β-ODAP, and cyanogenic glucosides.
- **Dehulling (Physical Seed Coat Removal):** It was found to reduce the concentrations of protease inhibitors in soybean and peanut as well as β-ODAP in grass pea.
- **Grating/Pounding:** When followed by drying or roasting, has been found to reduce cyanogenic glucosides in cassava.
- **Irradiation:** Using γ-irradiation up to 10 kGy can inactivate protease inhibitors, lectins, phytic acids, and oligosaccharides.
- **Germination:** Results in the hydrolysis of seed storage proteins in many legumes and can help reduce amylase and trypsin inhibitors as well as tannins.

5.2. Dietary Nutrients and Supplements

In many cases, it is possible to mitigate or even counteract the deleterious effects of antinutrients using dietary supplements or foods enriched with minerals, iodine, and sulfur-containing amino acids. Calcium-rich foods can offset soluble oxalate absorption. Mineral supplements containing Zn^{2+}, Cu^{2+}, Ca^{2+}, Fe^{2+}, Mg^{2+}, and Mn^{2+} can compensate for the reduction in their bioavailability by phytates, oxalates, tannins, and saponins. Fish, seafood, and edible seaweed (Romaris-Hortas et al., 2011) as well as iodine salt, can counteract the action of goitrogens by iodine supplementation. Finally, food enriched with sulfur containing amino acids such as animal proteins (e.g., meat, eggs, and milk), nuts, seeds, grains, legumes, allium vegetables (e.g., garlic, leeks, onions, scallions, and shallots), cruciferous vegetables and leafy green vegetables, can all reduce the risk of cyanide poisoning by cyanogenic glycosides and β-ODAP (Hillocks and Maruthi, 2012).

5.3. Plant Breeding

Breeding has been used to generate cultivars with reduced antinutrient content in several food crops, such as cyanogenic glycosides in cassava (Carvalho et al., 2015; Malik et al., 2020), β-ODAP in grass-pea (Dixit et al., 2016; Hillock and Maurthi, 2012), phytic acid in soybean (Gossner et al., 2019), saponin in quinoa (Ward, 2000), and trypsin inhibitor in mung bean (Das et al., 2015) and soybean (Bernard and Hymowitz, 1986). In some cases, such as oxalate in spinach (Shi et al., 2016), the genes responsible for antinutrient biosynthesis have recently been identified, and in others, such as C-glycosylflavones in pearl millet grains, breeding attempts are in progress (Boncompagni et al., 2018).

5.4. Genetic Editing and Silencing

A number of gene silencing and editing techniques have been used to reduce the levels of antinutrients in food crops. The three most prominent ones are gene and RNAi silencing and gene editing using CRISPR/cas9. RNAi silencing was used to reduce the levels of phytic acid in rice and soybean (Kumar et al., 2019; Kamarkar et al., 2020), saponin in soybean (Takagi et al., 2011) and α-amylase/trypsin inhibitor in wheat (Kalunke et al., 2020). Gene silencing was used to reduce the levels of phytic acid in corn and alkaloids in potatoes and tomatoes (Cardenas et al., 2015). CRISPR-Cas9 gene editing was used target α-amylase/trypsin inhibitor genes in wheat (Camerlengo et al., 2020), phytic acid in soybean (Song et al., 2022), saponin in ginseng (Choi et al., 2022), and cyanogenic glycosides in cassava (Juma et al., 2022; Gomez et al., 2023).

In cases where the metabolic pathway leading to antinutrient biosynthesis has not been entirely elucidated, gene editing can be used to target precursor metabolites required for synthesizing the antinutrient. For example, the antinutrient β-ODAP is synthesized in grass pea from its precursor's oxalic acid and L-DAPA (Malathi et al., 1970; Goldsmith et al., 2022). In order to degrade oxalic acid and, subsequently, reduce the concentration of β-ODAP, a constitutively expressed fungal oxalate decarboxylase gene was introduced into grass pea (Kumar et al., 2017). As a result, a reduction of up to 75% in oxalate levels and 73% in β-ODAP levels were obtained in the seeds of the transgenic grass pea relative to the wild-type plant (Kumar et al., 2016). The metabolic pathway responsible for β-ODAP biosynthesis in grass pea has recently been elucidated (Goldsmith et al., 2022), yet this knowledge is yet to be applied in the development of new cultivars devoid of β-ODAP.

6. Challenge: Reducing Antinutrients in Food Crops

Several factors impede the progress of attempts to breed out antinutrients from food crops. (1) The linkage of genes controlling the biosynthesis of antinutrients to others that is essential for the plant. (2) The long time and efforts required for successful breeding, the large number of antinutrient variants possible in a single crop (e.g., >40 different saponins in quinoa, El Hazzam et al., 2020). (3) Properties such as high heterozygosity, allopolyploidy, low fertility, unsynchronized flowering as well as tradeoffs between desirable and undesirable traits (Parmar et al., 2017; Ward, 2000). In some cases, such as oxalate in spinach (Shi et al., 2016), the concentrations of antinutrients in the plant are determined by multiple genes, each having a minor effect. This makes breeding and even genetic engineering efforts aiming at their reduction quite difficult. Additionally, the lack of commercial incentive for research and development of breeding programs (Rajsic et al., 2023) and the ability to eliminate antinutrients by other methods, such as food processing, decrease the number of attempts to breed out antinutrients.

The complex stress-associated regulation of antinutrient biosynthesis, e.g., drought-induced β-ODAP (Jiao et al., 2011) highlights the need to identify and target key metabolic enzymes in the biosynthesis pathway of antinutrients. The underutilization of crops like grass pea, millets, and cassava has also led to a lack of basic research efforts to elucidate the genes and metabolic pathways. Increasing the number of basic research efforts into the biosynthesis of antinutrients is a prerequisite to engineering cultivars devoid of them.

7. Conclusion

Antinutrients such as cyanogenic glucosides, β-ODAP, and goitrogens constitute a health risk for both humans and animals if consumed in large amounts, as primary diet components, over prolonged periods of time. Much of the knowledge regarding the negative impact of antinutrients on human health came from epidemics that followed events of food shortages or famines, at which survival depended on the consumption of plant foods like cassava tubers, grass pea seeds, and millet grains containing these antinutrients. Unfortunately, due to their antinutrients these stress tolerant crops remain underdeveloped as food crops on a global scale. At low doses, antinutrients have been found to have beneficial anticarcinogenic, antidiabetic, anti-inflammatory, antimicrobial, and immunostimulatory properties. Yet, crops that contain high concentrations of them are still considered potential health risks. Attempts to reduce or eliminate antinutrients in food crops by plant breeding are complicated by genetic linkages, and the number of antinutrients that can reside in a single plant. However, the advent of multiple omics technologies has facilitated the identification and targeting of the regulatory and biosynthetic pathways involved in antinutrient production enabling generation of novel cultivars with reduced antinutrients. Such crops are likely to become a valuable addition to the collection of food crops that feed the world in general, and during climate change conditions in particular.

References

Adamczyk, B., J. Simon, V. Kitunen, S. Adamczyk, and A. Smolander. 2017. Tannins and their complex interaction with different organic nitrogen compounds and enzymes: Old paradigms versus recent advances. *Chemistry Open*, 6: 610–614.

Adamolekun, B. 2010. Etiology of Konzo, epidemic spastic paraparesis associated with cyanogenic glycosides in cassava: Role of thiamine deficiency. *J. Neurol. Sci.*, 296: 30–33.

Attalla, K., S.B. De, and M. Monga, 2014. Oxalate content of food: A tangled web. *Urology*, 84: 555–559.

Baguma, M., F. Nzabara, G.M. Balemba, E.B. Malembaka, C. Migabo, G. Mudumbi, V. Bito, J. Cliff, J.M. Rigo, and J.N. Chabwine. 2021. Konzo risk factors, determinants, and etiopathogenesis: What is new? A systematic review. *Neurotoxicology*, 85: 54–67.

Barre, A., E. Damme, M. Simplicien, H. Benoist, and P. Rouge. 2020. Are dietary lectins relevant allergens in plant food allergy? *Foods*, 9(12): 1724.

Bendre, A.D., S. Ramasamy, and C.G. Suresh. 2018. Analysis of Kunitz inhibitors from plants for comprehensive structural and functional insights. *Int. J. Biol. Macromol.*, 113: 933–943.

Bernard, R.L., and T. Hymowitz. 1986. Registration of L81-4590, L81-4871, and L83-4387 soybean germplasm lines lacking the kunitz trypsin-inhibitor. *Crop Sci.*, 26: 650–651.

Boncompagni, E., G. Orozco-Arroyo, E. Cominelli, P.I. Gangashetty, S. Grando, T.T. Kwaku Zu, M.G. Daminati, E. Nielsen, and F. Sparvoli. 2018. Antinutritional factors in pearl millet grains: Phytate and goitrogens content variability and molecular characterization of genes involved in their pathways. *PLoS One*, 13: 30.

Bradbury, J.H., and F. Lambein, 2011. Konzo and neurolathyrism: Similarities and dissimilarities between these crippling neurodegenerative diseases of the poor. *Food Chem. Toxicol.*, 49: 537–538.

Bule, M., F. Khan, M.F. Nisar, K. Niaz, S. Nabavi, M. Saeedi, and A. Sanches Silva. 2020. Tannins (hydrolysable tannins, condensed tannins, phlorotannins, flavono-ellagitannins). *Recent Advances in Natural Products Analysis*, 132–146. Elsevier.

Camerlengo, F., A. Frittelli, C. Sparks, A. Doherty, D. Martignago, C. Larré, R. Lupi, F. Sestili, and S. Masci 2020. CRISPR-Cas9 Multiplex editing of the α-amylase/trypsin inhibitor genes to reduce allergen proteins in durum wheat. *Front. Sustain. Food Syst.* (July), 4.

Canivenc-Lavier, M.C., and C. Bennetau-Pelissero. 2023. Phytoestrogens and health effects. *Nutrients*, 15(2).

Cárdenas, P.D., P.D. Sonawane, U. Heinig, S.E. Bocobza, S. Burdman, and A. Aharoni. 2015. The bitter side of the nightshades: Genomics drives discovery in Solanaceae steroidal alkaloid metabolism. *Phytochem.*, 113: 24–32.

Carvalho, S.P., J.N.D. Caetano, and I.S. Pereira. 2015. *8th International Symposium on in vitro Culture and Horticultural Breeding*, 541–544.

Chai, W.W., and M. Liebman. 2005. Oxalate content of legumes, nuts, and grain-based flours. *J. Food Compos. Anal.*, 18: 723–729.

Chamberlain, C.A., M. Hatch, and T.J. Garrett. 2020. Oxalobacter formigenes produces metabolites and lipids undetectable in oxalotrophic bifidobacterium animalis. *Metabolomics*, 16: 1–10.

Chatterjee, S., Y. Rhee, P.S. Chung, R.F. Ge, and J.C. Ahn. 2018. Sulforaphene enhances the efficacy of photodynamic therapy in anaplastic thyroid cancer through Ras/RAF/MEK/ERK pathway suppression. *J. Photochem. Photobiol.*, B 179: 46–53.

Cheok, S.S. 1978. Acute cassava poisoning in children in Sarawak. *Trop. Doct.*, 8: 99–101.

Choi, H.S., H.B. Koo, S.W. Jeon, J.Y. Han, J.S. Kim, K.M. Jun, and Y.E. Choi. 2022. Modification of ginsenoside saponin composition via the CRISPR/Cas9-mediated knockout of protopanaxadiol 6-hydroxylase gene in Panax ginseng. *J. Ginseng Res.*, 46(4): 505–514.

Cid-Gallegos, M.S., L.J. Corzo-Rios, C. Jimenez-Martinez, and X.M. Sanchez-Chino. 2022. Protease inhibitors from plants as therapeutic agents: A review. *Plant Foods Hum. Nutr.*, 77: 20–29.

Das, S., Dewanjee, A. Maji, and K.K. Sarkar. 2015. Breeding mungbean *Vigna radiata* (L.) Wilczek for development of high yielding protein rich lines with low trypsin inhibitor. *Legume Res.*, 38: 447–451.

De Coninck, T., and E.J.M. Van Damme. 2021. Review: The multiple roles of plant lectins. *Plant Sci.*, 313: 111096.

De Jesus, N.Z.T., H. de Souza Falcão, I.F. Gomes, T.J. de Almeida Leite, G.R. de Morais Lima, J.M. Barbosa-Filho, J.F. Tavares, M.S.D. Silva, P.F. de Athayde-Filho, and L.M. Batista. 2012. Tannins, peptic ulcers, and related mechanisms. *Int. J. Mol. Sci.*, 13: 3203–3228.

Dey, S., A. Saxena, Y. Kumar, T. Maity, and A. Tarafdar. 2022. Understanding the antinutritional factors and bioactive compounds of Kodo Millet (*Paspalum scrobiculatum*) and Little Millet (*Panicum sumatrense*). *J. Food Qual.*, 2022, Article 1578448.

Di Dalmazi, G., and C. Giuliani. 2021. Plant constituents and thyroid: A revision of the main phytochemicals that interfere with thyroid function. *Food Chem. Toxicol.*, 152: 112158.

Diwan, F.H., I.A. Abdel-Hassan, and S.T. Mohammed. 2000. Effect of saponin on mortality and histopathological changes in mice. *East Mediterr. Health J.*, 6: 345–351.

Dixit, G.P., A.K. Parihar, A. Bohra, and N.P. Singh. 2016. Achievements and prospects of grass pea (*Lathyrus sativus* L.) improvement for sustainable food production. *Crop J.*, 4: 407–416.

Dixon, R.A., and N.L. Paiva.1995. Stress-induced phenylpropanoid metabolism. *Plant Cell*, 7: 1085–1097.

Elekofehinti, O.O., O. Iwaloye, F. Olawale, and E.O. Ariyo. 2021. Saponins in cancer treatment: Current progress and future prospects. *Pathophysiology*, 28: 250–272.

El Hazzam, K., J. Hafsa, M. Sobeh, M. Mhada, M. Taourirte, K. El Kacimi, and A. Yasri. 2020. An insight into saponins from Quinoa (*Chenopodium quinoa* Willd): A review. *Molecules*, 25(5): 1059.

Elnour, A., L. Hambraeus, M. Eltom, M. Dramaix, and P. Bourdoux, 2000. Endemic goiter with iodine sufficiency: A possible role for the consumption of pearl millet in the etiology of endemic goiter. *Am. J. Clin. Nutr.*, 71: 59–66.

Faizal, A., and D. Geelen, 2013. Saponins and their role in biological processes in plants. *Phytochem. Rev.*, 12: 877–893.

Falade, K.O., and J.O. Akingbala. 2010. Utilization of cassava for food. *Food Rev. Int.*, 27: 51–83.

Fearon, C., B. Murray, and H. Mitsumoto. 2015. Disorders of upper and lower motor neurons. *Bradley's Neurology in Clinical Practice*, 1484–1518.

Fu, L.L., C.C. Zhou, S. Yao, J.Y. Yu, B. Liu, and J.K. Bao. 2011. Plant lectins: Targeting programmed cell death pathways as antitumor agents. *Int. J. Biochem. Cell Biol.*, 43(10): 1442–1449.

Fuster, J.M.B., P.S. Cortés, J.P. Bestard, and F.G. Freixedas. 2017. Plant phosphates, phytate and pathological calcifications in chronic kidney disease. *Nefrología* (English Edition), 37: 20–28.

Gaitan, E., R.C. Cooksey, J. Legan, and R.H. Lindsay. 1995. Antithyroid effects *in vivo* and *in vitro* of vitexin: A C-glucosylflavone in millet. *J. Clin. Endocrinol. Metab.*, 80: 1144–1147.

Gaitan, E., N.C. Nelson, and G.V. Poole. 1991. Endemic goiter and endemic thyroid-disorders. *World J. Surg.*, 15: 205–215.

Gaitan, E., R.H. Lindsay, R.D. Reichert, S.H. Ingbar, R.C. Cooksey, J. Legan, E.F. Meydrech, J. Hill, and K. Kubota. 1989. Antithyroid and goitrogenic effects of millet: Role of C-glycosylflavones. *J. Clin. Endocrinol. Metab.*, 68: 707–714.

Getahun, H., F. Lambein, M. Vanhoorne, and P. Van der Stuyft. 2002. Pattern and associated factors of the neurolathyrism epidemic in Ethiopia. *Trop. Med. Int. Health*, 7: 118–124.

Gleadow, R.M., and B.L. Møller. 2014. Cyanogenic glycosides: Synthesis, physiology, and phenotypic plasticity. *Annu. Rev. Plant Biol.*, 65: 155–185.

Goldsmith, M., S. Barad, M. Knafo, A. Savidor, S. Ben-Dor, A. Brandis, T. Mehlman, Y. Peleg, S. Albeck, O. Dym, and E. Ben-Zeev. 2022. Identification and characterization of the key enzyme in the biosynthesis of the neurotoxin beta-ODAP in grass pea. *J. Biol. Chem.*, 298: 101806.

Gomez, M.A., K.C. Berkoff, B.K. Gill, A.T. Iavarone, S.E. Lieberman, J.M. Ma, A. Schultink, N.G. Karavolias, S.K. Wyman, R.D. Chauhan, and N.J. Taylor. 2023. CRISPR-Cas9-mediated knockout of CYP79D1 and CYP79D2 in cassava attenuates toxic cyanogen production. *Front. Plant Sci.*, 13: 1079254.

Gossner, S., F. Yuan, C. Zhou, Y. Tan, Q. Shu, and K.H. Engel. 2019. Stability of the metabolite signature resulting from the MIPS1 mutation in low phytic acid soybean (*Glycine max* L. Merr.) mutants upon cross-breeding. *J. Agric. Food Chem.*, 67: 5043–5052.

Hamsalakshmi, Alex, A.M., M.A. Marappa, S. Joghee, and S.B. Chidambaram. 2022. Therapeutic benefits of flavonoids against neuroinflammation: A systematic review. *Inflammopharmacology*, 30: 111–136.

Haque, A., M. Hossain, G. Wouters, and F. Lambein. 1996. Epidemiological study of lathyrism in northwestern districts of Bangladesh. *Neuroepidemiology*, 15: 83–91.

Hillocks, R.J. and M.N. Maruthi. 2012. Grass pea (*Lathyrus sativus*): Is there a case for further crop improvement? *Euphytica*, 186: 647–654.

Jiao, C.J., J.L. Jiang, L.M. Ke, W. Cheng, F.M. Li, Z.X. Li, and C.Y. Wang. 2011. Factors affecting beta-ODAP content in *Lathyrus sativus* and their possible physiological mechanisms. *Food Chem. Toxicol.*, 49: 543–549.

Juma, B.S., A. Mukami, C. Mweu, M.P. Ngugi, and W. Mbinda. 2022. Targeted mutagenesis of the CYP79D1 gene via CRISPR/Cas9-mediated genome editing results in lower levels of cyanide in cassava. *Front. Plant Sci.* (Oct.), 13.

Kalunke, R.M., S. Tundo, F. Sestili, F. Camerlengo, D. Lafiandra, R. Lupi, C. Larré, S. Denery-Papini, S. Islam, W. Ma, S. D'Amico, and S. Masci. 2020. Reduction of allergenic potential in bread wheat RNAi transgenic lines silenced for CM3, CM16 and 0.28 ATI genes. *Int. J. Mol. Sci.*, 21(16): 5817.

Karlund, A., I. Paukkonen, C. Gomez-Gallego, and M. Kolehmainen. 2021. Intestinal exposure to food-derived protease inhibitors: Digestion physiology and gut health-related effects. *Healthcare*, 9: 1002.

Karmakar, A., S. Bhattacharya, S. Sengupta, N. Ali, S.N. Sarkar, K. Datta, and S.K. Datta. 2020. RNAi-mediated silencing of ITPK gene reduces phytic acid content, alters transcripts of phytic acid biosynthetic genes, and modulates mineral distribution in rice seeds. *Rice Sci.*, 27(4): 315–328.

Kaspchak, E., L.I. Mafra, and M.R. Mafra. 2018. Effect of heating and ionic strength on the interaction of bovine serum albumin and the antinutrients tannic and phytic acids, and its influence on in vitro protein digestibility. *Food Chem.*, 252: 1–8.

Kaur, P., S.S. Purewal, K.S. Sandhu, M. Kaur, and R.K. Salar. 2019. Millets: A cereal grain with potent antioxidants and health benefits. *J. Food Meas. Charact.*, 13: 793–806.

Kumar, A., V. Kumar, V. Krishnan, A. Hada, A. Marathe, M. Jolly, and A. Sachdev. 2019. Seed targeted RNAi-mediated silencing of GmMIPS1 limits phytate accumulation and improves mineral bioavailability in soybean. *Sci. Rep.*, 9: 7744.

Kumar, S., P. Gupta, S. Barpete, A. Sarker, A. Amri, P.N. Mathur, and M. Baum. 2013. *Genetic and Genomic Resources of Grain Legume Improvement: 11. Grass Pea.* Elsevier Science.

Kumar, V., A. Chattopadhyay, S. Ghosh, M. Irfan, N. Chakraborty, S. Chakraborty, and A. Datta. 2016. Improving nutritional quality and fungal tolerance in soya bean and grass pea by expressing an oxalate decarboxylase. *Plant Biotechnol. J.*, 14(6): 1394–1405.

Kumar, V., A.K. Sinha, H.P. Makkar, and K. Becker. 2010. Dietary roles of phytate and phytase in human nutrition: A review. *Food Chem.*, 120: 945–959.

Lambein, F., S. Travella, Y.H. Kuo, M. Van Montagu, and M. Heijde. 2019. Grass pea (*Lathyrus sativus* L.): Orphan crop, nutraceutical, or just plain food? *Planta*, 250: 821–838.

Lesjak, M., and K.S. Srai, S. 2019. Role of dietary flavonoids in iron homeostasis. *Pharmaceuticals*, 12: 119.

Li, S., Y. Cui, Y. Zhou, Z. Luo, J. Liu, and M. Zhao. 2017. The industrial applications of cassava: Current status, opportunities, and prospects. *J. Sci. Food Agric.*, 97: 2282–2290.

Lin, B., X. Qi, L. Fang, L. Zhao, R. Zhang, J. Jing, ... and P. Xue. 2021. In vivo acute toxicity and mutagenic analysis of crude saponins from *Chenopodium quinoa* Willd husks. *RSC Adv.*, 11(8): 4829–4841.

Liu, Y., X. Jin, H.G. Hong, L. Xiang, Q. Jiang, Y. Ma, Z. Chen, L. Cheng, Z. Jian, Z. Wei, and J. Ai. 2020. The relationship between gut microbiota and short chain fatty acids in the renal calcium oxalate stones disease. *FASEB J.*, 34(8): 11200–11214.

Logue, B.A., D.M. Hinkens, S.I. Baskin, and G.A. Rockwood. 2010. The analysis of cyanide and its breakdown products in biological samples. *Crit. Rev. Anal. Chem.*, 40: 122–147.

Long, Y.C., Y.H. Ye, and Q.Y. Xing. 1996. Studies on the neuroexcitotoxin beta-N-oxalo-L-alpha, beta-diaminopropionic acid and its isomer alpha-N-oxalo-L-alpha, beta-diaminopropionic acid from the root of Panax species. *Int. J. Pept. Protein Res.*, 47: 42–46.

Lopez-Moreno, M., M. Garces-Rimon, and M. Miguel. 2022. Antinutrients: Lectins, goitrogens, phytates, and oxalates, friends or foe? *J. Funct. Food*, 89.

Malathi, K., G. Padmanaban, and P.S. Sarma. 1970. Biosynthesis of beta-n-oxalyl-l-alpha, beta-diaminopropionic acid, *Lathyrus-Sativus* neurotoxin. *Phytochem.*, 9: 1603.

Malik, A.I., P. Kongsil, V.A. Nguyễn, W. Ou, P. Srean, L.A.B. López-Lavalle, Y. Utsumi, C. Lu, P. Kittipadakul, H.H. Nguyễn, and H. Ceballos. 2020. Cassava breeding and agronomy in Asia: 50 years of history and future directions. *Breed. Sci.*, 70: 145–166.

Messina, M. 2016. Soy and health update: Evaluation of the clinical and epidemiologic literature. *Nutrients*, 8: 754.

Mishra, A., A. Behura, S. Mawatwal, A. Kumar, L. Naik, S.S. Mohanty, D. Manna, P. Dokania, A. Mishra, S.K. Patra, and R. Dhiman. 2019. Structure-function and application of plant lectins in disease biology and immunity. *Food Chem. Toxicol.*, 134: 110827.

Mitchell, T., P. Kumar, T. Reddy, K.D. Wood, J. Knight, D.G. Assimos, and R.P. Holmes. 2019. Dietary oxalate and kidney stone formation. *Am. J. Physiol. Renal Physiol.*, 316(3): F409–F413.

Montagnac, J.A., C.R. Davis, and S.A. Tanumihardjo. 2009. Nutritional value of cassava for use as a staple food and recent advances for improvement. *Compr. Rev. Food. Sci. Food Saf.*, 8: 181–194.

Muramoto, K. 2017. Lectins as bioactive proteins in foods and feeds. *Food Sci. Technol. Res.*, 23: 487–494.

Ngudi, D.D., Y.H. Kuo, M. Van Montagu, and F. Lambein. 2012. Research on motor neuron diseases konzo and neurolathyrism: Trends from 1990 to 2010. *PLoS Negl. Trop. Dis.*, 6: e1759.

Nguyen, V.D., H.L.T. Nguyen, L.C. Do, V. Van Tuan, P.T. Thuong, and T.N. Phan. 2018. A New saponin with anti-HIV-1 protease activity from Acacia pennata. *Nat. Prod. Commun.*, 13: 411–414.

Otubogun, F.M., R.O. Akinyemi, and A.O. Ogunniyi. 2019.Tropical ataxic neuropathy: Findings of a neuroepidemiological survey of Odeda, southwest Nigeria. *J. Neurol. Sci.*, 405: 116434.

Panda, P.K., P.P. Naik, P.P. Praharaj, B.R. Meher, P.K. Gupta, R.S. Verma, T.K. Maiti, M.K. Shanmugam, A. Chinnathambi, S.A. Alharbi, and G. Sethi. 2018. Abrus agglutinin stimulates BMP-2-dependent differentiation through autophagic degradation of β-catenin in colon cancer stem cells. *Mol. Carcinog.*, 57(5): 664–677.

Panghal, A., C. Munezero, P. Sharma, and N. Chhikara. 2021. Cassava toxicity, detoxification and its food applications: A review. *Toxin. Rev.*, 40: 1–16.

Parmar, A., B. Sturm, and O. Hensel. 2017. Crops that feed the world: Production and improvement of cassava for food, feed, and industrial uses. *Food Secur.*, 9: 907–927.

Paśko, P., P. Zagrodzki, K. Okoń, E. Prochownik, M. Krośniak, and A. Galanty. 2022. Broccoli sprouts and their influence on thyroid function in different *in vitro* and *in vivo* models. *Plants*, 20: 2750.

Patto, M.V., B. Skiba, E.C.K. Pang, S.J. Ochatt, F. Lambein, and D. Rubiales. 2006. Lathyrus improvement for resistance against biotic and abiotic stresses: From classical breeding to marker assisted selection. *Euphytica*, 147: 133.

Petrine, J.C.P., and B. Del Bianco-Borges. 2021. The influence of phytoestrogens on different physiological and pathological processes: An overview. *Phytother. Res.*, 35: 180–197.

Petroski, W., and D.M. Minich. 2020. Is there such a thing as "Anti-Nutrients"? A narrative review of perceived problematic plant compounds. *Nutrients*, 12.

Pramitha, J.L., S. Rana, P.R. Aggarwal, R. Ravikesavan, A.J. Joel, and M. Muthamilarasan, 2021. Diverse role of phytic acid in plants and approaches to develop low-phytate grains to enhance bioavailability of micronutrients. *Adv. Genet.*, 107: 89–120.

Prieto, M.A., C.J. Lopez, and J. Simal-Gandara. 2019. Glucosinolates: Molecular structure, breakdown, genetic, bioavailability, properties, and healthy and adverse effects. *Adv. Food Nutr. Res.*, 90: 305–350.

Quereshi, M.Y., D.J. Pilbeam, C.S. Evans, and E.A. Bell. 1977. The neurolathyrogen, α-amino-β-oxalylaminopropionic acid in legume seeds. *Phytochem.*, 16(4): 477–479.

Rajsic, P., R. Gray, A. Weersink, and I. Rajcan. 2023. Are there adequate incentives for research and innovation in the plant breeding supply chain? *In:* Stanton, John and Rosa Caiazza (Eds.), *Agricultural Value Chains: Some Selected Issues. Sustainable Development*. IntechOpen.

Ratnavathi, C.V. 2017. Nutritional qualities and value addition of millets. *In:* J.V. Patil (Ed.), *Millets and Sorghum: Biology and Genetic Improvement*. John Wiley & Sons Ltd.

Romarís-Hortas, V., C. García-Sartal, M. del Carmen Barciela-Alonso, R. Domínguez-González, A. Moreda-Piñeiro, and P. Bermejo-Barrera. 2011. Bioavailability study using an in-vitro method of iodine and bromine in edible seaweed. *Food Chem.*, 124: 1747–1752.

Saleem, S., N.U. Mushtaq, W.H. Shah, A. Rasool, K.R. Hakeem, and R.U. Rehman. 2021. Morpho-physiological, biochemical and molecular adaptation of millets to abiotic stresses: A review. *Phyton*, 90(5): 1363.

Sanchis, P., R. Rivera, F. Berga, R. Fortuny, M. Adrover, A. Costa-Bauza, F. Grases, and L. Masmiquel. 2018. Phytate decreases formation of advanced glycation end-products in patients with type II diabetes: Randomized crossover trial. *Sci. Rep.*, 8(1): 9619.

Sartelet, H., S. Serghat, A. Lobstein, Y. Ingenbleek, R. Anton, E. Petitfrere, Aguie-G. Aguie, L. Martiny, and B. Haye. 1996. Flavonoids extracted from fonio millet (*Digitaria exilis*) reveal potent antithyroid properties. *Nutrition*, 12(2): 100–106.

Serrano, J., R. Puupponen-Pimiä, A. Dauer, A.M. Aura, and F. Saura-Calixto. 2009. Tannins: Current knowledge of food sources, intake, bioavailability, and biological effects. *Mol. Nutr. Food Res.*, 53: S310–S329.

Shan, S., R. Yin, J. Shi, L. Zhang, F. Liu, Q. Qiao, and Z. Li. 2022. Bowman-Birk major type trypsin inhibitor derived from foxtail millet bran attenuate atherosclerosis via remodeling gut microbiota in ApoE–/–mice. *J. Agric. Food Chem.*, 70: 507–519.

Sharma, A. 2021. A review on traditional technology and safety challenges with regard to antinutrients in legume foods. *J. Food Sci. Technol.*, 58: 2863–2883.

Sharma, D., P. Singh, and S.S. Singh. 2018. Beta-N-oxalyl-L-alpha,beta-diaminopropionic acid induces wound healing by stabilizing HIF-1alpha and modulating associated protein expression. *Phytomedicine*, 44: 9–19.

Sharon, N. 2007. Lectins: Carbohydrate-specific reagents and biological recognition molecules. *J. Biol. Chem.*, 282: 2753–2764.

Sharon, N. and H. Lis. (Eds.) 2007. Detection, occurrence, and isolation. *Lectins*, pp. 33-61. Dordrecht: Springer.

Shi, A., B. Mou, and J.C. Correll. 2016. Association analysis for oxalate concentration in spinach. *Euphytica*, 212: 17–28.

Shi, J., K. Arunasalam, D. Yeung, Y. Kakuda, G. Mittal, and Y. Jiang. 2004. Saponins from edible legumes: Chemistry, processing, and health benefits. *J. Med. Food*, 7(1): 67–78.

Song, J.H., G. Shin, H.J. Kim, S.B. Lee, J.Y. Moon, J.C. Jeong, H.K. Choi, I.A. Kim, H.J. Song, C.Y. Kim, and Y.S. Chung. 2022. Mutation of GmIPK1 gene using CRISPR/Cas9 reduced phytic acid content in soybean seeds. *Int. J. Mol. Sci.*, 23(18): 10583.

Sun, H.X., Y. Xie, and Y.P. Ye. 2009. Advances in saponin-based adjuvants. *Vaccine*, 27: 1787–1796.

Sunil, S., and K. Leena. 2022. Contributing factors of hypothyroidism: A systematic review. *J. Clin. Diagnostic Res.*, 16(1).

Tagorti, G., B. Yalcin, M. Gunes, A.Y. Kursun, and B. Kaya. 2023. Genotoxic and genoprotective effects of phytoestrogens: A systematic review. *Drug Chem. Toxicol.*, 1–13.

Takagi, K., K. Nishizawa, A. Hirose, A. Kita, and M. Ishimoto. 2011. Manipulation of saponin biosynthesis by RNA interference-mediated silencing of β-amyrin synthase gene expression in soybean. *Plant Cell Rep.*, 30: 1835–1846.

Tan, R.Y., G.Y. Xing, G.M. Zhou, F.M. Li, W.T. Hu, F. Lambein, J.L. Xiong, S.X. Zhang, H.Y. Kong, H. Zhu, and Z.X. Li. 2017. Plant toxin β-ODAP activates integrin β1 and focal adhesion: A critical pathway to cause neurolathyrism. *Sci. Rep.*, 7(1): 40677.

Taylor, E.N., and G. Curhan. 2007. Oxalate intake and the risk for nephrolithiasis. *J. Am. Soc. Nephrol.*, 18: 2198–2204.

Taylor, J.R., and J. Awika, (Eds.). 2017. *Gluten-free Ancient Grains: Cereals, Pseudocereals, and Legumes: Sustainable, nutritious, and Health-Promoting Foods for the 21st Century*. Woodhead Publishing.

Tripp, R. 1992. Dietary consequences of changes in trade and agriculture for the nankane of northern ghana. *Ecol. Food Nutr.*, 27: 249–258.

Ward, S.M. 2000. Response to selection for reduced grain saponin content in quinoa (*Chenopodium quinoa* Willd.). *Field Crop. Res.*, 68: 157–163.

Westfall, S., and G.M. Pasinetti. 2019. The gut microbiota links dietary polyphenols with management of psychiatric mood disorders. *Front. Neurosci.*, 13: 1196.

Whitaker, J.R., 1997. Protease and α-amylase inhibitors of higher plants. Ch. 2. *In:* F. Shahidi (Ed.), *Antinutrients and Phytochemicals in Food*, 10–30. ACS Symposium Series.

Xu, Q., F. Liu, P. Chen, J.M. Jez, and H.B. Krishnan. 2017. Beta-N-Oxalyl-l-alpha,beta-diaminopropionic acid (beta-ODAP) content in *Lathyrus sativus*: The integration of nitrogen and sulfur metabolism through beta-cyanoalanine synthase. *Int. J. Mol. Sci.*, 18.

Yan, Z.Y., P.S. Spencer, Z.X. Li, Y.M. Liang, Y.F. Wang, C.Y. Wang, and F.M. Li. 2006. *Lathyrus sativus* (grass pea) and its neurotoxin ODAP. *Phytochem.*, 67(2): 107–121.

Zaheer, K., and M. Humayoun Akhtar. 2017. An updated review of dietary isoflavones: Nutrition, processing, bioavailability, and impacts on human health. *Crit. Rev. Food Sci. Nutr.*, 57: 1280–1293.

Zimmermann, M.B., and K. Boelaert. 2015. Iodine deficiency and thyroid disorders. *Lancet Diabetes Endocrinol.*, 3: 286–295.

Zimmermann, M.B. 2011. The role of iodine in human growth and development. *Semin. Cell Dev. Biol.*, 22: 645–652.

CHAPTER

14

Phytic Acid's Innovative Antinutrient Reduction

Wusirika Ramakrishna*[0000-0001-5378-0037], **Pankaj Ror**[0000-0001-7138-0088] **and Rahul Beniwal**[0009-0004-1595-9852]

Department of Biochemistry, Central University of Punjab, Ghudda, Bathinda - 151401, India

1. Introduction

Phytic acid (PA) is a naturally occurring compound found predominantly in the whole grains of cereals, legumes, nuts, and oilseeds. It is the primary storage form of phosphorus in plants and acts as a primary energy source for germinating seeds (Loewus, 2001). Its chemical structure is derived from inositol, a cyclic six-carbon sugar alcohol compound, with six hydroxy groups replaced by the phosphate groups forming inositol hexaphosphate ($C_6H_{18}O_{24}P_6$) (Figure 1). Inositol hexaphosphate ($InsP_6$ or IP_6), known by various names, phytin, phytate or phytic acid, was first identified in 1855 (Graf, 1983). The phosphate groups of PA are negatively charged at the optimal pH range, allowing them to interact with positively charged mineral atoms and other biomolecules. Cereal grains and their by-products are generally used worldwide to provide essential nutrients and energy in human and animal diets. The presence of PA in these plant-derived foods often leads to poor bioavailability of minerals such as zinc, iron, calcium, etc., because it tends to form chelates with these cations making them unavailable for absorption in the intestinal tract (Sarkhel and Roy, 2022). Therefore, due to its food micronutrient chelating ability, PA is classified as an antinutrient compound.

PA has 12 ionizable protons, giving it a unique structure responsible for its characteristic properties. It forms chelates with multivalent metal ions, forming water-insoluble complexes called phytate salts. The affinity of PA towards cations

*Corresponding author: rk.wusirika@cup.edu.in, wusirika@gmail.com.

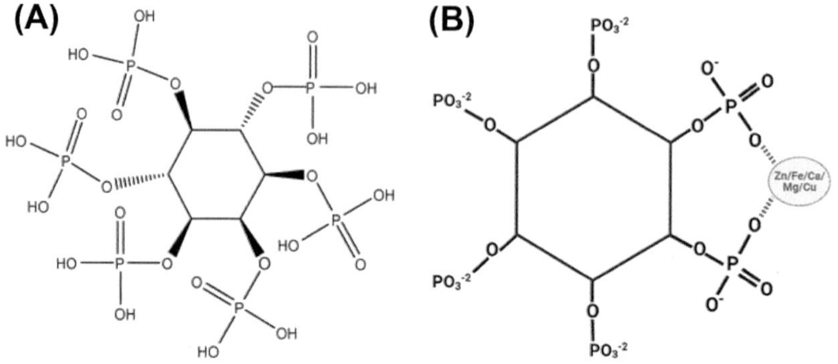

Figure 1: (A) Chemical structure of phytic acid ($C_6H_{18}O_{24}P_6$) and (B) a phytate salt.

decreases as follows: $Cu^{2+} > Zn^{2+} > Ni^{2+} > Co^{2+} > Mn^{2+} > Fe^{3+} > Ca^{2+}$ (Graf and Eaton, 1990). These and other metal ions like Mg^{2+}, Na^+, and K^+ bind with the phosphate group(s) to form complexes of varying solubility. Zinc is reported to be one of the most affected minerals as it forms the most insoluble and stable complex. Diets composed of high PA proportion and poor trace elements lead to higher nutritional impact whereas in well-balanced diets, a lower anti-nutritional effect of PA is observed (Bloot et al., 2021). Therefore, the micronutrient to PA molar ratio is an essential determinant in the nutritional value of any diet. The diets of the Asian population are characterized by a higher proportion of cereals and therefore have a high content of phytic acid, which is considered as a main reason for mineral deficiencies. Other components of food such as polyphenols, fats, organic acids, and complexing agents may also interact with minerals, and thus reduce their bioavailability (Silva and Bracarense, 2016; Zhang et al., 2020). Therefore, the effect on mineral absorption should also consider all dietary components.

Apart from chelating mineral ions, the antinutritional effect of PA is also observed in inhibiting the activity of digestive enzymes like pepsin, amylase, and trypsin (Kumar et al., 2020). PA binds to these enzymes and other proteins via electrostatic interactions. It can also interact with starch via hydrogen bonding. Humans do not produce phytases needed to break down phytates and liberate phosphorus. This results in about half the amount of ingested minerals (Fe, Zn, Ca) passing out of the body unabsorbed. However, a healthy population of intestinal microbiota aids in meeting the antinutritional effects of PA by producing microbial phytases (Famularo et al., 2005; González-Córdova et al., 2016).

Traditionally, many methods are adopted and widely used to reduce the content of PA and enhance the nutritional value of food before consumption. These methods include soaking grains, legumes, nuts and seeds in water or steeping for several hours, cooking, fermenting, and sprouting or germination (Wu et al., 2022). Combining these methods can reduce the phytate content substantially

(Lemmens et al., 2019). Novel processing methods like irradiation, ultrasound, and high pressure are also being employed (Grgić et al., 2023; Sarkhel and Roy, 2022). Another method is the explicit use of PA degrading enzyme phytase during food preparation. Phytases can be derived from microbial or endogenous sources. Endogenous phytases are found in most PA containing seeds of higher plants. Cereal phytases are more cost-effective and highly accepted among consumers than microbial phytases (Vashishth et al., 2017). Endogenous phytases are also less allergenic, therefore, more research is needed to identify new cereal phytases, characterize their respective genes and develop engineered crops for better absorption of micronutrients. Genes involved in PA biosynthesis and associated vacuolar transport have been characterized and genetically modulated to produce less PA in plants (Sahu et al,. 2023).

2. Biosynthesis and Abundance of Phytic Acid

The biosynthesis of inositol hexaphosphate starts a few days after flowering and continues during seed development till maturity (Rasmussen et al., 2010). The phytic acid, its salts, and complexes are stored in the protein bodies. Biosynthesis occurs through a pathway involving several enzymes. The starting point of the pathway is glucose-6-phosphate. It is converted by the enzyme *myo*-inositol-3-phosphate synthase (MIPS) into *myo*-inositol-3-phosphate (Ins-3-P). Further, the enzyme inositol tris/tetrakisphosphate kinase (ITPK) converts $InsP_3$ and $InsP_4$ into *myo*-inositol-1,3,4,5,6-5-phosphate [$Ins(1,3,4,5,6)P_5$]. ITPK is also known as ITP5/6K. Genes encoding this enzyme show multi-kinase activities, suggesting the presence of some intermediate pathways leading to the synthesis of $Ins(1,3,4,5,6)P_5$ (Suzuki et al., 2007). In the last step, the enzyme inositol pentakisphosphate 2-kinase (IPK1) phosphorylates $Ins(1,3,4,5,6)P_5$ resulting in the formation of phytic acid ($InsP_6$). The above-described lipid-independent pathway is the most common pathway and is found in cereals and legumes. In the lipid-dependent pathway, phospholipase C, inositol 1,4,5-trisphosphate kinase (IPK2), and IPK1 enzymes are involved in synthesizing PA (Suzuki et al., 2007). Both pathways are illustrated in Figure 2. Whereas all plant tissues use the lipid-dependent pathway, the lipid-independent pathway occurs primarily in the seeds (Frittelli et al., 2023). The biosynthesis of inositol hexaphosphate can be influenced by several factors, such as the availability of inositol, plant genotype, plant hormones, climatic conditions like soil and temperature, use of fertilizer, the activity of the enzymes involved in the biosynthesis pathway, and the presence of other molecules that can regulate the pathway (Bloot et al., 2021).

The prevalence of PA among various organisms, biochemistry and interaction pathways are documented in the NCBI PubChem database with compound identifier 890 (https://pubchem.ncbi.nlm.nih.gov/compound/Phytic-acid). Several plants synthesize PA to utilize excess phosphorus accumulated in developing seeds over and above what is needed for nominal cellular functions (Raboy, 2009).

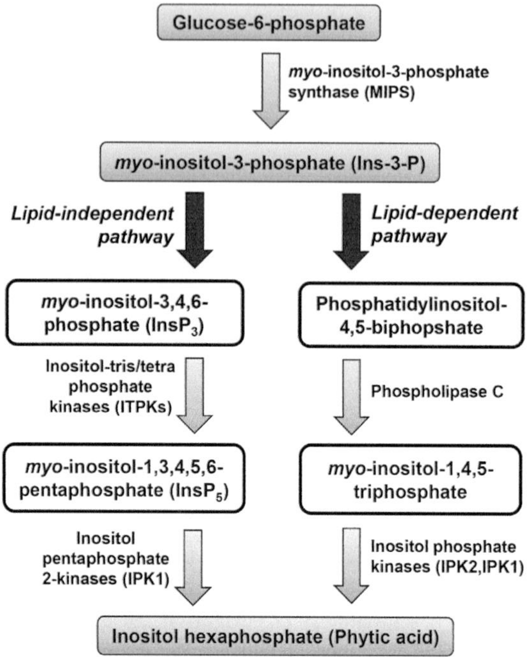

Figure 2: Phytic acid biosynthesis pathway.

Therefore, 50–80% of seed phosphorus is stored as PA (Thavarajah et al., 2010). Fruits and vegetables contain a lower amount of phytic acid. The PA content of some commonly used foods is summarized in Table 1. PA constitutes about 1–5% of dry weight in most cereals, legumes, oilseeds and nuts (Graf and Eaton, 1990). The distribution of PA also varies among various parts or layers of whole grain in different cereal plant species. For example, in corn, PA is distributed as follows: germ (88%), endosperm (3.2%), and hull (0.4%); whereas in wheat, the germ portion contains ~13% of total PA with 2.2% in endosperm and 87% in aleurone. Similarly, 80% of organic phosphorus in rice bran is contained in aleurone and pericarp layers with germ and endosperm containing only 7.6% and 1.2%, respectively (O'Dell et al., 1972). Therefore, grain processing methods also play a vital role to alter the nutritional value of consumed food as the products and by-products of processing contain differential proportions of PA along with other nutritional factors such as carbohydrates, proteins, lipids, minerals, and antioxidants.

3. Beneficial Properties of Phytic Acid

Although PA is an anti-nutritional factor, it also finds applications in medicine, food, and many other industries. It is important here to briefly summarize the positive roles of PA. It is considered a natural antioxidant compound

Table 1: Phytic acid content in various foods

	Food Products	Phytic Acid % (w/w)
Cereals	Wheat (*Triticum* spp.)	0.39-1.35
	Wheat bran	2.1-7.3
	Rice (*Oryza sativa/glaberrima*)	0.06-1.08
	Rice bran	2.56-8.7
	Maize (*Zea mays*)	0.72-2.22
	Maize germ	6.39
	Oat (*Avena sativa*)	0.42-1.16
	Barley (*Hordeum vulgare*)	0.38-1.16
	Millets (*Pennisetum* sp., etc)	0.18-1.67
	Sorghum (*Sorghum* spp.)	0.57-3.35
Legumes	Peanut (*Arachis hypogaea*)	0.17-4.47
	Lentils (*Lens culinaris*)	0.27-1.51
	Peas (*Pisum sativum*)	0.22-1.22
	Chickpeas (*Cicer arietinum*)	0.28-1.60
	Soybeans (*Glycine max*)	1.0-2.22
	Soy concentrate	10.7
	Navy bean	0.75-1.58
Nuts	Almonds (*Prunus dulcis*)	0.35-9.42
	Cashew nuts (*Anacardium occidentale*)	0.19-4.98
	Walnuts (*Juglans regia*)	0.20-6.69
	Pistachios (*Pistachia vera*)	0.29-2.83
Oilseeds	Rapeseed (*Brassica napus*)	2.50
	Rapeseed protein concentrate	5.3-7.5
	Linseed (*Linum usitatissimum*)	2.15-3.69
	Sesame seed (*Sesmun indicum*)	1.44-5.36
	Sunflower meal (*Helianthus annuus*)	3.9-4.3
Fruits	Banana (*Musa* sp.)	0.017-0.036
	Mango (*Mangifera indica*)	0.14
	Avocado (*Persea americana*)	0.05
Vegetables	Carrot (*Daucus carota*)	0.09
	Spinach (*Spinacia oleracea*)	0.07

Adapted from Schlemmer et al. (2009) and Bloot et al. (2021)

(Kumar et al., 2021) and can inhibit lipid peroxidation. PA is used in food preservatives to extend the shelf life of fruits, vegetables, and their oils. At high PA concentrations, the browning and putrefaction of fruits and vegetables is prevented due to inhibition of the enzyme polyphenol oxidase (Du et al., 2012). It also inhibits the iron-catalysed hydroxyl radical (•OH) formation owing to its iron binding ability and formation of completely inactive chelate. PA can be used to reduce the allergenicity of some foods due to its ability to form complexes with proteins thus reducing their solubility and altering their function (Wang and Guo, 2021). Dephosphorylated forms of phytic acid ($IP_{1\text{-}5}$) play an important role in cellular signal transduction whereas some function as secondary messengers and neuromodulators. $IP_{1\text{-}5}$ can regulate cell division and differentiation, endocytosis, and exocytosis (Kumar et al., 2021). PA has been found to lower serum cholesterol and triglycerides, and anti-platelet aggregation thereby reducing risk factors for cardiovascular disorders (Pujol et al., 2023). $IP_{1\text{-}6}$ also acts as a potent inhibitor of calcium oxalate crystallization and serves as a therapeutic agent for chronic kidney disease (Ekramzadeh et al., 2023; Grases et al., 2022).

Additionally, PA exhibits antineoplastic properties, i.e., inhibition of tumor cell growth and differentiation (Fox and Eberl, 2002). Significant anticancer activity of PA has been reported in both *in vivo* and *in vitro* studies in hepatocarcinoma, rhabdomyosarcoma, colon cancer, and early breast cancer (Kato et al., 2019; Lan et al., 2022; Vucenik, 2019). Other potential roles of PA include modulation of insulin secretion and sensitivity which is affected in diabetic patients (Sanchis et al., 2023), prevention of fatty liver (Katayama, 1999), modulation of immune response (Wee et al., 2021), antimicrobial effect on pathogenic fungi and bacteria (Asensio et al., 2022; Li et al., 2023), and inhibition of HIV-1 virus replication (Otake et al., 1999). PA has industrial applications, which include its role as an antimicrobial agent in beverages and milk products, a metal coating agent to prevent rusting and corrosion, an excellent flame-retardant component in cotton, polyester, and silk fabrics, anti-explosive additive for aviation gasoline, and heavy-metal contaminated waste treatment (Bloot et al., 2021).

4. Phytic Acid Reduction and Nutrient Biofortification

Phytic acid is stored mainly in rice bran, where it chelates cations and causes cation deficiency in consumers (Kishor et al., 2019). *Myo*-inositol-1,2,3,4,5,6-hexakisphosphate (PA) is the plant's most abundant form of phosphorous, but humans cannot digest it. Low phytic acid 1 (*lpa1*) is a mutation that interrupts the supply of *myo*-inositol during seed development. Non-lethal allelic variants of *lpa1* have been developed in rice (*lpa1-1*), maize, soybean, wheat, and barley. The use of hydropriming and increase in anthocyanin content can enhance seed germination in the wild type but not in the *lpa1-1* mutant. Phosphorous (P) that is reduced from seeds as PA in rice is replaced by inorganic phosphate. High inorganic P is typically present in all rice *lpa* allelic variants. Rice *lpa1* homozygote showed no altered germination and viability; however, total production is comparatively lower (Kishor et al., 2019). Genes targeted for low PA production are summarized in Table 2.

Table 2: Genes targeted in different plants to produce low phytic acid mutants

Sr. No.	Plant Name	Gene Targeted	Mutation/ Transgenic Method Used	Target Gene Function	% PA Reduced	Reference
1.	Durum wheat	*TdMRP3*	TILLING	Vacuolar transport	85	Frittelli et al., 2023
2.	Soybean	*GmIPK1*	CRISPR/Cas9	PA biosynthesis	≤25	Song et al., 2022
3.	Rice	*OsITPK6*	CRISPR/Cas9	PA biosynthesis	≤32	Jiang et al., 2019
4.	Maize	*MIPS1S*	Ethyl-methanesulfonate (EMS) treatment	PA biosynthesis	90	Pilu et al., 2003
5.	Maize	*ZmMRP4*	EMS	Membrane transporter	≤36	Abhijith et al., 2020
6.	Rice	*OsMIK*	Sodium azide and Methylnitrosourea	PA biosynthesis	75	Kim et al., 2008
7.	Oilseed rape	*BnITPK1* and *BnITPK4*	CRISPR-Cas9	PA biosynthesis	≤35	Sashidhar et al., 2020
8.	Wheat	*TaIPK1*	CRISPR/Cas9	PA biosynthesis	70	Ibrahim et al., 2022
9.	Wheat	*TaABCC13*	RNAi	PA transporter	22-34	Bhati et al., 2016
10.	Soybean	*GmMIPS1*	Antisense and RNAi	PA biosynthesis	38-41	Kumar et al., 2019

Depending on the mutation target, all *lpa* mutants can be divided into the following categories:

(a) Targeting the first step of PA biosynthesis, where glucose-6-phosphate is converted to *myo*-inositol-3-phosphate by *myo*-inositol-3-phosphate synthase.
(b) Intermediate phosphorylation steps by inositol-tris/tetraphosphate kinases (ITPKs) to produce inositol-pentaphosphate (InsP5).
(c) Last step of PA biosynthesis where InsP5 is phosphorylated at the second position by IPK1 to synthesize PA.
(d) Targeting MRP transporters which are responsible for the storage and vacuolar transport of PA.

Various approaches used to target genes related to PA metabolism and transport in plants are shown in Figure 3.

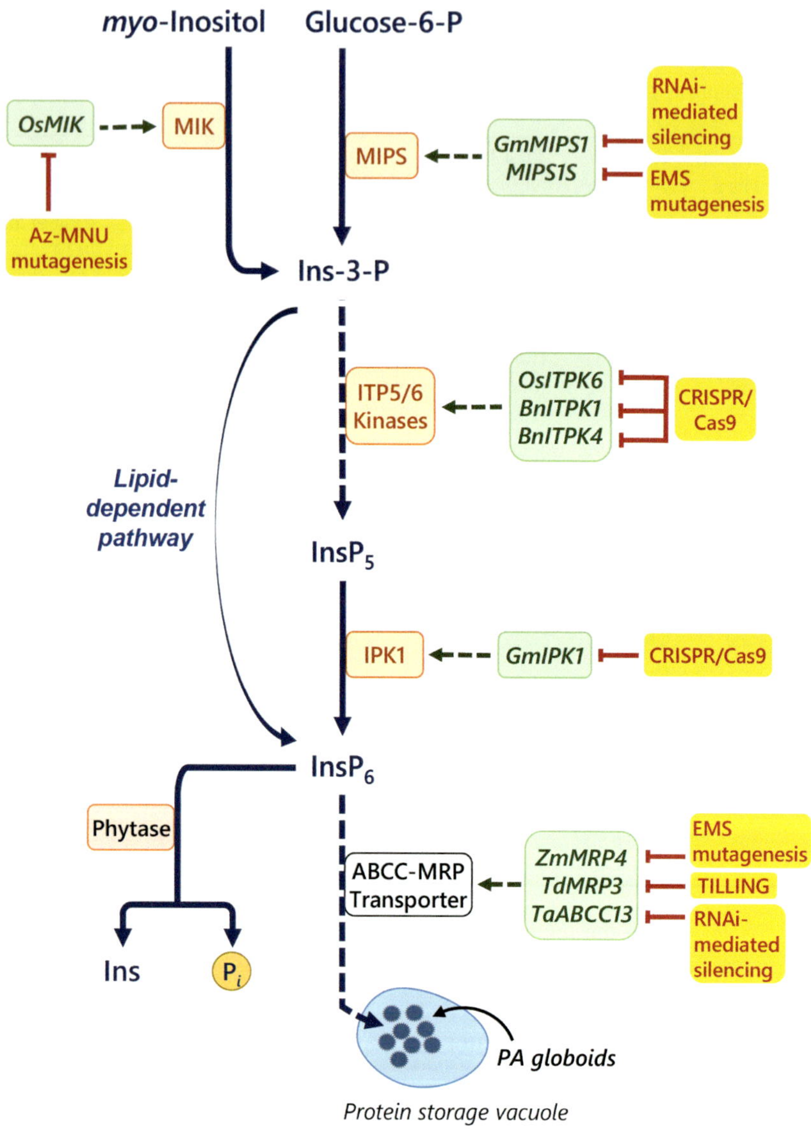

Figure 3: Illustration of various methods to reduce the phytic acid synthesis and accumulation in plants.

Abbreviations: Ins: Inositol, PA: Phytic acid, EMS: Ethyl-methanesulfonate, AzMNU: Azide-methylnitrosourea, TILLING: Targeting induced local lesions in genomes, MIK: *myo*-inositol kinase, MIPS: *myo*-inositol-3-phosphate synthase, ITP5/6K: Inositol tris/tetraphosphate-5,6-kinase, IPK1: Inositol-pentaphosphate-2-kinase, ABCC-MRP: ATP binding cassette type C-multidrug resistance-associated protein, CRISPR: Clustered regularly interspaced short palindromic repeats (Credit: BioRender.com).

Different crops and mutants produced for low PA content are discussed below.

4.1. Rice

Low PA rice with recessive allele *lpa1-3* called Sanggol has a point mutation. Sanggol was developed from the japonica rice cultivar and has 49% lower PA than the wild type. Sanggol takes a longer time for 50% flowering and has chalkier grains than the wild type. *OsLpa1* is localized on chromosome 2 and expresses three splice variants (Kishor et al. 2019). PA biosynthesis mutants have shown adverse effects like low yield, delayed germination, and lower seed set rates (Sashidhar et al., 2020). In rice seeds, 80% of the total PA is stored in aleurone and maternal teguments (Frittelli et al., 2023). Multidrug-resistant protein (MRP5) is a member of the ATP binding cassette (ABC) superfamily responsible for tonoplast transport and has a high affinity for PA. In the rice vacuole, in addition to MRP5, other transporters like SULTR-like phosphorus distribution transporter (SPDT) also contribute in P transport to grain, thereby increasing PA levels. Another study where *OsITPK6* was targeted using CRISPR/Cas9 produced four mutant lines from which two had tremendously low seed set (*ositpk6_3* and *ositpk6_4*) while *ositpk6_1* plant showed the highest rate of germination among the four mutant lines. PA phosphorus of *ositpk6_1* and *ositpk6_2* was decreased by 10.1% and 32.1%, while inorganic phosphorous increased by 4.12- and 5.18-fold, respectively. The growth of *ositpk6_1* was found to be inferior to Xidao 1 under salt and osmotic stress (Jiang et al., 2019). N15-375 and N15-186 mutated lines developed from the Japonica cultivar Nipponbare showed PA content of 43 and 75%, respectively, while no change in the total phosphorous content was observed compared to the wild type (Kim et al., 2008).

4.2. Maize

lpa1-7, *lpa1-241*, *lpa1-1*, *lpa 2-1*, and *lpa1-5525* are five low PA mutants identified and characterized in maize. Lpa1-1 has a C-T transition in *ZmMRP4*, which is a membrane transporter for PA (Abhijith et al., 2020). PA was reduced up to 66% in the *lpa1-7* mutant. The seeds of this mutant age faster, progressively reducing germination and tolerance to unfavorable storage conditions (Colombo et al., 2023). *lpa1-241* and *lpa241* are allelic to *lpa1-1* (Badone et al., 2010). The *lp1-241* homozygous mutant's kernel has a 25% higher anthocyanin concentration than the wild type, which is also an indicator of a plant under stress (Badone et al., 2010). *lpa1-241* scutellum color is blue rather than red like wild type which is due to the mislocalization of anthocyanin in the cytoplasm. *lpa1-1* also shows an increase in anthocyanin level but only a 0.1-fold increase. *lpa2-1* shows a 50% reduction in PA and an increase in inorganic phosphorous (Colombo et al., 2023). An A-G transition mutation was observed in *ZmITPK* in *lpa2-1*. Maize mutant line with kernel PA amount less than 20% showed a decreased germination rate (Pilu et al., 2003). A maize mutant (*lpa241*) with low PA was developed by treating pollen with ethyl-methanesulfonate (EMS), which has almost 10-fold more free phosphate and 90% less PA. However, seed germination was lowered by 30% in *lpa241*, but there was no reduction in plant development and seed maturation.

4.3. Wheat

Multidrug-Resistance associated Proteins 3 (TdMRP3) are responsible for accumulating PA inside the vacuole of durum wheat. Disrupting *TdMRP3* using Targeting Induced Local Lesions in Genomes (TILLING) leads to increased seed micronutrient accumulation. A double null mutant of *TdMRP3* enhanced iron, Zn, and Mn content. Iron accumulation in the double mutant is observed in the scutellum, aleuronic layer, and endosperm, while *TdMRP3* single mutant micronutrient profile is non-uniform. In the double null mutant, micronutrient Mg and S increased by 38.4% and 32.6%, respectively. The kernel weight and spike length do not show significant difference between control and double mutant. Spikelet per spike (–17.7%) and the number of kernels per spike (–34.5%) decreased in the double mutant, while the single kernel weight is more in the double mutant than control. Root diameter increased by 45.8% in the mutant *TdMRP3*, but the number of root tips, length, volume, and surface area decreased by 83.5, 61.2, 16.8, and 40.6%, respectively (Frittelli et al., 2023). Wheat (*Triticum aestivum* L.) mutant *Js-12-LPA* generated by treatment with 2% EMS showed a 43% reduction in bran PA and a four-fold increase in inorganic phosphorus. Globoids have smaller sizes and increased numbers per aleurone cell in *Js-12-LPA*. This decrease in size is hypothesized to be due to reduced PA content, which may be responsible for cross-linking smaller globoids and forming larger ones. When the total volume of wild type vs. *Js-12-LPA* globoid volume is compared, the increased number still has less volume than the number of globoids in the wild type (Joyce et al. 2005). Targeted mutation in wheat (*Triticum aestivum*) by CRISPR-Cas9 in inositol 1,3,4,5,6-pentakisphosphate 2-kinase (*TaIPK1*) gene resulted in reduced PA content and up to 2.1 and 1.9-fold increase in Fe and Zn. Plant height and germination were not altered in the *TaIPK1* mutant, but spike length and spikelet number decreased (Ibrahim et al., 2022).

A forward selection approach was used to select for natural low PA variants in wheat by performing genome-wide association studies that identified haplotypes that additively account for a 10% reduction in PA (Wen et al., 2022). This study also developed a reliable method for high-throughput PA quantification to facilitate the phenotyping and screening of large breeding populations.

4.4. Barley

PA content is influenced by genetic and environmental factors, and it affects the protein content in barley (Dai et al., 2007). *lpa* 1-1 mutant barley contain almost twice non-phytate P than the wild type, dispite the total P content is same in both (Li et al., 2001). Dietary absorption of P is therefore enhanced in *lpa* 1-1 barley as compared to the nutrition from non-mutant variety. Four barley mutants derived by sodium azide treatment shows PA reduction of 95%, 65%, 50% and 40% in M955, *lpa3-1*, *lpa1-1*, and *lpa2-1*, respectively (Bregitzer & Raboy, 2006). A non-phytate P increase accompanies a decrease in PA content. The inorganic phosphate content of mutants is ranked as *lpa2-1* ≤ *lpa1-1* < *lpa3-1* < M955, wherein *lpa1-1* shows the least loss in crop yield. The PA content in *lpa1-1* was lowered specifically in aleurone while PA was almost unchanged in the germ layer. A decrease in the expression of H^+/pyrophosphatase, a sugar transporter, and starch synthesis gene is observed in M955 (Bowen et al., 2007). H^+/pyrophosphatase is present in membrane of globoid where it generates electrochemical gradient by proton pumping. The expression of *brittle1*, which encodes ADP-glucose transporter providing substrate for starch synthesis decreased in M955. The combined decrease in these genes may be a reason for reduced yield, seed weight and different globloid structure found in M955.

4.5. Other Crops

MRP5 mutants from the common bean and soybean show up to a 90% reduction in PA content (Sashidhar et al., 2020). Inositol 1,3,4,5,6-pentakisphosphate 2-kinase encoded by *GmIPK1* converts *myo*-inopsitol-1,3,4,5,6-pentakisphosphate (IP5) to *myo*-inositol-1,2,3,4,5,6-hexakisphosphate (IP6) in soybean (*Glycine max* L.). Soybean has three homologs for the *IPK1* enzyme located on chromosomes 14, 6, and 4. Targeting *GmIPK1* using CRISPR-Cas9 reduces the PA content up to 25% without having any adverse agronomic effects in the transgenic plant.

In oilseed rape, three paralogs of ITPK were mutated, and there was about a 35% reduction in PA content. Oilseed rape underwent whole genome duplication multiple times. Even more interesting is that new mutations appeared in the T_2 generation that were not present in T_1. Similar to other studies targeting the reduction of phytic acid, oilseed rape also showed an increase in inorganic phosphorous content as PA decreased.

5. Heterologous Expression of Phytase

Heterologous expression of phytase is supposed to boost nutrient bioavailability by degrading PA in seeds. Introducing the phytase gene (*phyA*) from *Aspergillus japonicas* in wheat increased phytase activity by up to 99% and decreased phytate by up to 76% in seeds (Abid et al., 2017). Reduced PA in seed does not change the nutrient content in wheat; however, it does increase the bioavailability of Fe and Zn by 14.28 and 4%, respectively, compared to non-transgenic wheat. Fungal phytase has higher activity at higher temperatures, so it is more active in chapatti than in dough. Similarly, transgenic rice and maize have 130- and 50-fold higher phytase activity, respectively. *Aspergillus* phytase (*phyA*) can be expressed in wheat and used for non-ruminant animal feed to improve phytate phosphorous digestibility (Brinch-Pedersen et al., 2000). A cassette designed with a signal peptide and phytase under ubiquitin promoter increased phytase activity 4-fold upon introduction in wheat using the particle bombardment method. The signal sequence targets the *phyA* to the endoplasmic reticulum and ensures glycosylation. Leaves and developing seeds showed the production of microbial phytase. Heterologous *Aspergillus niger* phytase expressed with endosperm-specific promoter hydrolysed 86% InsP6 and removed all lower inositol phosphates (Brinch-Pedersen et al., 2003). Phytase provides a less expensive alternative to non-renewable inorganic phosphorus in feeds, reducing water pollution by limiting phosphorus run-off in manure. Xenogenic expression of the thermostable *mappA* gene in soybean (*Glycine max*) increased phytase activity and can be used in feed additives (Zhao et al., 2019). However, xenogenic expression of the mappA gene does reduce 100-grain weight by 4%.

6. Challenges and Future Prospects

Firstly, PA is an antioxidant, and plants that have reduced PA go through higher oxidative stress. One alternative is to replace PA with another antioxidant, but as shown by Colombo et al. (2023) anthocyanins do not reduce the effect of lowered PA. Other options to replace PA as an antioxidant needs to be explored. Secondly, lower production by PA mutants is a big issue, but increased nutrients of these mutants may persuade people to use them. It still needs to be established if lowered production of PA mutant plants is due to their antioxidative properties or if genes responsible for PA have other metabolic and developmental roles. The biosynthesis of PA involves alternate and multiple enzymatic pathways which pose a challenge for transgenic efforts as the function of silenced or mutated gene is often compensated by other members of the gene family. The disruption of PA biosynthetic pathway genes at later steps (i.e., IPK1 and ABCC/MRP transporter) using reverse genetics would be more beneficial as it would have negligible pleiotropic effects.

7. Conclusion

Phytic acid is considered an antinutrient as monogastric animals cannot digest it thereby reducing the absorption of nutrients. PA serves as a source of phosphorous for plants and also acts as an antioxidant. Targeting PA biosynthesis genes to generate cereals, nuts, and legumes with lower PA is an approach to enhance their absorption in the intestines. However, some drawbacks which include reduced germination, seed viability, and production, faster seed aging, chalky grain and increased time for 50% flowering were observed in *lpa* mutants. Therefore, a balanced and holistic approach is needed to regulate PA to reduce its antinutrient quality by considering multiple approaches that involve studying the interaction of PA with other biomacromolecules.

References

Abhijith, K.P., V. Muthusamy, R. Chhabra, S. Dosad, V. Bhatt, G. Chand, S.K. Jaiswal, R.U. Zunjare, S. Vasudev, D.K. Yadava, and F. Hossain. 2020. Development and validation of breeder-friendly gene-based markers for lpa1-1 and lpa2-1 genes conferring low phytic acid in maize kernel. *3 Biotech*, 10: 1–12.

Abid, N., A. Khatoon, A. Maqbool, M. Irfan, A. Bashir, I. Asif, M. Shahid, A. Saeed, H. Brinch-Pedersen, and K.A. Malik. 2017. Transgenic expression of phytase in wheat endosperm increases bioavailability of iron and zinc in grains. *Transgenic Res.*, 26: 109–122.

Asensio, G., A.M. Hernández-Arriaga, M. Martín-del-Campo, M.A. Prieto, L. Rojo, and B. Vázquez-Lasa. 2022. A study on Sr/Zn phytate complexes: Structural properties and antimicrobial synergistic effects against Streptococcus mutans. *Sci. Rep.*, 12: 20177.

Badone, F.C., E. Cassani, M. Landoni, E. Doria, D. Panzeri, C. Lago, F. Mesiti, E. Nielsen, and R. Pilu. 2010. The low phytic acid1-241 (lpa1-241) maize mutation alters the accumulation of anthocyanin pigment in the kernel. *Planta*, 231: 1189–1199.

Bhati, K.K., A. Alok, A. Kumar, J. Kaur, S. Tiwari, and A.K. Pandey. 2016. Silencing of ABCC13 transporter in wheat reveals its involvement in grain development, phytic acid accumulation and lateral root formation. *J. Exp. Bot.*, 67: 4379–4389.

Bloot, A.P.M., D.L. Kalschne, J.A.S. Amaral, I.J. Baraldi, and C. Canan. 2021. A review of phytic acid sources, obtention, and applications. *Food Rev. Int.*, 39: 73–92.

Bowen, D.E., E.J. Souza, M.J. Guttieri, V. Raboy, and J. Fu. 2007. A low phytic acid barley mutation alters seed gene expression. *Crop Science*, 47: S149-S159.

Bregitzer, P., and V. Raboy. 2006. Effects of four independent low-phytate mutations on barley agronomic performance. *Crop Science*, 46(3): 1318–1322.

Brinch-pedersen, H., A. Olesen, S.K. Rasmussen, and P.B. Holm. 2000. Generation of transgenic wheat (*Triticum aestivum* L.) for constitutive accumulation of an Aspergillus phytase. *Mol. Breed.*, 6: 195–206.

Brinch-pedersen, H., F. Hatzack, L.D. Sørensen, and P.B. Holm. 2003. Concerted action of endogenous and heterologous phytase on phytic acid degradation in seed of transgenic wheat (*Triticum aestivum* L.). *Transgenic Res.*, 1: 649–659.

Colombo, F., A. Pagano, S. Sangiorgio, A. Macovei, A. Balestrazzi, F. Araniti, and R. Pilu. 2023. Study of seed ageing in lpa1-1 maize mutant and two possible approaches to restore seed germination. *Int. J. Mol. Sci.*, 24: 732.

Dai, F., J. Wang, S. Zhang, Z. Xu, and G. Zhang, G. 2007. Genotypic and environmental variation in phytic acid content and its relation to protein content and malt quality in barley. *Food Chem.*, 105(2): 606–611.

Du, Y., S. Dou, and S. Wu. 2012. Efficacy of phytic acid as an inhibitor of enzymatic and non-enzymatic browning in apple juice. *Food Chem.*, 135: 580–582.

Ekramzadeh, M., L.W. Moore, K. Kalantar-Zadeh, and J.D. Kopple. 2023. Phytate and kidney health: The roles of dietary phytate in inhibiting intestinal phosphorus absorption and intravenous phytate in decreasing soft tissue calcification. *J. Ren. Nutr.*, 33: 225–227.

Famularo, G., C. De Simone, V. Pandey, A.R. Sahu, and G. Minisola. 2005. Probiotic lactobacilli: An innovative tool to correct the malabsorption syndrome of vegetarians? *Med. Hypotheses*, 65: 1132–1135.

Fox, C.H., and M. Eberl. 2002. Phytic acid (IP6), novel broad spectrum anti-neoplastic agent: A systematic review. *Complement Ther. Med.*, 10: 229–234.

Frittelli, A., E. Botticella, S. Palombieri, S. Masci, S. Celletti, M.C. Fontanella, S. Astolfi, P. De Vita, M. Volpato, and F. Sestili. 2023. The suppression of TdMRP3 genes reduces the phytic acid and increases the nutrient accumulation in durum wheat grain. *Front. Plant Sci.*, 14: 1079559.

González-Córdova, A.F., L.M. Beltrán-Barrientos, L. Santiago-López, H.S. Garcia, B. Vallejo-Cordoba, and A. Hernandez-Mendoza. 2016. Phytate-degrading activity of probiotic bacteria exposed to simulated gastrointestinal fluids. *LWT - Food Sci. Technol.*, 73: 67–73.

Graf, E., and J.W. Eaton. 1990. Antioxidant functions of phytic acid. *Free Radic. Biol. Med.*, 8: 61–69.

Graf, E. 1983. Applications of phytic acid. *J. Am. Oil Chem. Soc.*, 60(11): 1861–1867.

Grases, F., A. Costa-Bauzá, P. Calvó, F. Julià, J. Dietrich, R.M. Gomila, G. Martorell, and P. Sanchis. 2022. Phytate dephosphorylation products also act as potent inhibitors of calcium oxalate crystallization. *Molecules*, 27: 5463.

Grgić, T., Z. Pavišić, N. Maltar-Strmečki, B. Voučko, N. Čukelj Mustač, D. Ćurić, A. Le-Bail, and D. Novotni. 2023. Ultrasound-assisted modification of enzymatic and antioxidant activities, functional and rheological properties of oat and barley bran. *Food Bioproc. Technol.*, 1: 1–14.

Ibrahim, S., B. Saleem, N. Rehman, S.A. Zafar, M.K. Naeem, and M.R. Khan. 2022. CRISPR/Cas9 mediated disruption of Inositol Pentakisphosphate 2-Kinase 1 (TaIPK1) reduces phytic acid and improves iron and zinc accumulation in wheat grains. *J. Adv. Res.*, 37: 33–41.

Jiang, M., Y. Liu, Y. Liu, Y. Tan, J. Huang, and Q. Shu. 2019. Mutation of inositol 1,3,4-trisphosphate 5/6-kinase6 impairs plant growth and phytic acid synthesis in rice. *Plants*, 8: 1–12.

Joyce, C., A. Deneau, K. Peterson, I. Ockenden, V. Raboy, and J.N.A. Lott. 2005. The concentrations and distributions of phytic acid phosphorus and other mineral nutrients in wild-type and low phytic acid Js-12-LPA wheat (*Triticum aestivum*) grain parts. *Can. J. Bot.*, 83: 1599–1607.

Katayama, T. 1999. Hypolipidemic action of phytic acid (IP6): Prevention of fatty liver. *Anticancer Res.*, 19: 3695–3698.

Kato, A., Y. Hirakawa, and W. Hiraoka. 2019. Inositol hexaphosphate-induced cellular response in myeloid leukemia cells is mediated by nicotinamide adenine dinucleotide phosphate oxidase activation. *Fujita Med. J.*, 5: 98–103.

Kim, S.I., C.B. Andaya, J.W. Newman, S.S. Goyal, and T.H. Tai. 2008. Isolation and characterization of a low phytic acid rice mutant reveals a mutation in the rice orthologue of maize MIK. *Theor. Appl. Genet.*, 117: 1291–1301.

Kishor, D.S., C. Lee, D. Lee, J. Venkatesh, J. Seo, J.H. Chin, Z. Jin, S.K. Hong, J.K. Ham, and H.J. Koh. 2019. Novel allelic variant of Lpa1 gene associated with a significant reduction in seed phytic acid content in rice (*Oryza sativa* L.). *PLoS One*, 14: 1–20.

Kumar, A., B. Singh, P. Raigond, C. Sahu, U.N. Mishra, S. Sharma, and M.K. Lal. 2021. Phytic acid: Blessing in disguise, a prime compound required for both plant and human nutrition. *Food Res Int.*, 142: 110193.

Kumar, A., C. Sahu, P.A. Panda, M. Biswal, R.P. Sah, M.K. Lal, M.J. Baig, P. Swain, L. Behera, K. Chattopadhyay, and S. Sharma. 2020. Phytic acid content may affect starch digestibility and glycemic index value of rice (*Oryza sativa* L.). *J. Sci. Food Agric.*, 100: 1598–1607.

Kumar, A., V. Kumar, V. Krishnan, A. Hada, A. Marathe, C. Parameswaran, M. Jolly, and

A. Sachdev. 2019. Seed targeted RNAi-mediated silencing of GmMIPS1 limits phytate accumulation and improves mineral bioavailability in soybean. *Sci. Rep.*, 9: 7744.

Lan, T.T., Y. Song, X.H. Liu, C.P. Liu, H.C. Zhao, Y.S. Han, C.H. Wang, N. Yang, Z. Xu, M. Tao, and H. Li. 2022. IP6 reduces colorectal cancer metastasis by mediating the interaction of gut microbiota with host genes. *Front. Nutr.*, 9: 979135.

Lemmens, E., N. De Brier, P. Goos, E. Smolders, and J.A. Delcour. 2019. Steeping and germination of wheat (*Triticum aestivum* L.). I. Unlocking the impact of phytate and cell wall hydrolysis on bioaccessibility of iron and zinc elements. *J. Cereal Sci.*, 90: 102847.

Li, N., Y.X. Wu, Y.D. Zhang, S.R. Wang, G.C. Zhang, and J. Yang. 2023. Phytic acid is a new substitutable plant-derived antifungal agent for the seedling blight of Pinus sylvestris var. mongolica caused by Fusarium oxysporum. *Pestic. Biochem. Physiol.*, 191: 105341.

Li, Y.C., D.R. Ledoux, T.L. Veum, V. Raboy, K. Zyla, and A. Wikiera. 2001. Bioavailability of phosphorus in low phytic acid barley. *Journal of Applied Poultry Research*, 10(1): 86–91.

Loewus, F.A. 2001. Biosynthesis of phytate in food grains and seeds. *In:* N. Rukma Reddy, S.K. Sathe (Eds.), *Food Phytates*, 69–78. Florida, USA: CRC Press.

O'Dell, B.L., A.R. De Boland, and S.R. Koirtyohann. 1972. Distribution of phytate and nutritionally important elements among the morphological components of cereal grains. *J. Agric. Food Chem.*, 20: 718–723.

Otake, T., H. Mori, M. Morimoto, K. Miyano, N. Ueba, I. Oishi, N. Kunita, and T. Kurimura. 1999. Anti-HIV-1 activity of myo-inositol hexaphosphoric acid (IP6) and myo-inositol hexasulfate (IS6). *Anticancer Res.*, 19: 3723–3726.

Pilu, R., D. Panzeri, G. Gavazzi, S.K. Rasmussen, G. Consonni, and E. Nielsen. 2003. Phenotypic, genetic and molecular characterization of a maize low phytic acid mutant (lpa241). *Theor. Appl. Genet.*, 107: 980–987.

Pujol, A., P. Sanchis, F. Grases, and L. Masmiquel. 2023. Phytate intake, health and disease: "Let thy food be thy medicine and medicine be thy food." *Antioxidants*, 12: 146.

Raboy, V. 2009. Approaches and challenges to engineering seed phytate and total phosphorus. *Plant Sci.*, 177: 281–296.

Rasmussen, S.K., C.R. Ingvardsen, and A.M. Torp. 2010. Mutations in genes controlling the biosynthesis and accumulation of inositol phosphates in seeds. *Biochem. Soc. Trans.*, 38: 689–694.

Sahu, A., R. Verma, U. Gupta, S. Kashyap, and I. Sanyal. 2023. An overview of targeted genome editing strategies for reducing the biosynthesis of phytic acid: An anti-nutrient in crop plants. *Mol. Biotechnol.*, 65: 722.

Sanchis, P., P. Calvo, A. Pujol, R. Rivera, F. Berga, R. Fortuny, A. Costa-Bauza, F. Grases, and L. Masmiquel. 2023. Daily phytate intake increases adiponectin levels among patients with diabetes type 2: A randomized crossover trial. *Nutr. Diabetes*, 13: 2.

Sarkhel, S., and A. Roy. 2022. Phytic acid and its reduction in pulse matrix: Structure-function relationship owing to bioavailability enhancement of micronutrients. *J. Food Process Eng.*, 45: e14030.

Sashidhar, N., H.J. Harloff, and C. Jung. 2020. Knockout of multi-drug resistant protein 5 genes lead to low phytic acid contents in oilseed rape. *Front. Plant Sci.*, 11: 603.

Schlemmer, U., W. Frølich, R.M. Prieto, and F. Grases. 2009. Phytate in foods and significance for humans: Food sources, intake, processing, bioavailability, protective role, and analysis. *Mol. Nutr. Food Res.*, 53: S330–S375.

Silva, E.O., and A.P.F.R.L. Bracarense. 2016. Phytic acid: From antinutritional to multiple protection factor of organic systems. *J. Food Sci.*, 81: R1357–R1362.

Song, J.H., G. Shin, H.J. Kim, S.B. Lee, J.Y. Moon, J.C. Jeong, H.K. Choi, I.A. Kim, H.J. Song, C.Y. Kim, and Y.S. Chung. 2022. Mutation of GmIPK1 gene using CRISPR/Cas9 reduced phytic acid content in soybean seeds. *Int. J. Mol. Sci.*, 23: 10583.

Suzuki, M., K. Tanaka, M. Kuwano, and K.T. Yoshida. 2007. Expression pattern of inositol phosphate-related enzymes in rice (*Oryza sativa* L.): Implications for the phytic acid biosynthetic pathway. *Gene*, 405: 55–64.

Thavarajah, D., P. Thavarajah, C.T. See, and A. Vandenberg. 2010. Phytic acid and Fe and Zn concentration in lentil (*Lens culinaris* L.) seeds is influenced by temperature during seed filling period. *Food Chem.*, 122: 254–259.

Vashishth, A., S. Ram, and V. Beniwal. 2017. Cereal phytases and their importance in improvement of micronutrients bioavailability. *3 Biotech*, 7: 42.

Vucenik, I. 2019. Anticancer properties of inositol hexaphosphate and inositol: An overview. *J. Nutr. Sci. Vitaminol.*, 65: S18–S22.

Wang, R., and S. Guo. 2021. Phytic acid and its interactions: Contributions to protein functionality, food processing, and safety. *Compr. Rev. Food Sci. Food Saf.*, 20: 2081–2105.

Wee, Y., C. Yang, S.K. Chen, Y.C. Yen, and C.S. Wang. 2021. Inositol hexaphosphate modulates the behavior of macrophages through alteration of gene expression involved in pathways of pro- and anti-inflammatory responses, and resolution of inflammation pathways. *Food Sci. Nutr.*, 9: 3240–3249.

Wen, Z., P. Juliana, H.S. Dhugga, M. Pacheco, U.I. Martínez, A. Aguilar, M.I. Ibba, V. Govindan, R.P. Singh, and K.S. Dhugga. 2022. Genome-wide association study of phytic acid in wheat grain unravels markers for improving biofortification. *Front Plant Sci.*, 13: 830147.

Wu, N.N., R. Li, Z.J. Li, and B. Tan. 2022. Effect of germination in the form of paddy rice and brown rice on their phytic acid, GABA, γ-oryzanol, phenolics, flavonoids and antioxidant capacity. *Food Res. Int.*, 159: 111603.

Zhang, Y.Y., R.N.K. Stockmann, and S. Ajlouni. 2020. Revisiting phytate-element interactions: Implications for iron, zinc, and calcium bioavailability, with emphasis on legumes. *Crit. Rev. Food Sci. Nutr.*, 62: 1696–1712.

Zhao, Y., L. Zhu, C. Lin, Z. Shen, and C. Xu. 2019. Transgenic soybean expressing a thermostable phytase as substitution for feed additive phytase. *Sci. Rep.*, 9: 14390.

CHAPTER

15

Glucosinolates in Brassicas: Manipulating Content and Composition

Sanjula Sharma[a], Heena Rani[b], Shivani[c], Mehak Gupta[a], Manju Bala[c], Gurdeep Kaur[a] and Gurpreet Kaur[a*]

[a] Department of Plant Breeding and Genetics, Punjab Agricultural University, Ludhiana - 141004, Punjab, India

[b] ORISE, Cereal Crops Research Unit, USDA-ARS, 502 Walnut Street, Madison, WI, 53726, USA

[c] Food Grains and Oilseeds Processing Division, ICAR-Central Institute of Post-Harvest Engineering and Technology, Ludhiana - 141004, Punjab, India

1. Introduction

Glucosinolates (GLSs), sulphur and nitrogen-containing secondary metabolites are abundantly present in the genus *Brassica*. The genus Brassica belonging to the Brassicaceae family encompasses a diverse range of crops, including vegetables (*Brassica oleracea*), oilseeds (*Brassica rapa, Brassica napus*, and *Brassica juncea*), and condiments (*Brassica carinata, Brassica nigra*, and *Brassica juncea*) (Dixon, 2007). About 150 GLSs have been characterized in plants (Blažević et al., 2020), of which 17 have been reported in *Brassica* crops (Miao et al., 2017; Wu et al., 2021). GLSs are inactive until they have been hydrolyzed by the enzyme β-thioglucoside glucohydrolase (EC 3.2.3.1, myrosinases) to yield a variety of degradation products like isothiocyanates, thiocyanates, indoles, sinigrin, nitriles, etc. GLSs and their various hydrolyzed products contribute to human/animal health and also act as defensive compounds. They have been found to possess anticarcinogenic (Zhou et al., 2019), anti-inflammatory (Sun et al., 2019), and antimicrobial (Borges et al., 2015) properties. GLSs and their

*Corresponding author: gurpreetkaur@pau.edu

degraded products also defend against generalist and specialized pests (Chhajed et al., 2020). Some of the GLSs, progoitrin, and sinigrin have a negative impact on meal quality (Nambiar et al., 2021). Based on animal studies, several countries have established regulatory limits for rapeseed-mustard varieties with GLSs <30 µmoles/g in defatted meal. The low GLSs genotypes (<30 µmoles/g of defatted meal) along with low erucic acid (<2%) are registered under the trade names 'canola' or '00' quality in rapeseed-mustard. GLSs exhibit potential benefits and certain negative effects, as increasing the content of beneficial GLSs composition and reducing the adverse ones is desirable. Therefore, the levels and profile of GLSs have gained significant attention and are regarded as a valuable trait in Brassica breeding programs. Due to its many biological functions, researchers are actively exploring breeding and molecular genetics techniques to modulate GLSs content, aiming to increase the advantageous GLSs and reduce those with anti-nutritional effects. The information in this chapter will provide valuable insights into GLSs biosynthesis, hydrolysis, and hydrolytic products. Furthermore, it explores various breeding and metabolic engineering techniques to manage the balance of GLSs in Brassica crops effectively.

2. Glucosinolates Biosynthesis

Glucosinolates biosynthesis is a complex process, and using the model plant *Arabidopsis* has significantly contributed to our understanding of its different steps (Sønderby et al., 2010). GLSs biosynthesis is categorized into three distinct phases. The initial elongation phase involves the addition of methylene group to aliphatic and aromatic amino acids. The second phase consists of restructuring amino acid moiety to form the core structure of GLSs followed by its secondary modification in the third phase (Figure 1).

3. Glucosinolates Structure, its Hydrolysis and Hydrolytic Products

Structurally, GLSs are anions composed of thiohydroximates linked with three moieties: a sulfate, β-glucopyranosyl, and a variable side chain derived from amino acid resulting into about 150 different types of GLSs in vegetative tissues and seed of Brassicaceae (Figure 2). Based on precursor amino acids, these compounds are classified into three categories; aliphatic GLSs (from Ala, Ile, Leu, Met, and Val), aromatic GLSs (from Tyr and Phe), and indolic GLSs (from Trp) (Blažević et al., 2020; Sharma et al., 2022).

The biological activities of GLSs primarily come from their hydrolysates produced by β-thioglucosidase enzyme called myrosinase (E.C 3.2.3.1), which is present in specific idioblasts called myrosin cells, a separate compartment from its substrate. In some instances, both may be compartmentalized within a single cell, GLSs being in vacuoles and myrosinases in the cytosol or endoplasmic reticulum

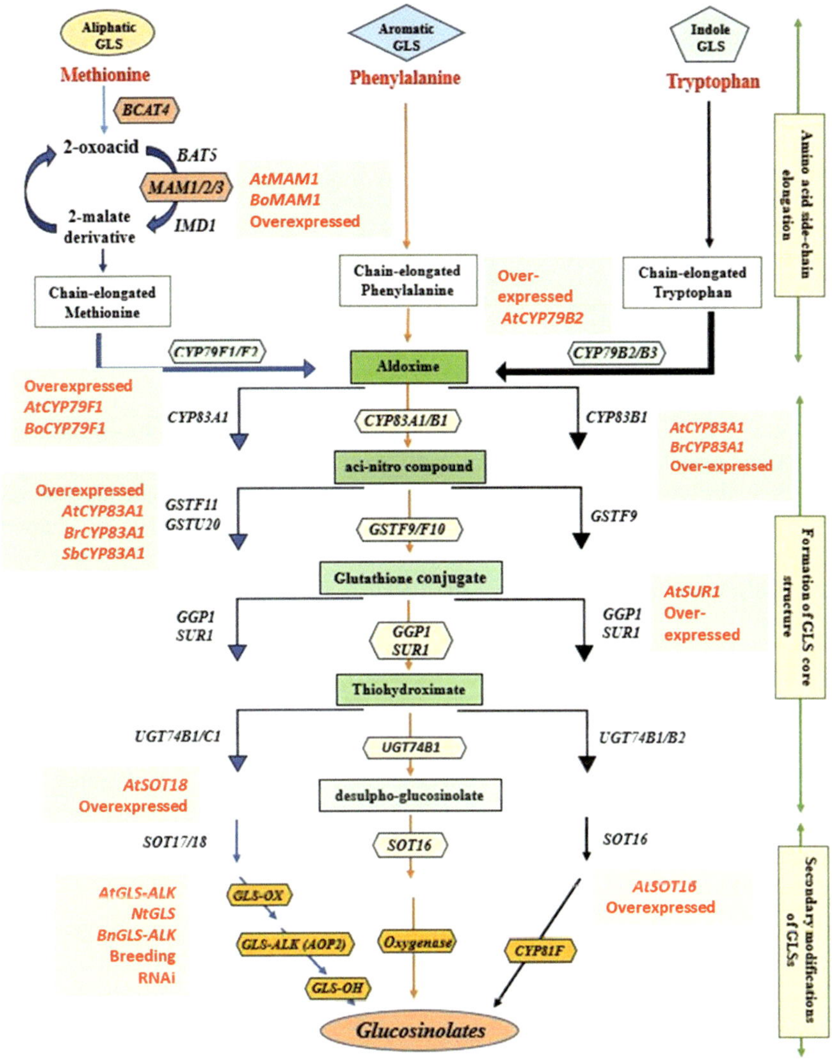

Figure 1: Illustration of glucosinolate pathway manipulation in plants through various techniques.

BCAT4: Branched chain amino acid aminotransferases 4; *BAT5*: Bile acid transporter 5; *MAM 1/2/3*: Methylthioalkylmalate synthase 1; *IMD1*: Isopropylmalate dehydrogenase 1 are involved in methionine side chain elongation; *CYP79F1/F2 & CYP79B2/B3* (Cytochrome P450 family genes): Involved in aldoxime formation; *CYP83A1/B1, GSTF9/10/11 & GSTU20*: Glutathione-S-transferase family genes; *GGP1*: γ-glutamyl peptidase; *SUR1*: Superroot1 *UGT74B1/B2/C1*, UDP-glucosyltransferase genes; *SOT16/17/18*: Sulphotransferases genes are involved in GLSs core formation step; Secondary modifications step involves *GLS-OX*, oxygenase; *GLS-ALK* (*AOP2*): Alkenylhydroxalkyl producing 2; *GLS-OH*: 2-oxoglutarate-dependent dioxygenase.

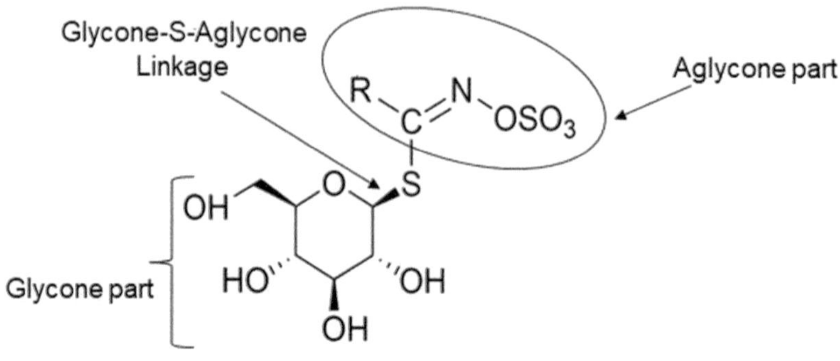

Figure 2: The basic structure GLSs.

(Mitreiter and Gigolashvili, 2021). Plant tissue damage caused by events like wounding, insect, or pathogen attack, activates the GLS-myrosinase system. This process hydrolytically cleaves the thioglycosidic bond in GLSs, producing glucose and an unstable aglycone moiety (thiohydroxamate-o-sulfonate). The breakdown products of GLSs depend on the side chain structure and the presence and type of specifier proteins. Hydrolyzed products of various GLSs are summarized in Table 1.

Table 1: The most abundant GLS of rapeseed-mustard meal: Their names, class, hydrolytic product along with structural formulae

Trivial Name	Systematic Name	Class	Hydrolysis Product
Sinigrin	2-propenyl	Aliphatic	ITC, nitriles
Gluconapin	3-butenyl	Aliphatic	ITC, nitriles
Glucobrassicanapin	4-pentenyl	Aliphatic	ITC, nitriles
Progoitrin	2-hydroxy-3-butenyl	Aliphatic	Oxazolidine-2-thiones
Gluconapoleiferin	2-hydroxy-4-pentenyl	Aliphatic	Oxazolidine-2-thiones
Glucosinalbin	4-hydroxybenzyl	Aromatic	Isothiocyanate
Glucobrassicin	3-indolylmethyl	Indolic	Indolyl-3-carbinol
4-hydroxy glucobrassicin	4-hydroxy-3-indolylmethyl	Indolic	Thiocyanate
4-methoxy glucobrassicin	4-methoxy-3-indolylmethyl	Indolic	Auxins

4. Glucosinolates Profile of Brassica

Glucosinolates are found in different parts of the plants, but their profiles and concentrations vary. Each species has multiple GLS types: four to five major

Table 2: Principal GLSs identified in common Brassica crops

Crop	Aliphatic GLSs										Indolic GLSs				Aromatic GLSs
	3 C side chains			4 C side chains				5 C side chains							
	GVN	GIB	SIN	GER	GRA	GNA	PRO	GAL	GBN	GNL	GBS	4OHGBS	4OMGBS	NGBS	GNST
Brassica rapa															
Chinese cabbage[a]	A	P	P	A	A	P	P	P	P	A	P	A	P	P	P
Turnip[a]	P	P	A	P	P	P	P	P	P	P	P	P	P	P	P
Brassica oleracea															
White cabbage[a]	P	P	P	P	P	P	P	A	P	P	P	P	P	P	P
Red cabbage[a]	P	P	P	P	P	P	P	P	P	P	P	P	P	P	P
Broccoli[a,b]	P	P	P	P	P	P	P	P	P	P	P	P	P	P	P
Cauliflower[a]	P	P	P	P	P	P	P	P	P	A	P	P	P	P	P
Brassica juncea															
Mustard seeds[c]	A	P	P	A	A	P	P	A	A	A	P	P	P	P	P
Brassica napus															
Oilseed rape[a]	A	A	A	P	P	P	P	P	P	P	P	P	P	P	P
Swede[a]	P	P	P	P	P	P	P	P	P	P	P	P	P	P	P

GVN: Glucoiberverine, GIB: Glucoiberin, SIN: Sinigrin, GER: Glucoerucin, GRA: Glucoraphanin, GNA: Gluconapin, PRO: Progoitrin, GAL: Glucoallysin, GBN: Glucobrassicanapin, GNL: Gluconapoleiferin, GBS: Glucobrassicin, 4OHGBS: 4hydroxyglucobrassicin, 4OMGBS: 4methoxyglucobrassicin, NGBS: Neoglucobrassicin, GNST: Gluconasturtiin.

Data Source: [a]Miao et al., 2021; [b]Miao et al., 2017; [c]Zhang et al., 2022.

ones (>80%) and several minor ones (<1%) (Mocniak et al., 2023). GLSs profile can change based on growth stages, plant parts, pest attack, nutrient status, and environment. Reproductive tissues contain 10–40 times higher concentrations than vegetative ones (Clarke, 2010). Numerous reviews and book chapters have discussed the distribution of GLSs in Brassica crops as summarized in Table 2.

5. Glucosinolates Contribution to Animals/Human Health

The dual role of GLSs as both beneficial and detrimental compounds has made them a double-edged sword in nutrition. While they have been shown to have anti-cancer and antiinflammatory properties, their antinutritional effects may lead to deficiencies of iodine, calcium, and iron in animals. Thus, it is recommended to avoid excessive use of feed from Brassicaceae despite being an excellent source of nutrition. Achieving a balance in the intake of GLS rich food with other dietary factors becomes crucial to maximize the benefits while minimizing their negative effects. A detail about the dual role of GLSs is mentioned below.

5.1. Benefits of GLSs

Hydrolyzed products of GLSs have been reported as anticarcinogenic (Zhou et al., 2019). Allyl-isothiocyanates (AITC), benzyl-ITC and sulforaphane suppressed the progression of cell cycle by stimulating the accumulation of reactive oxygen species, downregulating or upregulating different cyclin-Cdk complexes and inducing apoptosis (Grundemann and Huber, 2018). Glucoiberin, glucoraphanin, glucobrassicin, sinigrin, gluconapin, and total indole GLSs are also found to have positive correlation with antioxidant abilities that contribute towards lowering cancer risk. Hydrolytic products such as sulforaphane and indole-3-carbinol, have demonstrated significant anti-inflammation activity (Mohammadi et al., 2018; Sun et al., 2019). Derivatives of sinigrin and gluconasturtiin exhibited antimicrobial properties against both gram-negative and gram-positive bacteria, even at low levels (Borges et al., 2015).

5.2. Adverse Effect on Growth Performance of Animals

Hydrolyzed products of GLSs cause detrimental effects on animals like reduced fertility, thyroid dysfunctions, hepatic and renal dysfunction (Collett et al., 2014; Felker et al., 2016). The formation of DNA adducts has been reported from the degradation product of 1-methoxyindol-3-ylmethyl GLSs (Gerber et al., 2011). Oxazolidine-2-thione is a breakdown product of progoitrin, which causes goitre, stunted growth, inadequate production of eggs, and liver hypertrophy. Goitrin and thiocyanates cause hypothyroidism and goitre in ruminants and poultry (Felker et al., 2016). GLS-derived nitrile and epithionitrile compounds in the Brassica seed meal cause hepatic and renal dysfunction (Collett et al., 2014).

6. Strategies to Manipulate Content and Composition of Glucosinolates

Various strategies, including traditional breeding, molecular approaches, and genetic engineering techniques, have been used to manipulate the content and composition of GLSs. Several processing techniques have also been exploited to modify/reduce the GLSs level.

6.1. Breeding for Reduced Glucosinolates Content

Traditional breeding and selection techniques have been used to develop plant varieties with low GLSs content. In the first phase, efforts have been made to reduce GLSs content to acceptable levels. Genetic variation for GLSs content was very low in the rapeseed and mustard genotypes. However, a low amount of GLSs was found in the Polish *B. napus* cultivar 'Bronowski'. It contained only 10% GLSs compared to traditional genotypes (15 μmol/g defatted meal). The low GLSs content of Bronowski was transferred into low erucic acid germplasm of summer and winter rapeseed through conventional cross-breeding and pedigree selection. This led to release of two high-yielding double low canola quality genotypes named Tower (1973) and Erglu (1974) in Canada and Germany, respectively. In Indian mustard, success has been achieved following the development of the first low GLSs mustard genotype, BJ 1058 (Love et al., 1990). Since then many canola-quality varieties have been developed globally utilizing these sources of low GLSs and erucic acid in Indian mustard (Kaur et al., 2022).

6.2. Mapping for Glucosinolate Content and Composition

Various mapping techniques, including linkage maps, associative transcriptomics, and genome-wide association studies (GWAS), have been used to map genes of GLSs in rapeseed mustard, subsequently developing into markers for marker-assisted selection. Liu et al. (2020a) mapped a major QTL, *qGLS-C2,* on chromosome explaining 30.88–72.87% of the phenotypic variation. This QTL harboured the candidate gene, *BnaC2.MYB28/HAG1*, orthologous to the *AtMYB28,* which is responsible for the synthesis of aliphatic GLSs. The gene *MYB28,* for GLSs content, was also mapped through associative transcriptomics analysis in *B. napus* (Harper et al., 2012; Kittipol et al., 2019). GWAS analysis could map different copies of *AtMYB28* for the low seed GLSs content (Liu et al., 2020b). Ramchiary et al. (2007) could map six 'true' QTLs involved in synthesizing aliphatic GLSs in the F_1DH and BC_4DH populations of *B. juncea.* Subsequently, these QTLs were fine-mapped using a candidate gene approach. Out of which, three, i.e., QTLs, *J2GLS1, J3GLS2,* and *J9GLS3* could be used for marker-assisted selection (Bisht et al., 2009).

6.3. Metabolic Engineering of Glucosinolates

Several approaches, including manipulation of genes involved in the biosynthesis of GLSs, inhibition of transport processes of GLSs, and manipulation of transcription factors involved in the synthesis of GLSs have been employed for engineering the GLSs content.

6.3.1. Engineering of Genes involved in the Biosynthesis of Glucosinolates

Genes, *MAM1, CYP79F1,* and *CYP83A1* are involved in synthesizing aliphatic GLSs. All individual aliphatic GLSs could be increased by overexpression of *CYP83A1*, while 3-butenyl and 4-pentenyl GLSs content could be elevated in the transgenic line of *MAM1* in the Chinese cabbage (Zang et al., 2008b). Similarly, the role of genes, *MAM1* and *CYP83A1* has also been demonstrated in *B. napus* where synthesis of aliphatic GLSs elevated (Zhang et al., 2015). Brassica species, including *B. rapa, B. napus*, and *B. juncea*, have been found to contain high levels of 3-butenyl GLS and progoitrin with low levels of desirable anticarcinogen glucoraphanin. Silencing of *AOP2/GLS-ALK* gene could enhance levels of glucoraphanin in all parts of the *B. juncea* (Augustine and Bisht, 2015) and *B. napus* plant (Liu et al., 2012). Antisense for the gene *AOP2* could enhance the content of glucoraphanin in Chinese Kale (Qian et al., 2015). Genes *CYP79B2, CYP79B3*, and *CYP83B1* are involved in synthesizing indole GLSs. Transgenic for *CYP79B2* or *CYP79B3* along with *CYP83B1* resulted in higher levels of indolic GLSs in Chinese cabbage (Zang et al., 2008a).

6.3.2. Engineering of Genes involved in the Transportation of GLSs

Two transporters, *GTR1* and *GTR2*, engaged in the transport of synthesized GLSs *via* phloem to seeds in *A. thaliana*, might be genetically altered to remove GLSs from seeds while maintaining them in leaves (Nour-Eldin et al., 2012). The double mutant *gtr1 gtr2* of *A. thaliana* failed to accumulate GLSs in its seeds (Nour-Eldin et al., 2012). Subsequently, a mutation in one of seven and four of 12 orthologs of GTR using TILLING helped in achieving a low concentration of GLSs by 60–70% in both *B. rapa* and *B. juncea* (Nour-Eldin et al., 2017). Similarly, low levels of GLSs have been achieved through gene editing of transporter *GTR2* in *B. napus* (He et al., 2022). Gene editing of multiple homologs of *GTR1* and *GTR2* in *B. juncea* have been used to design plants with high GLSs in the leaves with acceptable content of GLSs in the seed while maintaining the defense and yield level (Mann et al., 2023). Other transporters, such as members of the usually multiple amino acids, move in and out transporter (*UMAMIT*) family – *UMAMIT 29, UMAMIT 30*, and *UMAMIT 31* – in *A. thaliana* function as exporters of GLSs. Tripple mutant for *UMAMIT 29*, *UMAMIT 30*, and *UMAMIT 31* accumulated deficient levels of GLSs (Xu et al., 2023).

Table 3: Alteration in GLS content of different Brassica species through processing techniques

S. No.	Target Crop	GLS Content in Control Experiments	Processing Technology	Reference
1.	*Brassica napus*	18.61 μ moles/g in rapeseed meal	Steam explosion using high steam pressure (2.4, 3.2 or 4.0 MPa) at different processing times: 30 sec: >90% reduction of GLSs 150 sec: 99% reduction of GLSs	Gu et al., 2011
2.	*Brassica rapa*	12.99 μ moles/g DW	Different cooking methods: Steam:11.8 μ moles/g DW Boil: 4.66 μ moles/g DW High pressure: 4.64 μ moles/g DW	Francisco et al., 2010
3.	*Brassica juncea*	36.20 ± 0.86 μ moles/g in untreated seeds	Microwave treatment at frequency of 2450 MHz, power 900 W for 6 min; 2 min: 28.04 ± 5.33 μ moles/g; 4 min: 21.93 ± 0.98 μ moles/g; 6 min: 15.00 ± 1.50 μ moles/g (Postharvest yellowing at different temperatures i.e., 20°C, 10°C and 0°C)	Verma et al., 2019
4.	*Brassica oleracea*	Fresh Broccoli (floret) tissue At 20°C: 3.200±0.07 μ moles/g DW At 10°C: 3.201±0.16 μ moles/g DW At 0°C: 3.203±0.079 μ moles/g DW	At 20°C: 2.363±0.052 μ moles/g DW At 10°C: 2.266±0.021 μ moles/g DW At 0°C: 2.579±0.037 μ moles/g DW (Slight yellowing) At 20°C: 1.227±0.016 μ moles/g DW At 10°C: 1.186±0.006 μ moles/g DW At 0°C: 1.275±0.015 μ moles/g DW (Serious yellowing)	Yang et al., 2022

6.3.3. Engineering of Genes Governing Transcription Factors

Transcription factors (TF) are involved in regulating the biosynthesis of both aliphatic and indole GLSs. The biosynthesis of aliphatic GLSs is regulated by TFs, *MYB28, MYB29*, and *MYB76* (Gigolashvili et al., 2007, 2008) whereas, *MYB34/ATR1* (*Altered tryptophan regulation 1*), *MYB51 (HIG1)*, *MYB122 (HIG2)* are associated with biosynthesis of indolic GLSs (Gigolashvili et al., 2007; Frerigmann and Gigolashvili, 2014). The double mutant for *MYB28* and *MYB29* could not synthesize aliphatic GLSs (Gigolashvili et al., 2007; Sønderby et al., 2007) whereas, overexpression of *MYB29* could enhance the accumulation of glucoraphanin and 2-propenyl GLS in *B. oleracea* (Zuluaga et al., 2019). GLS content was also reduced through RNAi of *MYB28* in *B. juncea* (Augustine et al., 2013) and silencing of *MYB28* in *B. juncea* and *B. oleracea* (Augustine and Bisht, 2019; Yin et al., 2017).

6.3.4. Other Techniques to Alter the Composition of Glucosinolates

Traditional and novel food processing techniques have been used to detoxify inherent undesirable GLSs. These include manipulation of temperature, light intensity, and fermentation etc. (Table 3).

7. Conclusion

In light of the growing interest in diet and health, it is necessary to have comprehensive knowledge about GLSs profiles and levels in plants. The studies presented in this chapter offer valuable information for developing promising Brassica varieties with optimal GLSs content, particularly those containing GLSs with beneficial effects on human health. Genetic engineering of key genes involved in GLSs biosynthesis and transport pathway holds great potential to reduce the GLSs content in seed meal. However, it is important to consider the potential implications of genetically modified organisms and address any associated safety concerns. Furthermore, research is required to understand the influence of environmental factors and agronomic practices on GLSs accumulation and hydrolysis. In conclusion, reducing GLSs level in Brassica crops requires a multifaceted approach involving traditional breeding methods, genetic engineering, and optimization of environmental and agronomic practices. This integrated approach holds great potential for enhancing the nutritional value of Brassica foods that benefit both human health and animal productivity.

References

Akbari, E., and M. Namazian. 2020. Sulforaphane: A natural product against reactive oxygen species. *Comput. Theor. Chem.*, 1183: 112850.

Augustine, R., and N.C. Bisht. 2019. Targeted silencing of genes in polyploids: Lessons learned from *Brassica juncea*-glucosinolate system. *Plant Cell Rep.*, 38: 51–57.
Augustine, R., and N.C. Bisht. 2015. Biotic elicitors and mechanical damage modulate glucosinolate accumulation by coordinated interplay of glucosinolate biosynthesis regulators in polyploid Brassica juncea. *Phytochemistry*, 117: 43–50.
Augustine, R., A. Mukhopadhyay, and N.C. Bisht. 2013. Targeted silencing of BjMYB28 transcription factor gene directs development of low glucosinolate lines in oilseed *Brassica juncea. Plant Biotechnol. J.*, 11(7): 855–866.
Baasanjav-Gerber, C., B.H. Monien, I. Mewis, M. Schreiner, J. Barillari, R. Iori, and H. Glatt. 2011. Identification of glucosinolate congeners able to form DNA adducts and to induce mutations upon activation by myrosinase. *Mol. Nutr. Food Res.*, 55(5): 783–792.
Bisht, N.C., V. Gupta, N. Ramchiary, Y.S. Sodhi, A. Mukhopadhyay, N. Arumugam, and A.K. Pradhan. 2009. Fine mapping of loci involved with glucosinolate biosynthesis in oilseed mustard (*Brassica juncea*) using genomic information from allied species. *Theor. Appl. Genet.*, 118: 413–421.
Blažević, I., S. Montaut, F. Burčul, C.E. Olsen, M. Burow, P. Rollin, and N. Agerbirk. 2020. Glucosinolate structural diversity, identification, chemical synthesis, and metabolism in plants. *Phytochemistry*, 169: 112100.
Borges, A., A.C. Abreu, C. Ferreira, M.J. Saavedra, L.C. Simões, and M. Simões. 2015. Antibacterial activity and mode of action of selected glucosinolate hydrolysis products against bacterial pathogens. *J. Food Sci. Technol.*, 52: 4737–4748.
Chhajed, S., I. Mostafa, Y. He, M. Abou-Hashem, M. El-Domiaty, and S. Chen. 2020. Glucosinolate biosynthesis and the glucosinolate–myrosinase system in plant defence. *Agronomy*, 10(11): 1786.
Clarke, D.B. 2010. Glucosinolates, structures, and analysis in food. *Anal. Methods*, 2(4): 310–325.
Collett, M.G., B.L. Stegelmeier, and B.A. Tapper. 2014. Could nitrile derivatives of turnip (*Brassica rapa*) glucosinolates be hepato- or cholangiotoxic in cattle? *Journal of Agricultural and Food Chemistry*, 62(30): 7370–7375.
Dixon, G.R. 2007. *Vegetable Brassicas and Related Crucifers* (No. 14). CABI. UK.
Felker, P., R. Bunch, and A.M. Leung. 2016. Concentrations of thiocyanate and goitrin in human plasma, their precursor concentrations in brassica vegetables, and associated potential risk for hypothyroidism. *Nutr. Rev.*, 74(4): 248–258.
Francisco, M., P. Velasco, D.A. Moreno, C. García-Viguera, and M.E. Cartea. 2010. Cooking methods of *Brassica rapa* affect the preservation of glucosinolates, phenolics and vitamin C. *Food Res. Int.*, 43(5): 1455–1463.
Frerigmann, H., and T. Gigolashvili. 2014. MYB34, MYB51, and MYB122 distinctly regulate indolic glucosinolate biosynthesis in Arabidopsis thaliana. *Mol. Plant*, 7(5): 814–828.
Gigolashvili, T., M. Engqvist, R. Yatusevich, C. Müller, and U.I. Flügge. 2008. HAG2/MYB76 and HAG3/MYB29 exert a specific and coordinated control on the regulation of aliphatic glucosinolate biosynthesis in Arabidopsis thaliana. *New Phytol.*, 177(3): 627–642.
Gigolashvili, T., R. Yatusevich, B. Berger, C. Müller, and U.I. Flügge. 2007. The R2R3-MYB transcription factor HAG1/MYB28 is a regulator of methionine-derived glucosinolate biosynthesis in Arabidopsis thaliana. *Plant J.*, 51(2): 247–261.

Gründemann, C., and R. Huber. 2018. Chemoprevention with isothiocyanates: From bench to bedside. *Cancer Lett.*, 414: 26–33.

Gu, X., W. Dong, and Y. He. 2011. Detoxification of rapeseed meals by steam explosion. *J. Am. Oil Chem. Soc.*, 88: 1831–1838.

Harper, A.L., M. Trick, J. Higgins, F. Fraser, L. Clissold, R. Wells, and I. Bancroft. 2012. Associative transcriptomics of traits in the polyploid crop species *Brassica napus*. *Nat. Biotechnol.*, 30(8): 798–802.

He, Y., Z. Yang, M. Tang, Q.Y. Yang, Y. Zhang, and S. Liu. 2022. Enhancing canola breeding by editing a glucosinolate transporter gene lacking natural variation. *Plant Physiol.*, 188(4): 1848–1851.

Kaur, G., S. Sharma, S. Langyan, J. Kaur, P. Yadava, and S.S. Banga. 2022. Advanced breeding for oil and oil cake quality in *Brassica juncea*. *The Brassica juncea Genome*, 413–438. Cham: Springer International Publishing.

Kittipol, V., Z. He, L. Wang, T. Doheny-Adams, S. Langer, and I. Bancroft. 2019. Genetic architecture of glucosinolate variation in *Brassica napus*. *J. Plant Physiol.*, 240: 152988.

Liu, Y., X. Zhou, M. Yan, P. Wang, H. Wang, Q. Xin, L. Yang, D. Hong, and G. Yang. 2020a. Fine mapping and candidate gene analysis of a seed glucosinolate content QTL, qGSL-C2, in rapeseed (*Brassica napus* L.). *Theor. Appl. Genet.*, 133: 479–490.

Liu, S., H. Huang, X. Yi, Y. Zhang, Q. Yang, C. Zhang, C. Fan, and Y. Zhou. 2020b. Dissection of genetic architecture for glucosinolate accumulations in leaves and seeds of *Brassica napus* by genome-wide association study. *Plant Biotechnol. J.*, 18(6): 1472–1484.

Liu, Z., A.H. Hirani, P.B.E. McVetty, F. Daayf, C.F. Quiros, and G. Li. 2012. Reducing progoitrin and enriching glucoraphanin in *Brassica napus* seeds through silencing of the GSL-ALK gene family. *Plant Mol. Biol.*, 79: 179–189.

Love, H.K., G. Rakow, J.P. Raney, and R.K. Downey. 1990. Development of low glucosinolate mustard. *Can. J. Plant Sci.*, 70(2): 419–424.

Mann, A., J. Kumari, R. Kumar, P. Kumar, A.K. Pradhan, D. Pental, and N.C. Bisht. 2023. Targeted editing of multiple homologues of GTR1 and GTR2 genes provides the ideal low-seed, high-leaf glucosinolate oilseed mustard with uncompromised defence and yield. *Plant Biotechnol. J.*, 21(11): 2182–2195.

Miao, H., W. Zeng, J. Wang, F. Zhang, B. Sun, and Q. Wang. 2021. Improvement of glucosinolates by metabolic engineering in Brassica crops. *aBIOTECH*, 2(3): 314–329.

Miao, H., J. Wang, C. Cai, J. Chang, Y. Zhao, and Q. Wang. 2017. Accumulation of glucosinolates in broccoli. *Glucosinolates*, 133–162. Cham: Springer.

Mitreiter, S., and T. Gigolashvili. 2021. Regulation of glucosinolate biosynthesis. *J. Exp. Bot.*, 72(1): 70–91.

Mocniak, L.E., K.R. Elkin, S.L. Dillard, R.B. Bryant, and K.J. Soder. 2023. Building comprehensive glucosinolate profiles for brassica varieties. *Talanta*, 251: 123814.

Mohammadi, S., A. Memarian, S. Sedighi, N. Behnampour, and Y. Yazdani. 2018. Immunoregulatory effects of indole-3-carbinol on monocyte-derived macrophages in systemic lupus erythematosus: A crucial role for aryl hydrocarbon receptor. *Autoimmunity*, 51(5): 199–209.

Nambiar, D.M., J. Kumari, R. Augustine, P. Kumar, P.K. Bajpai, and N.C. Bisht. 2021. GTR1 and GTR2 transporters differentially regulate tissue-specific glucosinolate contents and defence responses in the oilseed crop *Brassica juncea*. *Plant, Plant Cell Environ.*, 44(8): 2729–2743.

Nour-Eldin, H.H., S.R. Madsen, S. Engelen, M.E. Jørgensen, C.E. Olsen, J.S. Andersen, and B.A. Halkier. 2017. Reduction of antinutritional glucosinolates in Brassica oilseeds by mutation of genes encoding transporters. *Nat. Biotechnol.*, 35(4): 377–382.

Nour-Eldin, H.H., T.G. Andersen, M. Burow, S.R. Madsen, M.E. Jørgensen, C.E. Olsen, I. Dreyer, R. Hedrich, D. Geiger, and B.A. Halkier. 2012. NRT/PTR transporters are essential for translocation of glucosinolate defence compounds to seeds. *Nature*, 488(7412): 531–534.

Qian, H., B. Sun, H. Miao, C. Cai, C. Xu, and Q. Wang. 2015. Variation of glucosinolates and quinone reductase activity among different varieties of Chinese kale and improvement of glucoraphanin by metabolic engineering. *Food Chem.*, 168: 321–326.

Ramchiary, N., N.C. Bisht, V. Gupta, A. Mukhopadhyay, N. Arumugam, Y.S. Sodhi, D. Pental, and A.K. Pradhan. 2007. QTL analysis reveals context-dependent loci for seed glucosinolate trait in the oilseed *Brassica juncea*: Importance of recurrent selection backcross scheme for the identification of 'true' QTL. *Theor. Appl. Genet.*, 116: 77–85.

Sharma, S., M. Bala, G. Kaur, S. Tayyab, and S.R. Feroz. 2022. Chemical composition of oil and cake of *Brassica juncea*: Implications on human and animal health. *The Brassica juncea Genome*, 29–55. Cham: Springer International Publishing.

Sønderby, I.E., M. Burow, H.C. Rowe, D.J. Kliebenstein, and B.A. Halkier. 2010. A complex interplay of three R2R3 MYB transcription factors determines the profile of aliphatic glucosinolates in Arabidopsis. *Plant Physiology*, 153(1): 348–363.

Sønderby, I.E., B.G. Hansen, N. Bjarnholt, C. Ticconi, B.A. Halkier, and D.J. Kliebenstein. 2007. A systems biology approach identifies a R2R3 MYB gene subfamily with distinct and overlapping functions in regulation of aliphatic glucosinolates. *PLoS One*, 2(12): e1322.

Sun, B., Y.X. Tian, Q. Chen, Y. Zhang, Y. Luo, X.R. Wang, F. Zhang, and H.R. Tang. 2019. Variations in the glucosinolates of the individual edible parts of three-stem mustards (*Brassica juncea*). *R. Soc. Open Sci.*, 6(2): 182054.

Verma, A., A. Sharma, P.K. Rai, and D. Kumar. 2019. Effect of microwave pre-treatment on quality parameters in Indian mustard. *J. Food Sci. Technol.*, 56: 4956–4965.

Wu, X., F. Chen, X. Zhao, C. Pang, R. Shi, C. Liu, C. Sun, W. Zhang, X. Wang, and J. Zhang. 2021. QTL mapping and GWAS reveal the genetic mechanism controlling soluble solids content in *Brassica napus* shoots. *Foods*, 10(10): 2400.

Xu, D., N.C.H. Sanden, L.L. Hansen, Z.M. Belew, S.R. Madsen, L. Meyer, M.E. Jørgensen, P. Hunziker, D. Veres, C. Crocoll, and A. Schulz. 2023. Export of defensive glucosinolates is key for their accumulation in seeds. *Nature*, 1–7.

Yang, Q., Q. Zhou, X. Zhou, H. Fang, Y. Zhao, B. Wei, and S. Ji. 2022. Insights into profiling of glucosinolates and genes involved in its metabolic pathway accompanying post-harvest yellowing of broccoli. *Postharvest Biol. Technol.*, 185: 111780.

Yin, L., H. Chen, B. Cao, J. Lei, and G. Chen. 2017. Molecular characterization of MYB28 involved in aliphatic glucosinolate biosynthesis in Chinese kale (*Brassica oleracea* var. *alboglabra* Bailey). *Front. Plant Sci.*, 8: 1083.

Zang, Y.X., J.H. Kim, Y.D. Park, D.H. Kim, and S.B. Hong. 2008a. Metabolic engineering of aliphatic glucosinolates in Chinese cabbage plants expressing Arabidopsis MAM1, CYP79F1, and CYP83A1. *BMB Rep.*, 41(6): 472–478.

Zang, Y.X., M.H. Lim, B.S. Park, S.B. Hong, and D.H. Kim. 2008b. Metabolic engineering of indole glucosinolates in Chinese cabbage plants by expression of Arabidopsis CYP79B2, CYP79B3, and CYP83B1. *Mol. Cells*, 25(2).

Zhou, J.W., M. Wang, N.X. Sun, Y. Qing, T.F. Yin, C. Li, and D. Wu. 2019. Sulforaphane-induced epigenetic regulation of Nrf2 expression by DNA methyltransferase in human Caco-2 cells. *Oncol. Lett.*, 18(3): 2639–2647.

Zuluaga, D.L., N.S. Graham, A. Klinder, A.E. van Ommen Kloeke, A.R. Marcotrigiano, C. Wagstaff, R. Verkerk, G. Sonnante, and M.G. Aarts. 2019. Overexpression of the MYB29 transcription factor affects aliphatic glucosinolate synthesis in *Brassica oleracea*. *Plant Mol. Biol.*, 101: 65–79.

CHAPTER
16

Lectins: Antinutrients or Nutrients – Unveiling Their Dual Role

Manjeet Kaur Sangha[a*], Diksha Singla[b] and Manju Bala[c]

[a] Department of Biochemistry, Punjab Agricultural University, Ludhiana, Punjab, India

[b] Advanced Post-Harvest Technology Center, Centre for Applied Research, Innovation and Entrepreneurship, Lethbridge College, AB T1K 1L6, Alberta, Canada, 0000-0001-5188-748X

[c] ICAR-Central Institute of Post-Harvest Engineering and Technology, Ludhiana, Punjab, India

1. Introduction

Lectins are ubiquitous glycoproteins or carbohydrate-binding proteins. They are also known as agglutinins/hemagglutinins and present in all living organisms, from bacteria to human cells. According to the origin, lectins are classified into five categories: algal, bacterial, fungal, plant, and animal (Figure 1). They are further categorized based on structure and specification. Based on characterization and carbohydrate-binding capacity, lectins have been used to purify various glycoproteins (Mishra et al., 2019). Due to carbohydrate recognition domain (CRD) and reversible binding capacity, they can bind to any glycosyl ligand without affecting the covalent structure (André et al., 2015).

Based on the synthesis of lectins, they are categorized into classical and stress-inducible. These cannot be differentiated based on their three-dimensional structure. The synthesis and localization of classical lectins follow the signal secretory route pathway from the endoplasmic reticulum to the Golgi apparatus and finally to the vacuole. In contrast, in some cases, lectins are synthesized and located in the cytoplasm. They sometimes can enter the nucleus (Van Damme et al., 2008), called "nucleocytoplasmic lectins" (Lannoo et al., 2010).

*Corresponding author: manjeet_kaursangha@pau.edu

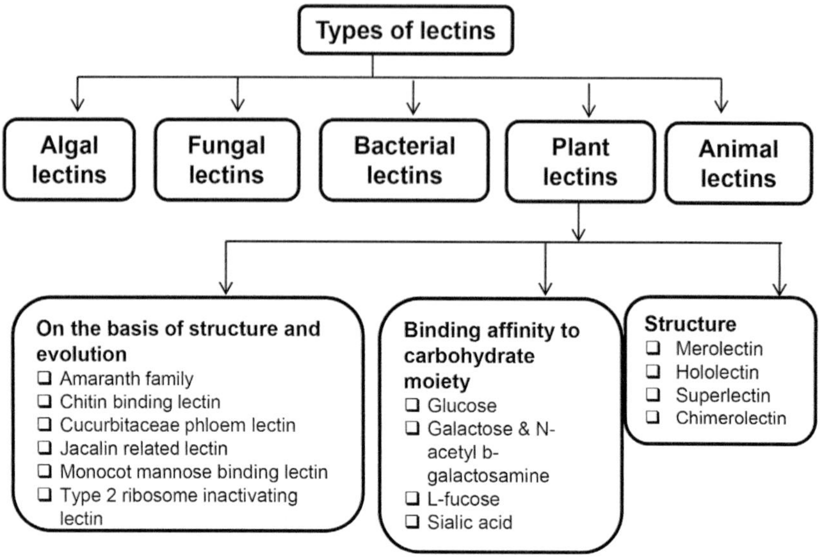

Figure 1: Classification of lectins.

2. Distribution of Lectins Based on Crop

The distribution of lectins in different crops can vary depending on the specific plant species and other factors such as growing conditions and environmental factors. Examples of different types of lectins found in various crops are wheat germ agglutinin (WGA) (van Buul and Bronus, 2014), soybean agglutinin (SBA) (Pan et al., 2023), concanavalin A (Con A) (Liu et al., 2022), peanut agglutinin (PNA) (Zhao et al., 2014), *Ulex europaeus* agglutinin (UEA) (Ordonez et al., 1987).

3. Biological Activities of Lectins in Humans

Lectins, multifunctional proteins found in various organisms, play a diverse role in biological activities. They can promote cell adhesion and aggregation by binding to specific carbohydrates on cell surfaces, crucial for processes like embryonic development, wound healing, and immune responses. Lectins also possess immunomodulatory powers, influencing immune cells and cytokine production, either suppressing or stimulating inflammatory responses. Some lectins exhibit antiviral and antibacterial properties by blocking pathogens from entering host cells, activating the immune system for pathogen eradication. These proteins can stimulate cell division and proliferation, particularly in immune cells like lymphocytes and macrophages, while enhancing macrophage phagocytic activity. Certain lectins, such as ricin and abrin, possess potent cytotoxic effects, inhibiting protein synthesis and inducing cell death. Hemagglutination, another property, enables lectins to clump red blood cells, serving applications in blood typing, pathogen detection, and glycoprotein purification. Remarkably, lectins also display anti-cancer activity, inducing apoptosis in cancer cells, as seen with wheat germ agglutinin and other lectins. In the realm of gut health, some lectins act as prebiotics, promoting the growth of beneficial gut bacteria that produce short-chain fatty acids with numerous health benefits.

4. Biological Activities of Lectins in Plant Stress Tolerance

Lectins have multiple applications in plant stress tolerance (Table 1). They exhibit anti-insect properties by impeding insect development and increasing mortality. This is achieved by altering enzyme activity in larvae, with esterase activity rising while acid phosphatase and alkaline phosphatase activities decline (Marothia et al., 2023). Genetically engineered lectins in crops like wheat, rice,

Table 1: Biological activities of lectin

Biological Activity	Source of Lectin	Reference
Antiviral	*Lablab purpureus* (L.) Sweet	Liu et al., 2020
	Triticum aestivum L.	Auth et al., 2021
	Nicotiana tabacum	Gordts et al., 2015
	Urtica dioica	Gordts et al., 2015
	Tamarindus indica	Kaur et al., 2019
	Musa acuminata Colla	Chan et al., 2022
	Momordica balsamina lectin	Powell and Gbodossou, 2020
Antiinflammatory	*Andira anthelmia*	Do Nascimento et al., 2021
	Parkia playcephala	Madhu and Sharada, 2019
	Machaerium acutifolium Vogel	Santos et al., 2019
	Tetracarpidium conophorum	Oladokun et al., 2019
	Schinus terebinthifolia	Ramos et al., 2020
	Bryothamnion triquetrum	Fontenelle et al., 2018
Antifungal	*Machaerium acutifolium* Vogel	Dias et al., 2020
	Moringa oleifera Lam.	Santos et al., 2020
	Portulaca elatior Mart. ex Rohrb	Silva et al., 2019
	Helianthus annuus L.	Del Rio et al., 2019
	Phaseolus vulgaris L.	Ang et al., 2014
	Canavalia brasiliensis Benth.	Klafke et al., 2013
	Mucuna pruriens (L.) DC.	Klafke et al., 2013

tobacco, and potatoes offer potential solutions for integrated pest management. Moreover, lectins possess anti-fungal abilities, indirectly hindering fungal growth by binding to carbohydrates on the fungal cell wall and disrupting chitin synthesis. Certain plant lectins, including hevein and chitin-binding polypeptides, inhibit fungal growth, while chimerolectins, like class I chitinases, possess fungicidal properties due to their catalytic domain (Lam and Ng, 2011).

5. Harmful Effects of Lectins in Humans

Although lectins are naturally present in many plant-based foods, they can pose certain potential drawbacks for human health. Digestive issues such as bloating, gas, diarrhea, and stomach cramps can arise in some individuals after consuming lectin-rich foods, as lectins have the ability to bind to the gut lining, potentially leading to inflammation and damage to the digestive tract, particularly impacting those with underlying digestive disorders like irritable bowel syndrome (IBS), inflammatory bowel disease (IBD), or leaky gut syndrome. Additionally, lectins may disrupt nutrient absorption in the digestive tract, for instance, certain lectins like PHA in raw kidney beans can inhibit the absorption of essential minerals such as iron and zinc. Lectin allergies are a concern for those with pre-existing food allergies, as individuals allergic to peanuts, soy, or gluten may also react to lectins found in beans, legumes, and grains, with symptoms ranging from mild discomfort to severe anaphylaxis. Furthermore, lectin-carbohydrate interactions can trigger immune responses that lead to inflammation and cell damage, potentially contributing to autoimmune diseases such as celiac disease and non-celiac gluten sensitivity. Antinutritional effects of lectins involve inhibiting digestive enzymes and interfering with nutrient absorption, occasionally causing gastrointestinal symptoms like nausea, vomiting, and diarrhea. Additionally, lectins can disrupt the effectiveness of medications by binding to them in the gut, possibly impacting insulin sensitivity and the balance of the gut microbiome. Overall, while lectins are a natural part of many plant-based foods, their potential adverse effects warrant consideration, especially for individuals with specific health conditions or sensitivities.

6. Conventional Methods of Lectin Reduction

Several conventional methods exist for reducing lectin levels in foods. Heat-based techniques, including cooking, soaking, sprouting, and fermenting, are effective in denaturing and breaking down certain lectins, rendering them more digestible. Prolonged boiling can significantly reduce lectin content in beans and legumes, while soaking nuts, beans, and legumes in water for several hours or overnight can aid in lectin removal (Thompson et al., 1983; Kumari, 2018). Enzyme treatments, such as using α-galactosidase to break down complex carbohydrates, can also diminish lectin levels in plant-based foods. Lectin-binding proteins, produced through genetic engineering or isolated from natural sources, can neutralize lectins. Additionally, fermentation, facilitated by microbes like Lactobacillus and Bifidobacterium, produces enzymes that effectively break down lectins (Reddy and Pierson, 1994). Peeling or removing the skins of certain plant-based foods, like potatoes, tomatoes, and peppers, which often contain high lectin content, before consumption or cooking can mitigate lectin exposure. Maintaining a diverse and balanced diet, encompassing various plant-based foods, can also help minimize the potential adverse effects of lectins by avoiding excessive exposure to any single type of lectin.

7. Genetic Manipulation of Lectins in Plants

Biological interventions offer a versatile approach to modulating lectin levels in plants, catering to specific objectives (Figure 2), as outlined below.

7.1. Transgenic Approach

The method of lectin biofortification in plants through genetic modification entails the creation of transgenic plants capable of producing and accumulating lectins. These proteins play a pivotal role in enhancing both the nutritional profile and resistance to insect pests, presenting an eco-friendly alternative to conventional pesticide use. Notably, mannose-binding plant lectins, such as *Galanthus nivalis* (snowdrop) agglutinin (GNA) and *Allium sativum* (garlic) leaf agglutinin (ASAL), have been extensively investigated for their efficacy in pest control. The introduction of GNA into transgenic lines of various crops, including grapevine, oilseed rape, potato, sweet potato, sugarcane, sunflower, rice, tobacco, and tomato, has proven effective in bolstering resistance against both Homopteran and Lepidopteran pests (Mi et al., 2017; Bharathi et al., 2008; Yarasi et al., 2008; Liu et al., 2015; Schuler et al., 1998). Transgenic rice plants expressing GNA have exhibited resistance to the rice brown planthopper (Rao et al., 1998), while ASAL-expressing transgenic rice has demonstrated high-level resistance against significant sap-sucking pests such as the brown planthopper. Potato plants engineered to produce Con A have displayed resistance to aphids and nematode pests (Griffiths, 2000). Extensive research has been conducted on lectin receptor-like kinases, referred to as chimeric lectins, for their potential to confer tolerance to various stresses (Liu et al., 2015). The locus *Bph3*, encompassing a cluster of three genes encoding lectin receptor kinases (*OsLecRK1-OsLecRK3*), has been instrumental in enhancing resistance to both brown and white-backed planthoppers when introgressed into susceptible rice varieties via transgenic strategies (Liu et al., 2015).

7.2. Breeding Approach

In the context of soybeans, researchers have harnessed traditional breeding techniques to develop cultivars with reduced lectin content (Pull et al., 1978; Staswick et al., 1983). For instance, the cultivar 'Enrei' was meticulously bred

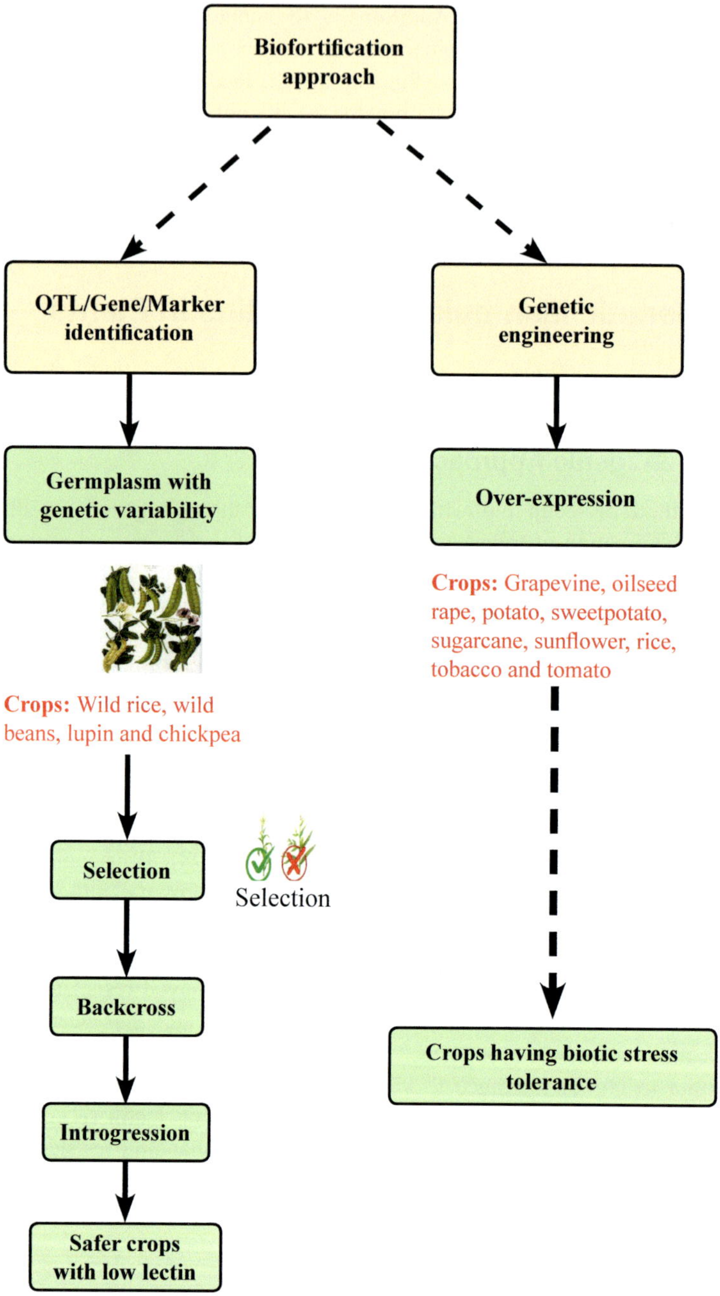

Figure 2: Diverse biofortification approaches to enhance lectin content in different crops.

over several generations to incorporate desired traits. Some wild type genotypes naturally produce lower amounts of lectins than their cultivated counterparts. Wild rice (*Zizania* spp.), wild beans, lupin (*Lupinus* spp.), lentil (*Lens culinaris* subsp. *orientalis*), and chickpea (*Cicer reticulatum*) contain lower levels of lectins than cultivated varieties (Koul et al., 2022). Researchers succeeded in transferring the genetic trait of lectin reduction from wild bean accessions into various bean cultivars (Campion et al., 2019). Marker-assisted breeding of the *Bph3* locus has yielded lines with enhanced resistance to both brown and white-backed planthoppers. Combining GLA and ASAL genes through marker-assisted breeding has led to the development of rice lines with superior pest resistance compared to individual gene-carrying transgenic lines (Bharathi et al., 2018).

8. Conclusion

Lectins are plant-based antinutrients with both potential health benefits, like anti-cancer and anti-inflammatory properties, and drawbacks, such as digestive issues and interference with nutrient absorption. Individual factors, including gut health and microbiome composition, influence how lectins affect our well-being. Proper preparation, like soaking, cooking, and sprouting, can enhance their digestibility. Lectin expression has been enhanced in several transgenic crops for insect and pest resistance. Breeding has been carried out to reduce their contents in edible parts of plant. Future perspectives on lectins involve genetic engineering and novel processing to reduce antinutrient effects. Personalized dietary recommendations based on individual lectin tolerance, potential treatments for autoimmune diseases, and enhancing health benefits through prebiotic lectin-rich functional foods are on the horizon. Biotechnological techniques for reducing lectin content in plants are still in the research stage. This chapter serves as a valuable resource for researchers, breeders, and nutritionists interested in improving the nutritional profile of plant foods and addressing antinutritional challenges associated with lectins.

References

André, S., H. Kaltner, J.C. Manning, P.V. Murphy, and H.J. Gabius. 2015. Lectins: Getting familiar with translators of the sugar code. *Molecules*, 20: 1788–1823.

Ang, A.S., R.C. Cheung, X. Dan, Y.S. Chan, W. Pan, and T.B. Ng. 2014. Purification and characterization of a glucosamine-binding antifungal lectin from *Phaseolus vulgaris cv.* Chinese pinto beans with anti-proliferative activity towards nasopharyngeal carcinoma cells. *Appl. Biochem. Biotechnol.*, 172(2): 672–686.

Auth, J., M. Fröba, M. Große, P. Rauch, N. Ruetalo, M. Schindler, M. Morokutti-Kurz, P. Graf, A. Dolischka, E. Prieschl-Grassauer, C. Setz, and U. Schubert. 2021. Lectin from *Triticum vulgaris* (WGA) inhibits infection with SARS-CoV-2 and its variants of concern alpha and beta. *International Int. J. Mol. Sci.*, 22(19): 10205.

Bharathi, Y., V.D. Reddy, K.V. Rao, and I.C. Pasalu. 2018. Effect of mannose specific lectins ASAL and GNA on the feeding behavior of BPH (*Nilaparvata lugens stal.*). *Int. J. Curr. Microbiol. App. Sci.*, 7(9): 1426–1436.

Campion, B., D. Perrone, I. Galasso, and R. Bollini. 2009. Common bean (*Phaseolus vulgaris* L.) lines devoid of major lectin proteins. *Plant Breed.*, 128(2): 199–204.

da Silva, J.D.F., S.P. da Silva, P.M. da Silva, A.M. Vieira, L.C.C. de Araújo, T. de Albuquerque Lima, A.P.S. de Oliveira, L.V. do Nascimento Carvalho, M.G. da Rocha Pitta, M.J.B. de Melo Rêgo, and I.O. Pinheiro. 2019. *Portulaca elatior* root contains a trehalose-binding lectin with antibacterial and antifungal activities. *Int. J. Biol. Macromol.*, 126: 291–297.

Chan, J.F., Y.J. Oh, S. Yuan, H. Chu, M.L. Yeung, D. Canena, C.C. Chan, V.K. Poon, C.C. Chan, A.J. Zhang, J.P. Cai, Z.W. Ye, L. Wen, T.T. Yuen, K.K. Chik, H. Shuai, Y. Wang, Y. Hou, C. Luo, W.M. Chan, Z. Qin, K.Y. Sit, W.K. Au, M. Legendre, R. Zhu, L. Hain, H. Seferovic, R. Tampé, K.K. To, K.H. Chan, D.G. Thomas, M. Klausberger, C. Xu, J.J. Moon, J. Stadlmann, J.M. Penninger, C. Oostenbrink, P. Hinterdorfer, K.Y. Yuen, and D.M. Markovitz. 2022. A molecularly engineered, broad-spectrum anti-coronavirus lectin inhibits SARS-CoV-2 and MERS-CoV infection *in vivo*. *Cell Rep. Med.*, 18: 3(10): 100774.

Del Rio, M., L. de la Canal, M. Pinedo, H.M. Mora-Montes, and M. Regente. 2019. Effects of the binding of a *Helianthus annuus* lectin to Candida albicans cell wall on biofilm development and adhesion to host cells. *Phytomedicine*, 58: 152875.

Dias, L.P., A.L. Santos, N.M. Araújo, R.R. Silva, M.H. Santos, R.R. Roma, B.A. Rocha, J.T. Oliveira, and C.S. Teixeira. 2020. *Machaerium acutifolium* lectin alters membrane structure and induces ROS production in *Candida parapsilosis*. *Int. J. Biol. Macromol.*, 163: 19–25.

Do Nascimento, F.L.F., A. de Freitas Pires, M.R.L. Mota, P.H.C. Isaias, D.F. de Araujo, M.G. de Queiroz Martins, C.G. Moreira, J.B. Cajazeiras, B.S. Cavada, K.S. do Nascimento, and A.M.S. Assreuy. 2021. The anti-inflammatory effect of andira anthelmia lectin in rats involves inhibition of the prostanoid pathway, TNF-a and lectin domain. *Mol. Biol. Rep.*, 49(9): 8847–8857.

Fontenelle, T.P.C., G.C. Lima, J.X. Mesquita, J.L.d.S. Lopes, T.V. de Brito, F.D.C. Vieira Júnior, A.B. Sales, K.S. Aragão, M.H.L.P. Souza, A.L.D.R. Barbosa, and A.L.P. Freitas. 2018. Lectin obtained from the red seaweed Bryothamnion triquetrum: Secondary structure and anti-inflammatory activity in mice. *Int. J. Biol. Macromol.*, 112: 1122–1130.

Gordts, S.C., M. Renders, G. Férir, D. Huskens, E.J.M. Van Damme, W. Peumans, J. Balzarin, and D. Schols. 2015. NICTABA and UDA, two GlcNAc-binding lectins with unique antiviral activity profiles. *J. Antimicrob. Chemother.*, 70(6): 1674–1685.

Griffiths, B.S., I.E. Geoghegan, and W.M. Robertson. 2000. Testing genetically engineered potato, producing the lectins GNA and Con A, on non-target soil organisms and processes. *J. Appl. Ecol.*, 37(1): 159–170.

Kaur, R., R. Mudgal, J. Jose, P. Kumar, and S. Tomar. 2019. Glycan-dependent chikungunya viral infection divulged by antiviral activity of NAG specific chi-like lectin. *Virol. J.*, 526: 91–98.

Klafke, G.B., G.M.S.G. Moreira, L.G. Monte, J.L. Pereira, T.M. Brandolt, M.O. Xavier, T. Santi-Gadelha, O.A. Dellagostin, and L. da Silva Pinto. 2013. Assessment of plant lectin antifungal potential against yeasts of major importance in medical mycology. *Mycopathologia*, 175: 147–151.

Koul, B., K. Sharma, V. Sehgal, D. Yadav, M. Mishra, and C. Bharadwaj. 2022. Chickpea (*Cicer arietinum* L.) biology and biotechnology: From domestication to biofortification and biopharming. *Plants*, 11(21): 2926.

Kumari, S. 2018. *The Effect of Soaking Almonds and Hazelnuts on Phytate and Mineral Concentrations*. Doctoral dissertation, University of Otago.

Lam, S.K., and T.B. Ng. 2011. Lectins: Production and practical applications. *Appl. Microbiol. Biotechnol.*, 89: 45–55.

Lannoo, N., and E.J.M. Van Damme. 2010. Nucleocytoplasmic plant lectins. *Biochim. Biophys. Acta*, 1800: 190–201.

Liu, Y., H. Hao, and T. Hou. 2022. Concanavalin A-induced autoimmune hepatitis model in mice: Mechanisms and future outlook. *Open Life Sci.*, 17(1): 91–101.

Liu, Y., H. Wu, H. Chen, Y. Liu, J. He, H. Kang, Z. Sun, G. Pan, Q. Wang, J. Hu, and F. Zhou. 2015. A gene cluster encoding lectin receptor kinases confers broad-spectrum and durable-insect resistance in rice. *Nat. Biotechnol.*, 33(3): 301–305.

Liu, Y.-M., M.D. Shahed-Al-Mahmud, X. Chen, T.-H. Chen, K.-S. Liao, J.M. Lo, Y.-M. Wu, M.-C. Ho, C.-Y. Wu, C.-H. Wong, J.-T. Jan, and C. Ma. 2020. A carbohydrate binding protein from the edible lablab beans effectively blocks the infections of influenza viruses and SARS-CoV-2. *Cell Rep.*, 32(6): 108016.

Madhu, C.S., and A.C. Sharada. 2019. Anti-inflammatory activity of partially purified lectin from *Praecitrullus fistulosus* phloem exudates. *Asian J. Pharm. Clin. Res.*, 12(1): 91–94.

Marothia, D., N. Kaur, C. Jhamat, I. Sharma, and P.K. Pati. 2023. Plant lectins: Classical molecules with emerging roles in stress tolerance. *Int. J. Biol. Macromol.*, 244: 125272.

Mi, X., X. Liu, H. Yan, L. Liang, X. Zhou, J. Yang, H. Si, and N. Zhang. 2017. Expression of the *Galanthus nivalis* agglutinin (GNA) gene in transgenic potato plants confers resistance to aphids. *C.R. Biol.*, 340(1): 7–12.

Mishra, A., A. Behura, S. Mawatwal, A. Kumar, L. Naik, S.S. Mohanty, D. Manna, P. Dokania, A. Mishra, S.K. Patra, and R. Dhiman. 2019. Structure-function and application of plant lectins in disease biology and immunity. *Food Chem. Toxicol.*, 134: 110827.

Oladokun, B.O., O.N. Omisore, O.A. Osukoya, and A. Kuku. 2019. Anti-nociceptive and anti-inflammatory activities of *Tetracarpidium conophorum* seed lectin. *Sci. Afr.*, 3: e00073.

Ordóñez, N.G., T. Brooks, S. Thompson, and J.G. Batsakis. 1987. Use of Ulex europaeus agglutinin I in the identification of lymphatic and blood vessel invasion in previously stained microscopic slides. *Am. J. Surg. Pathol.*, 11(7): 543–550.

Pan, L., T.E.C. Xu, X. Fan, J. Xia, Y. Liu, J. Liu, J. Zhao, N. Bao, Y. Zhao, H. Sun, G. Qin, and M.H. Farouk. 2023. The apoptotic effects of soybean agglutinin were induced through three different signal pathways by down-regulating cytoskeleton proteins in IPEC-J2 cells. *Sci. Rep.*, 13(1): 5753.

Powell, M.D., and E.V.A. Gbodossou. 2020. *Antiviral Compositions and Methods* (United States). WO2020132062A1.

Pull, S.P., S.G. Pueppke, T. Hymowitz, and J.H. Orf. 1978. Soybean lines lacking the 120,000-dalton seed lectin. *Science*, 200(4347): 1277-1279.

Ramos, D.d.B.M., M.T.d.M.F. Araújo, T.C.d.L. Araújo, Y.A. Silva, A.C.L.A. dos Santos, M.G. e Silva, P.M.G. Paiva, R.L. Mendes, and T.H. Napoleão. 2020. Antinociceptive activity of *Schinus terebinthifolia* leaf lectin (SteLL) in sarcoma 180-bearing mice. *J. Ethnopharmacol.*, 259: 112952.

Rao, K.V., K.S. Rathore, T.K. Hodges, X. Fu, E. Stoger, D. Sudhakar, ... and J.A. Gatehouse. 1998. Expression of snowdrop lectin (GNA) in transgenic rice plants confers resistance to rice brown planthopper. *Plant J.*, 15(4): 469–477.

Reddy, Neethu, and M.D. Pierson. 1994. Reduction in antinutritional and toxic components in plant foods by fermentation. *Food Res. Int.*, 27: 281–290.

Santos, L.M.M., P.M. Silva, M.C. Moura, A.R. Carvalho Junior, P.K. Amorim, T.F. Procópio, L.C.B.B. Coelho, L.C.N. Silva, P.M.G. Paiva, N.D.L. Santos, and T.H. Napoleão. 2021. Anti-Candida activity of the water-soluble lectin from Moringa oleifera seeds (WSMoL). *J. Mycol. Med.*, 31(2): 101074.

Santos, A.L.E., G.O. Leite, R.F. Carneiro, R.R. Roma, V.F. Santos, M.H.C. Santos, R.O. Pereira, R.C. Silva, C.S. Nagano, A.H. Sampaio, B.A.M. Rocha, P. Delatorre, A.R. Campos, and C.S. Teixeira. 2019. Purification and biophysical characterization of a mannose/N-acetyl-d-glucosamine-specific lectin from *Machaerium acutifolium* and its effect on inhibition of orofacial pain via TRPV1 receptor. *Arch. Biochem. Biophys.*, 664: 149–156.

Schuler, T.H., G.M. Poppy, B.R. Kerry, and I. Denholm. 1998. Insect-resistant transgenic plants. *Trends Biotechnol.*, 16(4): 168–175.

Staswick, P., and M.J. Chrispeels. 1983. Expression of lectin genes during seed development in normal and phytohemagglutinin-deficient cultivars of *Phaseulus vulgaris*. *J. Mol. Appl. Genet.*, 2: 525–535.

Thompson, Lilian U., L. Rea, Ramona, and J.A. Jenkins David. 1983. Effect of heat processing on hemagglutinin activity in red kidney beans. *J. Food Sci.*, 48: 235–236.

Van Buul, V.J., and F.J. Brouns. 2014. Health effects of wheat lectins: A review. *J. Cereal Sci.*, 59(2): 112–117.

Van Damme, E.J., N. Lannoo, and W.J. Peumans. 2008. Plant lectins. *Adv. Bot. Res.*, 48: 107–209.

Yarasi, B., V. Sadumpati, C.P. Immanni, D.R. Vudem, and V.R. Khareedu. 2008. Transgenic rice expressing *Allium sativum* leaf agglutinin (ASAL) exhibits high-level resistance against major sap-sucking pests. *BMC Plant Biol.*, 8: 102.

Zhao, Q., C.A. Duckworth, W. Wang, X. Guo, H. Barrow, D.M. Pritchard, J.M. Rhodes, and L.G. Yu. 2014. Peanut agglutinin appearance in the blood circulation after peanut ingestion mimics the action of endogenous galectin-3 to promote metastasis by interaction with cancer-associated MUC1. *Carcinogenesis*, 35(12): 2815–2821.

CHAPTER

17

Tackling Challenges, Seizing Future Opportunities in Biofortification

Himanshu Sharma, Aman Kumar and Parveen Chhuneja*[0000-0002-8599-9479]

School of Agricultural Biotechnology, Punjab Agricultural University, Ludhiana, Punjab, India

1. Introduction

Biofortified crops have immense potential to enhance nutritional quality of foods for malnourished population. Undoubtedly, biofortification through existing methodologies provides micronutrients to vulnerable populations in a cost-effective way (Qaim et al., 2007; Meenakshi et al., 2010) but biofortification of food crops faces various challenges. These challenges are related to the breeding approaches, agronomic practices, consumer acceptance, and regulatory approaches. Dealing with such challenges is a tedious task. We need to address challenges to overcome situations related to malnutrition and hidden hunger. Addressing hidden hunger through cost-effective breeding techniques (Gautam et al., 2021) requires overcoming various breeding challenges. Maintaining genetic stability in biofortified crops over an extended period is a key challenge. Additionally, these crops must exhibit resistance to both biotic and abiotic stresses (Koul et al., 2012).

Agronomic studies, while essential for developing biofortified crops adapted to specific climates and local practices, present their own challenges. Monitoring soil conditions, including temperature, pH, and nutrient availability, is critical. Identifying regions with nutrient deficiencies and tailoring crops to different demographic groups is crucial, though developing multi-nutrient biofortified crops catering to population heterogeneity poses a challenge (Nuthalapati et al., 2022). This approach enhances consumer health and simplifies diets, potentially reducing

*Corresponding author: pchhuneja@pau.edu

the cost of nutrient-rich crops. Yet, boosting nutrient content is insufficient; nutrient bioavailability factors must also be considered.

Given the financial constraints of farmers and consumers in developing countries, maintaining crop yields builds trust in biofortified crops. Effective policies and regulatory frameworks are essential for handling and distributing these crops. Raising awareness about biofortified crops ensures consistent demand and reinforces farmers' trust that their income won't be compromised (Rodas-Moya et al., 2022). Considering all aspects, future opportunities along with challenges in context to biofortified crops have been discussed in the current chapter. Future opportunities that can be used for the development of biofortified food crops are discussed in detail. In addition, the policies related to the outreach of biofortified crops to consumers are also discussed. This includes the awareness of plant scientists, farmers, suppliers, and consumers regarding collaborations, partnerships, and precision agriculture practices. Figure 1 highlights different challenges and opportunities and the same have been discussed in the following sections.

Figure 1: Challenges in nutrient biofortification of food crops.

2. Addressing Limited Genetic Diversity and Pre-breeding Challenges

Biofortified crops play a crucial role in enhancing nutritional value, but they must meet specific criteria, including high yields, disease resistance at target locations, and good grain quality (Welch and Graham, 2004). Conventional approaches are viable only when they can augment nutrient levels in food crops without compromising productivity and health benefits. However, modern food crops exhibit limited genetic diversity compared to landraces and wild ancestors (Connorton and Balk, 2019). Major staples like wheat, rice, and maize suffer from restricted genetic variability in micronutrient content (Zhao et al., 2009; Brar et al., 2011; Prasanna et al., 2020), hindering meaningful grain improvements. For instance, modern wheat contains less iron than older cultivars, and contemporary rice varieties have lower iron levels compared to traditional ones (Fan et al., 2008; Anandan et al., 2011). Overcoming this genetic diversity challenge necessitates exploring secondary and tertiary gene pools for biofortification.

The utilization of wild relatives in breeding programs offers untapped potential for crop biofortification. Many crop species possess vast unexplored wild germplasm, presenting opportunities for future biofortified varieties. Wild wheat species like *Triticum monococcum*, *Triticum boeoticum*, *Triticum dicoccoides*, and *Aegilops tauschii* exhibit a broader range of grain micronutrient density (Cakmak et al., 2000; Chhuneja et al., 2006; Ortiz-Monasterio and Graham, 2000). Researchers have initiated broad hybridization programs, developing amphiploids by crossing Triticum durum with Aegilops tauschii to transfer desired genes from the latter to bread wheat. Synthetic wheat varieties with significant variation in grain micronutrient content have been created (Calderini and Ortiz-Monasterio, 2003). Additionally, diversity panels and multiparent advanced generation inter-cross (MAGIC) populations have gained popularity as genetic resources for breeding lines with desirable traits. MAGIC plus populations, offering a broader genetic background, hold promise (Bandillo et al., 2013) and can be leveraged for biofortification. These resources facilitate QTL analysis and gene discovery with high mapping resolution.

However, utilizing secondary and tertiary gene pools presents numerous challenges, including crossability barriers, chromosome pairing issues, and linkage drag (Kumar et al., 2016; Sharma et al., 2021). Crossability barriers restrict the use of wide gene pools and can impact embryo fertility across generations (Ladizinsky and Abbo, 1993). Developing germplasm from these pools is time-consuming and complex, requiring rigorous screening and selection methods. Identifying specific genes responsible for nutrient traits and incorporating them into biofortification programs also poses challenges. Genomic segments from wild species may lead to poor agronomic traits, such as reduced yield, increased plant height, and subpar threshing characteristics. This often results in a trade-off between micronutrient content and yield, known as the "dilution effect" (Connorton and Balk, 2019). While utilizing wild germplasm for biofortification presents challenges, the potential benefits in terms of enhanced nutritional value and resilience to changing environmental conditions make it a compelling avenue for crop improvement. Realizing the full potential of wide germplasm demands the skilful integration of advanced genetic and breeding techniques.

3. Complex Genetic Control Influences Biofortification

Accumulation of micronutrients in specific parts of plants is a complex process which is controlled by a variety of processes such as nutrient uptake from soil, assimilation, and transformation within tissues and accumulation in sink. There are specific transporters involved in each step which are regulated by complex genetic mechanisms (Brumbarova et al., 2015; Connorton et al., 2017). The biggest challenge while developing biofortified crops is to channel the nutrients in specific tissues. Other challenges include enhancing nutrient uptake, facilitation of nutrient distribution, and increasing storage capacity of grains. The genetic mechanisms that regulate these processes are quite complex and not fully understood. The uptake of micronutrients requires a plethora of genes (Brumbarova et al., 2015) and identification of related genes to incorporate them into plants for biofortification is a difficult task (Connorton and Balk, 2019). *Ferritin* transporters are reported to be responsible for transport of iron in rice and wheat (Rehman et al., 2021). Iron biofortification in rice has been achieved by expressing combination of genes including *Ferritin, Nicotinamine Synthase (NAS),* and other associated genes (Boonyaves et al., 2016; Trijatmiko et al., 2016). Genes available in roots of crop plants are also prime factors involved in the biofortification. Ihemere, (2012) has established the role of *FEA1* (*Iron assimilatory* 1) gene in the starchy roots of Casava for iron transport. Vacuolar transport of micronutrients is influenced by the *VIT1* (Vacuolar iron transporter 1) gene (Narayanan et al., 2015). Phytosiderophores are involved in chelating micronutrients around rhizosphere and facilitating micronutrient uptake. Increased release of phytosiderophores in rhizosphere can result in increased uptake of micronutrients from the soil (Masuda et al., 2009; Lee et al., 2012). Understanding the mechanisms controlling phytosiderophore production in crops can offer new opportunities in enhancing the micronutrient uptake in plants. In cereal crops, the most important target in biofortification is increasing the micronutrient content in endosperm (Vasconcelos et al., 2017). Most of the micronutrients such as Fe, Zn, and anthocyanins in cereal crops are accumulated in the outer pericarp and aleurone layers (Garg et al., 2016; Singh et al., 2013). Identifying nutrient specific transporters and their regulatory mechanisms in facilitating movement from aleurone to endosperm will offer great opportunities to increase the micronutrient levels in grains.

4. Multinutrient Biofortification

Traditional breeding methods and recent molecular breeding techniques such as marker assisted selection have resulted in development of many biofortified varieties. Currently more than 340 varieties of 11 staple crops are being cultivated in more than 40 countries (Bouis, 2018). In India, currently 46 varieties of biofortified crops belonging to 7 crops have been released (HarvestPlus, 2023). Most of the biofortification programmes were designed solely for specific limiting nutrients such as Fe, Zn, provitamin A, lysine, and tryptophan (Bouis, 2018). However, the micronutrient deficiencies in majority of the population are complex. Especially in poor and developing areas, the lack of food diversity means that the micronutrient deficiencies cannot be overcome by a few biofortified crops. Therefore, the biggest challenge and opportunity in biofortification is stacking nutritional quality traits in a single cultivar. The future of biofortification lies in the development and implementation of strategies that target the enhancement of multiple essential nutrients simultaneously (Athar et al., 2020). Recent advances in biotechnology and genetic engineering offer opportunities for simultaneous enhancement of multiple nutrients in food crops (Van Der Straeten et al., 2020). By identifying and manipulating key genes involved in the synthesis and accumulation of various nutrients, crops with enhanced nutritional profiles can be bred.

5. Bioavailability of Nutrients in Food Crops

A significant challenge to fully harnessing the human health benefits of biofortified crops is the limited bioavailability of nutrients. Over the past decade, both conventional breeding and transgenic techniques have been employed to improve the contents of provitamin A, Fe, and Zn in various staple food crops (HarvestPlus, 2023). But the fraction of micronutrients that is bioavailable to be absorbed by human body remains a challenge. Elevated micronutrient levels in food do not necessarily ensure efficient absorption within the body. Numerous *in vitro* and *in vivo* studies have been conducted to investigate the factors that influence the bioavailability of micronutrients in crops (Bechoff et al., 2017). Antinutrients are major factors that affect the bioavailability of nutrients in host body therefore biofortified crops need to be developed in a manner to reduce the antinutrients. The common antinutritional factors are phytates, lectins, glucosinolates, oxalates, or polyphenols in food crops (House, 1999; La Frano et al., 2014). Such compounds are able to chelate micronutrients like iron and zinc to form insoluble complexes in gastrointestinal tracts, resulting in decreasing the bioavailability of the nutrients. Even a small reduction in antinutritional factors can improve bioavailability of nutrients significantly. But devising strategies to reduce the content of antinutrients in biofortified crops has remained a challenge since the mechanisms of production and genetic control of antinutrient elements in many crops are not completely understood.

Many antinutritional factors are integral part of the plant metabolism and provide resistance to plants from biotic and abiotic stresses (Bouis et al., 2010). Therefore, completely eliminating them from plants can have adverse effects on their basic physiology and survival. Finding a correct balance between the content of antinutritional factors and nutrients in food is also a big challenge. Antinutrient factors such as phytic acid cannot be completely eliminated from crops but strategies to lower their content can help in increasing the bioavailable nutrients. Phytic acid content in food crops can be reduced by identifying and targeting candidate genes. Using RNAi, *Inositol pentakisphosphate kinase-1* (*IPK1*) was studied in wheat (Aggarwal et al., 2018) and rice (Ali et al., 2013). Additionally, biomolecules such as ascorbic acid, cysteine can affect the bioavailability of iron (Bouis et al., 2010; La Frano et al., 2014). Emergence of new genome editing technologies such as CRISPR/Cas9 system have opened new opportunities in eliminating or reducing the content of these antinutrient factors to acceptable limits. This approach has been successful in rice, beans, and maize where significant reduction of phytic acid was observed (Shi et al., 2003; Panzeri et al., 2011; Pandey et al., 2021). Various promoters have been identified to increase the bioavailable micronutrients but due to limited knowledge about the genetic variability in food crops, effective increase in bioavailable nutrients is still a challenge (La Frano et al., 2014).

6. Agronomic Challenges for Biofortification of Food Crops

Agronomic parameters play a crucial role in the biofortification of food crops. There are several challenges related to agronomic practices for the development and utilization of biofortified crops. The prime challenge is to enhance the nutritional content by improving and maintaining the overall crop productivity. In addition to this, biofortified crops should have the ability to resist pathogens. The combination of disease resistance with nutritional quality is a complex task (Portilla et al., 2022). Biofortified food crops are also expected to adapt to diverse environmental conditions including air temperature, moisture availability, soil types, etc. Maintaining the nutritional status of crops under different agro-climatic conditions is a challenge (Sharma et al., 2023).

It is hard to use appropriate agronomic-related practices for biofortified crops including the appropriate irrigation methods, nutrient management, pest control strategies, and harvesting practices (Yadav et al., 2023). Agronomic management of food crops for seed production, harvesting, storage, and access is necessary to facilitate the widespread cultivation of biofortified varieties (De Valença et al., 2017). Agronomists also face challenges to design and handle biofortified crops with ideal maturation period, lodging resistance, and harvest index. Such challenges can be overruled by providing the site-specific guidelines for the development of food crops.

Moreover, grain nutrient content is also affected by the availability of nutrients in rhizosphere and affected by the clay content, ion exchange capacity, soil organic matter content, and soil pH (Marschner, 2012; Nguyen and Marschner, 2017). It was evident from studies that achievement of biofortified grains was also affected by the soil nutrient deficiency (Manyevere et al., 2017; Reza et al., 2017; Shukla et al., 2017). This challenge can be tackled by providing the suitable levels of nutrients in the soil during reproductive stage of plant development (Cakmak and Kutman, 2018). Physiochemical properties and nutrient content in soil are determining factors for the grain micronutrient content (Wissuwa et al., 2008). In addition, foliar spray in early grain-filling stage can be helpful in alleviating the grain micronutrient content.

7. Variable Geographic Demands and Consumer Awareness

Nutrient deficiencies in different populations can vary across geographic regions (Harding et al., 2018). Therefore, focus on the nutrient needs and preferences of local communities is required; this can lead to regionally adapted biofortified crops. Several studies were conducted in developing countries to upgrade the nutritional status of population using biofortified crops. Based on data and analytics, efforts have been made to provide biofortified varieties to developing countries.

Consumer awareness plays a crucial role in the successful utilization of biofortified food crops (Birol et al., 2015). These crops, enriched with enhanced levels of essential nutrients, have the potential to address malnutrition for the improvement of public health. Limited knowledge and awareness about biofortified crops among consumers can hinder their acceptance. Many consumers may not be familiar with the concept of biofortification and its potential benefits. Educational campaigns and information dissemination are crucial for creating awareness and improving acceptance (Pixley et al., 2013). To handle region-specific challenges, development of the targeted crop can be a solution. QPM-based maize crop was targeted for developing countries including African countries to provide nutritional benefits to the affected population. Development of QPM-based multinutrient maize with a high level of tryptophan is beneficial for a targeted population including infants and small children. QPM maize has a relatively higher vitamin B_3 content and bioavailability due to high tryptophan and low leucine content (Goredema-Matongera et al., 2021).

Also the sensory attributes of biofortified crops, including taste, texture, and overall quality, can impact consumer acceptance. Maintaining quality parameters for biofortified grains is a challenge. Consumers may be reluctant to biofortified crops with inferior taste or texture compared to traditional varieties (De Groote et al., 2014). Therefore, it is important to develop biofortified varieties that maintain or enhance sensory attributes while providing increased nutritional value.

Additionally, cultural and dietary preferences influence consumer acceptance of new food crops (Razzaq et al., 2021). Traditional food choices and cultural practices may limit the acceptance of biofortified crops that are unfamiliar or do not align with existing dietary habits. Incorporating biofortified crops into familiar dishes and considering cultural preferences can help increase acceptance.

8. Challenges to Successful Integration of Biofortified Crops in the Supply Chain

The main challenge to successfully integrate newly developed biofortified crops in the supply chain is acceptability by farmers. The initial investment required for purchasing bio-fortified seeds, implementing appropriate agronomic practices, and maintaining the genetic purity of these crops can be a barrier for many small-scale and resource-constrained farmers. Financial support, in the form of subsidies, grants, or access to affordable credit, can alleviate this financial burden and incentivize farmers to adopt biofortified crops. Incentivizing seed purchases can play a pivotal role in motivating farmers to transition to biofortified crop varieties. This can be achieved through subsidies, reduced seed costs, or bulk-buying discounts. By reducing the financial barriers associated with adopting new crops, farmers are more likely to experiment with biofortified varieties.

The opportunities in this area include equipping farmers with comprehensive knowledge about the nutritional benefits of these crops and their potential to improve both household diets and income. Awareness campaigns, workshops, and training sessions tailored to local contexts can effectively disseminate this information and build confidence among farmers. Moreover, showcasing successful biofortified crop cultivation through field demonstrations and the recognition of local champions can inspire and motivate other farmers to follow suit. Additionally, offering a guaranteed premium price through a minimum support price for biofortified grains in the market can incentivize farmers to cultivate more biofortified crops. It can offset the initial costs of adopting these crops, including investment in quality seeds and agronomic practices, encouraging more farmers to embrace biofortified varieties. Supporting farmers in cultivating biofortified crops is in harmony with wider global objectives related to sustainable agriculture, public health, and poverty reduction. In this context, financial support to farmers becomes a crucial tool not only for improving nutrition but also for advancing agricultural sustainability and social well-being.

9. Channelling Biofortified Crops to Safeguard Nutritional Integrity

Safeguarding the nutritional integrity of biofortified crops and distinguishing them in the market from other crops is another challenge in successful integration of biofortification in the system. To solve this purpose, dedicated channels are required for biofortified crops. Separate channelling of biofortified crops will involve establishing distinct supply chains and distribution channels specifically dedicated to biofortified crop varieties. Establishing a robust value-chain network that links seed multiplication centres, seed sellers, and farmers will be the foundational step. This network will ensure a seamless flow of high-quality biofortified seeds. The first crucial step in popularizing the biofortified varieties will involve reinforcing the seed-supply chain to ensure the production and distribution of high-quality seeds. Preserving genetic purity is of paramount importance to maintain the integrity of these quality traits, necessitating the identification of dedicated seed production areas. By creating dedicated channels, it will become possible to maintain the identity and quality of biofortified crops from production through to consumption, reducing the risk of contamination or dilution with conventional crops. This approach will also include proper labelling and marketing of biofortified products, allowing consumers to easily identify and access these nutritious options. As a consequence, separate channelling will help build consumer trust and generate awareness about the importance of biofortified crops in addressing malnutrition and improving public health.

10. Conclusion

Biofortified crops hold immense promise for improving dietary nutrition in regions plagued by malnutrition and nutrient deficiencies. This chapter delves into the challenges and potential pathways for advancing nutrient biofortification in food crops. Key challenges include limited genetic diversity, complex genetic control of biofortification traits, and the presence of antinutrient factors, all of which must be addressed during crop development. Agronomic practices and environmental factors can also significantly affect crop micronutrient content. Traditional breeding methods have limitations in targeting specific genes and biofortification parameters, necessitating advanced technological interventions. Advancements in genetic and breeding technologies offer solutions to these hurdles. Programmes focused on developing biofortified varieties should consider varying nutrient deficiencies across different regions and incentivize farmers to cultivate these crops. Creating distinct supply chains for biofortified products can maintain genetic purity. Public-private partnerships, from breeding to product distribution, are a promising approach. Raising consumer awareness and acceptance is a final challenge that can be tackled through awareness campaigns and quality maintenance. While biofortification faces challenges, the future holds substantial opportunities. Collaborative efforts among scientists, policymakers, and farmers can lead to sustainable adoption of biofortified varieties, ultimately improving the nutritional status of affected populations.

References

Aggarwal, S., A. Kumar, K.K. Bhati, G. Kaur, V. Shukla, S. Tiwari, and A.K. Pandey. 2018. RNAi-mediated downregulation of inositol Pentakisphosphate Kinase (IPK1) in wheat grains decreases Phytic acid levels and increases Fe and Zn accumulation. *Front. Plant Sci.*, 9: 259.

Ali, N., S. Paul, D. Gayen, S.N. Sarkar, K. Datta, and S.K. Datta. 2013. Development of low phytate rice by RNAi mediated seed-specific silencing of inositol 1,3,4,5,6-pentakisphosphate 2-kinase gene (IPK1). *PLoS One*, 8(7): e68161.

Anandan, A., G. Rajiv, R. Eswaran, and M. Prakash. 2011. Genotypic variation and relationships between quality traits and trace elements in traditional and improved rice (*Oryza sativa* L.) genotypes. *J. Food Sci.*, 76(4): H122–H130.

Athar, T., M.K. Khan, A. Pandey, F.G. Yilmaz, M. Hamurcu, E.E. Hakki, and S. Gezgin. 2020. Biofortification and the involved modern approaches. *J. Elem.*, 25(2): 717–731.

Bandillo, N., C. Raghavan, P.A. Muyco, M.A.L. Sevilla, I.T. Lobina, C.J. Dilla-Ermita, C.W. Tung, S. McCouch, M. Thomson, R. Mauleon, and R.K. Singh. 2013. Multi-parent advanced generation inter-cross (MAGIC) populations in rice: Progress and potential for genetics research and breeding. *Rice*, 6(1): 1–15.

Bechoff, A., and C. Dhuique-Mayer. 2017. Factors influencing micronutrient bioavailability in biofortified crops. *Annals of the New York Academy of Sciences*, 1390(1): 74–87.

Birol, E., J.V. Meenakshi, A. Oparinde, S. Perez, and K. Tomlins. 2015. Developing country consumers' acceptance of biofortified foods: A synthesis. *Food Security*, 7: 555–568.

Boonyaves, K., W. Gruissem, and N.K. Bhullar. 2016. NOD promoter controlled AtIRT1 expression functions synergistically with NAS and FERRITIN genes to increase iron in rice grains. *Plant Mol. Biol.*, 90: 207–215.

Bouis, H. 2018. Reducing mineral and vitamin deficiencies through biofortification: Progress under HarvestPlus. *Hidden Hunger: Strategies to Improve Nutrition Quality*, 118: 112–122. Karger Publishers.

Bouis, H.E., and R.M. Welch. 2010. Biofortification: A sustainable agricultural strategy for reducing micronutrient malnutrition in the global South. *Crop Sci.*, 50: S20–S32.

Brar, B., S. Jain, R. Singh, and R.K. Jain. 2011. Genetic diversity for iron and zinc contents in a collection of 220 rice (*Oryza sativa* L.) genotypes. *Indian J. Genet.*, 71(01): 67–73.

Brumbarova, T., P. Bauer, and R. Ivanov. 2015 Molecular mechanisms governing Arabidopsis iron uptake. *Trends Plant Sci.*, 20: 124–133.

Cakmak, I., and U.B. Kutman. 2018. Agronomic biofortifcation of cereals with zinc: A review. *Soil Sci.*, 69: 172–180.

Cakmak, I., H. Ozkan, H.J. Braun, R.M. Welch, and V. Romheld. 2000. Zinc and iron concentrations in seeds of wild, primitive, and modern wheats. *Food Nutr. Bull.*, 21: 401–403.

Calderini, D.F., and I. Ortiz-Monasterio. 2003. Grain position affects macro and micronutrient grain concentrations in cultivars and synthetic hexaploid wheats. *Crop Sci.*, 43: 141–151.

Chhuneja, P., H.S. Dhaliwal, N.S. Bains, and K. Singh. 2006. Aegilops kotschyi and Aegilops tauschii as sources for higher levels of grain iron and zinc. *Plant Breed.*, 125: 529–531.

Connorton, J.M., and J. Balk. 2019. Iron biofortification of staple crops: Lessons and challenges in plant genetics. *Plant Cell Physiol.*, 60(7): 1447–1456.

De Groote, H., C.K. Chege, K. Tomlins, and N.S. Gunaratna. 2014. Combining experimental auctions with a modified home-use test to assess rural consumers' acceptance of quality protein maize, a biofortified crop. *Food Qual.*, 3: 13.

De Valença, A.W., A. Bake, I.D. Brouwer, and K.E. Giller. 2017. Agronomic biofortification of crops to fight hidden hunger in sub-Saharan Africa. *Glob. Food Secur.*, 12: 8–14.

Fan, M.S., F.J. Zhao, S.J. Fairweather-Tait, P.R. Poulton, S.J. Dunham, and S.P. McGrath. 2008. Evidence of decreasing mineral density in wheat grain over the last 160 years. *J. Trace Elem. Med. Biol.*, 22: 315–324.

Garg, M., M. Chawla, V. Chunduri, R. Kumar, S. Sharma, N.K. Sharma, and S.P. Singh. 2016. Transfer of grain colours to elite wheat cultivars and their characterization. *J Cereal Sci.*, 71: 138–144.

Gautam, D., B. Shrestha, and B. Subedi. 2021. Biofortified maize: Cornerstone in plant breeding to combat hidden hunger in developing countries. *Arch. Agri. Environ. Sci.*, 6(1): 100–107.

Gemede, H.F., and N. Ratta. 2014. Antinutritional factors in plant foods: Potential health benefits and adverse effects. *International Journal of Nutrition and Food Sciences*, 3(4): 284–289.

Goredema-Matongera, N., T. Ndhlela, C. Magorokosho, C.N. Kamutando, A. van Biljon, and M. Labuschagne. 2021. Multinutrient biofortification of maize (*Zea mays* L.) in Africa: Current status, opportunities, and limitations. *Nutrients*, 13(3): 1039.

Harding, K.L., V.M. Aguayo, W.A. Masters, and P. Webb. 2018. Education and micronutrient deficiencies: An ecological study exploring interactions between women's schooling and children's micronutrient status. *BMC Public Health*, 18(1): 1–13.

HarvestPlus. 2023. *Database of Biofortified Crops Released*. Biocropshplus. https://bcr.harvestplus.org.

House La, W.A. 1999. Trace element bioavailability as exemplified by iron and zinc. *Field Crops Res.*, 60: 115–141.

Ihemere, U.E., N.N. Narayanan, and R.T. Sayre. 2012. Iron biofortification and homeostasis in transgenic cassava roots expressing the algal iron assimilatory gene, FEA1. *Front. Plant Sci.*, 3: 171.

Koul, B., K. Sharma, V. Sehgal, D. Yadav, M. Mishra, and C. Bharadwaj. 2022. Chickpea (*Cicer arietinum* L.) Biology and biotechnology: From domestication to biofortification and biopharming. *Plants*, 11(21): 2926.

Kumar, J., R. Kant, S.M. Quatadah, S. Kumar, and A. Sarker. 2016. Biofortification through breeding interventions in lentil. *Biofortification of Food Crops*, 147–156.

La Frano, M.R., F.F. de Moura, E. Boy, B. Lönnerdal, and B.J. Burri. 2014. Bioavailability of iron, zinc, and provitamin A-carotenoids in biofortified staple crops. *Nutr. Rev.*, 72(5): 289–307.

Ladizinsky, G., and S. Abbo. 1993. Cryptic speciation in Lens culinaris. *Genet. Resour. Crop Evol.*, 40: 1–5.

Lee, S., Y.S. Kim, U.S. Jeon, Y.K. Kim, and J.K. Schjoerring. 2012. Activation of rice nicotianamine synthase 2 (OsNAS2) enhances iron availability for biofortification. *Mol. Cells*, 33: 269–275.

Manyevere, A., P. Muchaonyerwa, P.N.S. Mnkeni, and I. Dhau. 2017. Spatial variability of selected soil micronutrients under smallholder crop production in Zanyokwe, eastern cape, South Africa. *South Afr. J. Plant Soil*, 34: 339–349.

Marschner, P. 2012. Rhizosphere biology, Chapter 15. *In: Mineral Nutrition of Higher Plants*, 369–388. 3rd Edn. Academic Press, London.

Masuda, H., K. Usuda, T. Kobayashi, Y. Ishimaru, Y. Kakei, M. Takahashi, K. Higuchi, H. Nakanishi, S. Mori, and N.K. Nishizawa. 2009. Overexpression of the barley nicotianamine synthase gene HvNAS1 increases iron and zinc concentrations in rice grains. *Rice*, 2: 155–166.

Meenakshi, J.V., N.L. Johnson, V.M. Manyong, H. DeGroote, J. Javelosa, D.R. Yanggen, ... and E. Meng. 2010. How cost-effective is biofortification in combating micronutrient malnutrition? An ex-ante assessment. *World Development*, 38(1): 64–75.

Narayanan, N., G. Beyene, R.D. Chauhan, E. Gaitán-Solis, M.A. Grusak, N. Taylor, and P. Anderson. 2015. Overexpression of Arabidopsis VIT1 increases accumulation of iron in cassava roots and stems. *Plant Sci.*, 240: 170–181.

Nguyen, T.T., and P. Marschner. 2017. Soil respiration, microbial biomass, and nutrient availability in soil after addition of residues with adjusted N and P concentrations. *Pedosphere*, 27(1): 76–85.

Ortiz-Monasterio, I., and R.D. Graham. 2000. Breeding for trace minerals in wheat. *Food Nutr. Bull.*, 21: 392–396.

Pandey, A.K., S. Aggarwal, V. Meena, and A. Kumar. 2021. Phytic acid reduction in cereal grains by genome engineering: Potential targets to achieve low phytate wheat. *Genome Engineering for Crop Improvement*, 146–156.

Panzeri, D., E. Cassani, E. Doria, G. Tagliabue, L. Forti, B. Campion, R. Bollini, C.A. Brearley, R. Pilu, E. Nielsen, and F. Sparvoli. 2011. A defective ABC transporter of the MRP family, responsible for the bean lpa1 mutation, affects the regulation of the phytic acid pathway, reduces seed myo-inositol and alters ABA sensitivity. *New Phytol.*, 191(1): 70–83.

Pixley, K., N.P. Rojas, R. Babu, R. Mutale, R. Surles, and E. Simpungwe. 2013. Biofortification of maize with provitamin A-carotenoids. *In*: Tanumihardjo (Ed.), *Carotenoids and Human Health*, 271–292. Springer Science.

Portilla, A.E., V.M. Mayor-Duran, H.F. Buendia, M.W. Blair, K. Cichy, and B. Raatz. 2022. Climbing bean breeding for disease resistance and grain quality traits. *Legume Sci.*, 4(2): e122.

Prasanna, B.M., N. Palacios-Rojas, F. Hossain, V. Muthusamy, A. Menkir, T. Dhliwayo, ... and X. Fan. 2020. Molecular breeding for nutritionally enriched maize: Status and prospects. *Front. Genet.*, 10: 1392.

Qaim, M., A.J. Stein, and J.V. Meenakshi. 2007. Economics of biofortification. *Agric. Econ.*, 37: 119–133.

Razzaq, A., Y. Tang, and P. Qing. 2021. Towards sustainable diets: Understanding the cognitive mechanism of consumer acceptance of biofortified foods and the role of nutrition information. *Int. J. Environ. Res.*, 18(3): 1175.

Rehman, A.U., S. Masood, N.U. Khan, M.E. Abbasi, Z. Hussainm, and I. Ali. 2021. Molecular basis of iron biofortification in crop plants: A step towards sustainability. *Plant Bree.*, 140(1): 12–22.

Reza, S.K., D.C. Nayak, S. Mukhopadhyay, T. Chattopadhyay, and S.K. Singh. 2017. Characterizing spatial variability of soil properties in alluvial soils of India using geostatistics and geographical information system. *Arch. Agron. Soil Sci.*, 63: 1489–1498.

Rodas-Moya, S., F.M. Giudici, B. Mudyahoto, E. Birol, S.R. Kodish, C. Lachat, and E.J. Feskens. 2022. Critical review of indicators, metrics, methods, and tools for monitoring and evaluation of biofortification programs at scale. *Front. Nutr.*, 9: 963748.

Sharma, H., S. Singh, M. Shamshad, A.K. Padhy, R. Kaur, L. Kashyap, ... and N.S. Bains. 2023. Variability in iron, zinc, phytic acid, and protein content in pre-breeding wheat germplasm under different water regimes. *Plant Growth Regul.*, 100(2): 531–543.

Sharma, S., R. Sharma, M. Govindaraj, R.S. Mahala, C.T. Satyavathi, R.K. Srivastava, ... and B. Kilian. 2021. Harnessing wild relatives of pearl millet for germplasm enhancement: Challenges and opportunities. *Crop Sci.*, 61(1): 177–200.

Shi, J., H. Wang, Y. Wu, J. Hazebroek, R.B. Meeley, and D.S. Ertl. 2003. The maize low-phytic acid mutant lpa2 is caused by mutation in an inositol phosphate kinase gene. *Plant Physiol.*, 131(2): 507–515.

Shukla, A.K., N.K. Sinha, P.K. Tiwari, C. Prakash, S.K. Behera, N.K. Lenka, V.K. Singh, B.S. Dwivedi, K. Majumdar, A. Kumar, P.C. Srivastava, S.P. Pachauri, M.C. Meena, B.L. Lakaria, and S. Siddiqui. 2017. Spatial distribution and management zones for sulphur and micronutrients in Shivalik Himalayan region of India. *Land Degrad. Dev.*, 28: 959–969.

Singh, S.P., K. Vogel-Mikuš, I. Arčon, P. Vavpetič, L. Jeromel, P. Pelicon, and R. Tuli. 2013. Pattern of iron distribution in maternal and filial tissues in wheat grains with contrasting levels of iron. *J. Exp. Bot.*, 64(11): 3249–3260.

Trijatmiko, K.R., C. Duenas, N. Tsakirpaloglou, L. Torrizo, F.M. Arines, C. Adeva, J. Balindong, N. Oliva, M.V. Sapasap, J. Borrero, and J. Rey. 2016. Biofortified indica rice attains iron and zinc nutrition dietary targets in the field. *Sci. Rep.*, 6(1): 19792.

Van Der Straeten, D., N.K. Bhullar, H. De Steur, W. Gruissem, D. MacKenzie, W. Pfeiffer, ... and H. Bouis. 2020. Multiplying the efficiency and impact of biofortification through metabolic engineering. *Nat. Commun.*, 11(1): 5203.

Vasconcelos, M.W., W. Gruissem, and N.K. Bhullar. 2017. Iron biofortification in the 21st century: Setting realistic targets, overcoming obstacles, and new strategies for healthy nutrition. *Curr. Opin. Biotechnol.*, 44: 8–15.

Welch, R.M., and R. Graham. 1999. A new paradigm for world agriculture: Meeting human needs – Productive, sustainable, nutritious. *Field Crop Res.*, 60: 1–10.

Wissuwa, M., A.M. Ismail, and R.J.P. Graham. 2008. Rice grain zinc concentrations as affected by genotype, native soil-zinc availability, and zinc fertilization. *Plant Soil*, 306: 37–48.

Yadav, A.K., A. Seth, V. Kumar, and A. Datta. 2023. Agronomic biofortification of wheat through proper fertilizer management to alleviate zinc malnutrition: A review. *Commun. Soil Sci. Plant Anal.*, 54(2): 154–177.

Zhao, F.J., Y.H. Su, S.J. Dunham, M. Rakszegi, Z. Bedo, and S.P. McGrath. 2009. Variation in mineral micronutrient concentrations in grain of wheat lines of diverse origin. *J. Cereal Sci.*, 49: 290–295.

CHAPTER
18

Omics in Biofortification: Unlocking Nutritional Potential

Apoorv Tiwari[a*][0000-0003-2294-0688]**, Era Chaudhary**[a]**, Lakshmi Giri**[a]**, Anjali Sharma**[a,b] **and Monika Garg**[a]

[a] National Agri-Food Biotechnology Institute, Sector 81, Sahibzada Ajit Singh Nagar - 140308, Punjab, India

[b] Central University of Punjab, Bathinda - 151401, Punjab, India

1. Introduction

Food production is becoming more demanding because the global population is constantly growing and due to climatic changes there is a decline in the production of quality food, thus making the development of nutrition rich, genetically modified crops a research priority (Malleshi et al., 2021; Raza et al., 2019). As a result, science is at a turning point and amid several milestones; the question is how to proceed and address these problems. Crop improvement is mainly requiring high crop yield, resistance to biotic as well as abiotic stresses, with high nutritional value (Gonzalez Guzman et al., 2022). Some contemporary and cutting-edge techniques and tools have been developed and utilized in crop plants to achieve that goal (Kaur et al., 2021).

The advancement of omics and next-generation high-throughput technologies promises trait improvement and increases the agronomic qualities of crops. Bioinformatics and large-scale omics techniques like genomics, transcriptomics, proteomics, metabolomics, ionomics, phenomics, and machine-learning techniques are being applied to the large volumes of data that are assisting us in understanding the mechanisms underlying crop features and improvements (Zhang et al., 2022). However, despite these developments, efforts to combine contrasting omics studies within common pathways and cellular networks of crop systems are still in their initial stages. Functional genomics, along with bioinformatics tools and

*Corresponding author: apoorvbio@gmail.com

databases, has helped in the identification, characterization, and annotation of nutritionally important genes and proteins for providing scientific rationales of nutritional importance. Proteomics assists researchers to understand the role of proteins in plant nutrient homeostasis (Bayat, 2002). It helps in monitoring protein fluctuations under varied developmental and environmental influences, as it is programmed by the genome and mediated by the transcriptome.

Breeding practices would improve through high-resolution germplasm fingerprinting or SNP genotyping. It will help select parents for making crosses and introgression of new alleles from exotic genotypes or by mapping of QTLs associated with desirable traits (Kushanov et al., 2021). Although transgenic work has a greater focus, the success rate and acceptability of reproductive activity are far greater. To address the issue of hunger, biofortified crops have a brighter future (Garg et al., 2018). Therefore, it is quite pertinent to identify the potential markers associated with high grain yield and nutrients trait and such markers would help in introgressing these traits into elite breeding lines which would accelerate the selection process and enhance the breeding efficiency. There is sufficient genomic information available in model crop plants like rice, wheat, maize and others, which makes QTL mapping an easy task. GBS (genotyping by sequencing) would play a fundamental role in increasing the number of SNP markers for QTL mapping. SNPs linked with genic, non-genic, and intragenic regions could be useful for the identification of potential genes, alleles, and regulatory regions which play a pivotal role in the improvement of nutritional traits. The role of such genomic regions could be useful not only in dissecting the complexity of traits but also in the creation of smart crops through genetic/genomic engineering (Shastry and protocols, 2009). In this chapter, we focus on different aspects and future prospects of omics techniques, including, genomics, proteomics, metabolomics, ionomics, phenomics, NGS data analysis, and recently developed computational algorithms for crop biofortification and sustainable agriculture. This chapter offers the updated views on large scale multidimensional omics-based approaches in the biofortification of crops to improve qualitative and quantitative traits in crop species.

2. Omics Approaches in Crop Biofortification

Omics technologies will undoubtedly gain a better understanding of the genetic makeup and various molecular networks in different crops (Figure 1). Similarly, genetically modified metabolic pathways hold considerable promise for accelerating the production of nutrient-rich food crops. As the most promising biofortification technique, 'omics' aided nutritional breeding is expected to be very helpful in establishing nutritional security for roughly two billion nutrient-deficient people worldwide (Jain et al., 2019) (Figure 2).

2.1. Genomics

In the last two decades, the development of a huge amount of sequencing

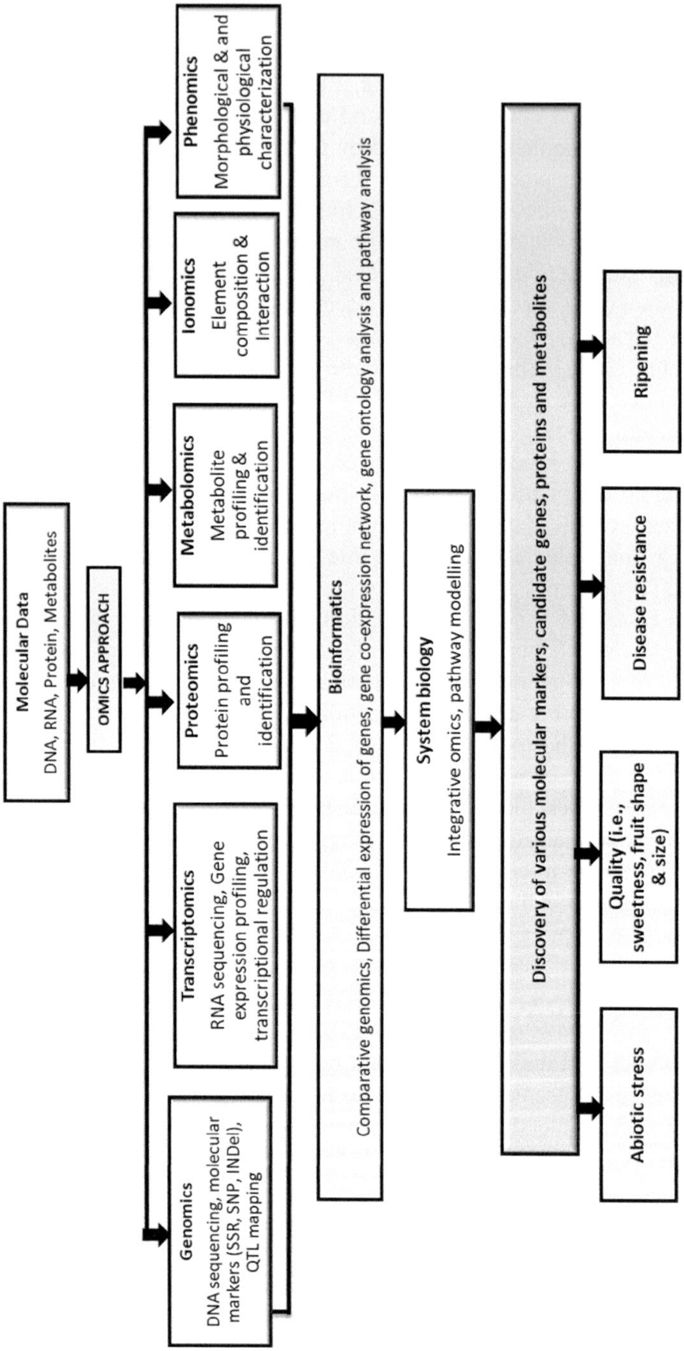

Figure 1: Bioinformatics, and systems biology approaches for a multi-omics study.

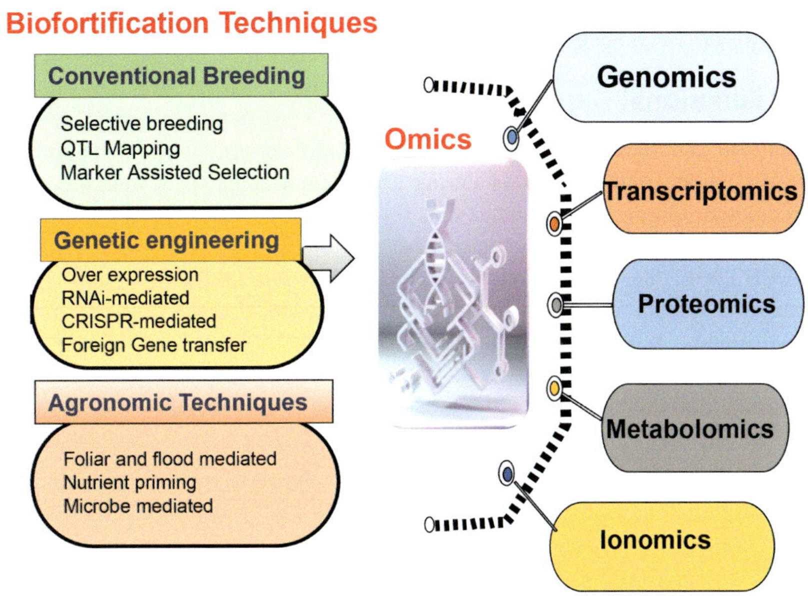

Figure 2: Different Omics approaches for crop biofortification.

information has been made possible by advancements in DNA sequencing technologies. NGS technology has revolutionized the influence of sequencing on our understanding of crop genomes and gene regulation (Behjati et al., 2013). These approaches are now showing significant promise in metagenomic and agronomic research, as well as in plant breeding of superior varieties. The field of genomics emphasizes structure, function, mapping, evolution, epigenomic, and genome editing. Genetic variation can be understood through genomics, which may improve crop breeding effectiveness and lead to the genetic advancement of agriculturally important nutrition-rich crops.

2.2. Structural Genomics

Structural genomics deals with molecular markers used to tag or map genes of interest for crop breeding advancement. The advancement of NGS technologies has made it feasible to detect and annotate SNPs easily. QTLs mapping can be performed using molecular markers like SNPs and AFLPs, which can then be connected to phenotypic information. Genome-wide association studies, which use SNPs in sequence data to detect associations between genetic variants and traits, can be performed on wide populations of crops (Srivastava et al., 2020). Genome-wide association studies have demonstrated a correlation between crops' structural variations and agronomic parameters. To increase crop quality and output, breeders can now use GBS and MAS to improve hybrid breeding. Analyzing associations between genotypes and phenotypes can lead to the discovery of QTLs, which can then be confirmed by functional genomics techniques (Tiwari et al., 2021; Ishikawa, 2017).

2.3. Functional Genomics

Functional genomics is concerned with the dynamic expression of gene products in a particular context, such as during a particular stage of development or a particular condition (Cano-Gamez and Trynka, 2020). Crops could be improved with the applications of genome engineering tools such as CRISPR/Cas9 and TALEN. With the use of TALEN and CRISPR/Cas9, scientists were able to generate a mutant variety of bread wheat with increased resistance to mildew (TaMlo) and a mutant variety of tomatoes (SlMlo) (Zhang et al., 2018). Most consumable crops such as wheat, soybean, rice, maize, and sorghum, have reportedly had their genomes edited using the CRISPR/Cas9 method. In maize, rice, wheat, and barley, comparative genomics has uncovered a wide variety of mutations associated with crop development, yield, and stress tolerance. To predict how mutation-derived varieties may influence food production and quality on a global scale, scientists have discovered a wide range of mutant features in crop plants. Accordingly, both functional genomics has proven to be effective in enhancing crop improvement, productivity with higher nutrients (Chaudhary et al., 2019).

2.4. Epigenomics

Heritable changes that do not involve in the change of the DNA sequence are called epigenetics and these epigenetic variations are reversible. They are mediated by a variety of processes, including DNA methylation and histone post-translational modification. The procedure of bisulfite sequencing allows the detection and analysis of methylation markers at a base-pair level of resolution. There has been extensive validation of the bisulfite sequencing approach about DNA methylation and chromatin-regulated genes in tomato, maize, and soybean crops (Miyata et al., 2017). When measuring DNA methylation in wheat and foxtail millet under salt stress, an MSAP method is typically used. The epigenome modifications linked to ripening in tomatoes and the tissue-cultured, stable epigenome variations in rice crops have both been the subject of DNA methylation investigations. These results point to the potential requirement of epigenomics applications in improving crop resistance to stress.

2.5. Pan Genomics

The pan-genome is made up of a whole set of genomes shared by a population of the species. It includes two sets of genomes, core and dispensable. The sets of essential genes in all individuals are known as a core genome, whereas dispensable genes, also called accessory genes, are unique to each or shared by some individuals in a population (Carlos Guimaraes et al., 2015). The variation of core genes is typically lower than that of ‘dispensable’ genes. Assembling and analyzing pan-genomes has demonstrated that relying on a single reference genome for a species can hinder our comprehension based on diverse traits. The genetic diversity present in a population is not accurately reflected by a single reference sequence for a species. Recent pangenomic researches are ongoing on various crops including rice, wheat, maize, soybean, *Brassica sp*. Using nine diverse morphologically cultivars of *Brassica oleracea* and a wild related variety. A pangenomic analysis indicated that numerous variable genes had been labeled with vital agronomic features, such as disease resistance, vitamin synthesis, and glucosinolate metabolism (Golicz et al., 2016). More variations like SNPs and InDels are present in non-essential genes rather than in the essential genome, established in *Brachypodium sp.*, rice, and soybean. Therefore, elite genes in crop wild relatives could be mined using pangenomic investigations to improve crops.

2.6. Transcriptomics

Transcriptomics is the study of the full set of RNA transcripts in an organism's genome in a cell or tissue that has gained popularity for examining how genes are expressed in response to various conditions over a prolonged time (Wang et al., 2009). This method aids in understanding a gene's primary function by examining the genes that are differentially expressed *in vitro*. Traditional profiling like differential display-PCR (DD-PCR), cDNAs-AFLP, and SSH were initially used to study transcriptomic expression, but had the disadvantage of lower resolution (Rajeevan et al., 2001; Taylor et al., 2017). Soon after, the development of new analysis methods such as microarrays, digital gene expression profiling, NGS, RNAseq, and SAGE were used for RNA expression profiling. Microarray analysis using an Affymetrix GeneChip array has made it better to identify the differentially expressed genes in crops during various developmental and reproductive stages. This method has been applicable in crops including maize, rice, wheat, and sorghum in response to extreme biotic and abiotic stress. Together, all these transcriptomic techniques may be extremely important for controlling the gene expression profile of different crop species (Lowe et al., 2017).

2.7. Proteomics

Proteomics, a method for profiling all of an organism's expressed proteins, is broken down into four different categories: structural, functional, sequence, and expression proteomics. In structural proteomics, the protein structure is studied to determine their supposed functions. Different techniques including computer-based modeling, electron microscopy, NMR, crystallization, and the X-ray diffraction of protein crystals, can be used to analyze structural proteomics (Maveyraud and Mourey, 2020). A protein's functions are identified through functional proteomics, and these functions are investigated using a variety of techniques, including yeast one/two hybrids and protein-based microarray.

The sequences of amino acids are typically identified by sequence proteomics. High-performance liquid chromatography, size exclusion chromatography, affinity chromatography, and ion exchange chromatography are some of the chromatography-based methods used in traditional proteomics (Coskun, 2016). However, more sophisticated methods like western blotting and enzyme-linked immunosorbent assay, 2D-gel electrophoresis, SDS-PAGE, and 2D-differential gel electrophoresis, could also be used. MS, IT-MS, MALDI-TOF, ESI, and collision-induced dissociation are used to determine the molecular mass of peptides (Artigues et al., 2016).

2.8. Metabolomics

The extensive study of all metabolites involved in various cellular processes in any biological system is known as metabolomics. The term 'metabolome' includes the entire set of metabolites produced by different metabolic interacting pathways in the system. Proteomics only revealed the final gene products; however, metabolomics can measure how proteins are expressed metabolically and pinpoint the key biochemical events that influence how genes function. The most popular methods in plant metabolomics are GC-MS, CE-MS, LC-MS, and NMR. Although mass spectrometry techniques (CE-MS, LC-MS, and GC-MS) are suitable, sensitive, and selective but time-consuming, but NMR is quick and selective. The β-carotene accumulation in the rice endosperm caused by modification in specific metabolic pathways has been proved to increase the nutritional quality in genetically modified rice. In tomato, metabolic engineering was used to increase the production of a secondary metabolite, anthocyanin. A comprehensive view ranging from the genome to the metabolome as well as phenotypic traits of various functions is thus provided by metabolomics in combination with other omics approaches (Chong et al., 2019).

2.9. Ionomics

Ionomics is an elemental profiling technique that examines total mineral nutrients in a living entity (Ali et al., 2021). Some elemental high-throughput profiling techniques have been used to measure ionome nutrient density in plants like ICP-OES, XFM, ICP-MS, and ED-XRF (Huang and Salt, 2016). The XRF localizes the elements more quickly and at a lower cost than both ICP-MS and OES. A few platforms like ionomic HUB or iHUB, have been created, making the public's access to ionomic data free for several species like Arabidopsis, rice, yeast, and soybean (Baxter et al., 2007). To identify the genetic diversification of mineral elements in the rice crop, the leaf and grain ionomes have been examined using the ICP-MS technique. The relationships of various traces elements, minerals nutrients, and metabolites under salt stress in barley were also explained by the ionomics study. Under water stress, micro- and macronutrients concentrations have also been observed in tomato cultivars using elemental profiling analysis. Ionomic profiling was also used to examine the nutrient availability in a variety of fruit varieties like apple, mango, orange, blueberries, and kiwi (Parent et al., 2013).

2.10. Phenomics

Phenomics is described as the process of study of various phenotypes associated with the wide collection of multi-array phenotypic data of crops. Reliable, efficient, multifunction, and high-throughput phenotypic techniques are becoming more significant instruments in breeding operations for accelerating genetic gain (Zhao et al., 2019). The phenome can be affected by genotype-phenotype-environment interactions. The collection of phenotypic available data is merely the initial stage in phenomics research. For the improvement in crops' genetics, phenotyping should be extremely correct for any target gene and QTL-based mapping of the desired traits also assists in this to identify their roles through forward and reverse phenomics applications (Pasala and Pandey, 2020). Automated high-throughput imaging technologies could identify the 'finest' or the 'excellent' germplasm lines. Techniques for biomass, color imaging: for canopy, far-infrared imaging and UV-Red, green, and blue imaging, LIDAR could be used to measure growth parameters for root system. Specifically, to examine the root system, MRIs have been used that enable quick phenotyping of the traits (Su, 2020). To determine crop phenotyping parameters, various studies have used RGB based on visible light imaging. This UV- based imaging was used to evaluate the *Triticum durum* root system grown in rhizoboxes-filled soil and also to assess the phenotyping of plant shoots under different stress responses. Plant phenomics currently faces a major challenge to organize the databases of collected phenotyping/traits so that they can be used further to produce new insights by merging phenomics with other approaches of omics to enhance crop management (Fasoula et al., 2020). Different omics based biological databases and tools widely used by researchers for crop improvement and breeding programs are listed in Tables 1 and 2.

3. Application of Omics Techniques in Crop Biofortification

The most susceptible groups to malnutrition are young children and pregnant women, which can result in stunted growth, health issues, and a variety of problems with childbirth. Although seven minerals like Fe, Zn, Cu, Ca, Mg, Se and several vitamins including vitamin A and B9 (folate) are frequently deficient in human diets, deficiencies of these vitamins are among the most serious (Tardy et al., 2020). Within the forthcoming years, the biofortification of staple crops through genetic modification, using an intelligent strategy of both traditional breeding and genetic engineering approaches, should progress significantly.

Bioinformatics employs computational tools for the management and evaluation of biological data (genomic, proteomic, metabolomics). Furthermore,

Table 1: Databases and tools for omics approach and their applications in crop improvement

S. No.	Database Name	Application	Web Address
1.	BBDG	It is a database of both sequences and gene expression. It enables researchers to connect expression patterns with gene function.	http://bioinformatics.towson.edu/ BBGD/
2.	BRAD	It is a database of brassica genomic sequences and primarily consists of genome assemblies, predicted genes, and their annotations.	http://brassicadb.org/brad/
3.	CerealsDB	It gives SNP information for the wheat genome.	http://www.cerealsdb.uk.net/ cerealgeno/mics/
4.	CottonFGD	It is a database which incorporates the genomic/transcriptomic data with research/analysis/visualization modules.	https://cottonfgd.org/
5.	CottonGen	It gives cotton genomic, genetic, and breeding resources for crop improvement.	https://www.cottongen.org/
6.	CottonQTLdb	Cotton crop QTL analysis.	http://www.cottonqtldb.org
7.	CropSNPdb	Multicrop (Brassica and Wheat) SNPs database.	http://snpdb.appliedbioinformatics.com.au/
8.	CSRDB	Small RNAs database of cereals.	http://sundarlab.ucdavis.edu/smrnas/
9.	Ensembl Plants	It is a database that organizes, analyzes, and visualizes genome-scale data for significant agricultural and model plants.	https://plants.ensembl.org/index.html
10.	GrainGenes	It is a *Triticeae* and *Avena* database which provides molecular and phenotypic information for wheat, barley, rye, and oat.	http://www.graingenes.org
11.	Gramene	It is a structured, interactive and integrated data repository for crop and model plant for functional genomics analysis.	http://www.gramene.org/

12.	IWGSC	International consortium of wheat producers, plant scientists, and public and private breeders working together.	https://www.wheatgenome.org/
13.	MaizeDIG	It is a maize genotypic-phenotypic database based on BioDIG.	https://maizedig.maizegdb.org/
14.	MaizeGDB	It is a database of biological information about the plant Zea mays, Multi omics.	https://www.maizegdb.org/
15.	MaizeSNPDB	It is SNP database of maize lines.	https://github.com/venyao/ MaizeSNPDB
16.	Oryzabase	It is a database of integrated biological and genomic resource of rice.	https://shigen.nig.ac.jp/rice/ oryzabase/
17.	PCD	It allows access to the annotation of existing transcriptome data, assisting pulse breeders and researchers in their projects.	https://www.pulsedb.org/
18.	Plant Reactome	It is an open-source, peer-reviewed route database that is manually curated.	http://plants.reactome.org/
19.	PlantGDB	It gives tools and resources for plant genomics.	http://www.plantgdb.org/
20.	PlantTFDB	Plant transcription factors are predicted using transcriptomics.	http://planttfdb.cbi.pku.edu.cn/
21.	Phytozome	This database is used to visualize, analyze the plant genomes, as well as genomes.	https://phytozome-next.jgi.doe.gov/
22.	PMN	It is a database of genomic sequence metabolism for plants and algae.	https://www.plantcyc.org/
23.	PPDB	It is a proteome database of Arabidopsis thaliana and maize.	http://ppdb.tc.cornell.edu/
24.	RAP-DB	Annotated genomics structure and function of rice.	http://rapdb.dna.affrc.go.jp/
25.	RGPDB	This database comprises over 1200 root-associated genes and their associated promoter sequences for different plants.	http://sysbio.unl.edu/RGPDB/
26.	Ricebase	It is an integrative genomic database for rice.	https://ricebase.org/
27.	RiceVarMap	A comprehensive database of rice genomic variants and functional annotation.	http:/ricevarmap.ncpgr.cn

(*Contd.*)

Table 1: (*Contd.*)

S. No.	Database Name	Application	Web Address
28.	RiceXPro	It is a library of gene expression profiles of rice.	http://ricexpro.dna.affrc.go.jp/
29.	SFGD	Genomics and transcriptomics.	https://www.sunflowergenome.org
30.	SGH	It is a web-based platform that provides consolidated access to sugarcane reference genome sequences and its genomic information.	https://sugarcane-genome.cirad.fr/
31.	SiFGD	Functional genomics database of *Setaria italica* (Foxtail Millet).	http://structuralbiology.cau.edu.cn/SIFGD/
32.	SorghumFDB	Genome data and functional annotations of sorghum database for prediction of gene functions.	http://structuralbiology.cau.edu.cn/ sorghum/index.html
33.	SorGSD	SNP database.	http://sorgsd.big.ac.cn/
34.	SoyBase	A comprehensive archive of professionally maintained genetics, genomics, and associated data resources for soybean.	https://www.soybase.org/
35.	SoyKB	A comprehensive web resource for soybean.	http://soykb.org/
36.	SoyNet	It is a co-expression and functional gene network and database of soybean.	https://www.inetbio.org/soynet/
37.	TFGD	Metabolomics, functional genomics, and integrated transcriptomics.	http://ted.bti.cornell.edu/
38.	URGI	This portal keep the databases and tools of genetic and genomic wheat.	https://wheat-urgi.versailles.inra.fr/About-us
39.	WGI	Genomics database of wheat.	http://wheatgenome.info/
40.	Wheat Initiative	A platform for the community of international wheat researchers as well as a source of information for the Wheat Initiative.	https://www.wheatinitiative.org/
41.	WheatGPE	Analysis of relationships between genotype, phenotype, and environment in wheat.	http://www.wheatdb.org/

42.	WheatIS	A single-access web-based system for data resources and bioinformatics tools is offered by the Wheat Information System.	http://www.wheatis.org/
43.	NCBI	Provides resources for biological research and development.	https://www.ncbi.nlm.nih.gov/
44.	UniProt	Offers extensive resources on protein sequences, containing functional and structural information.	https://www.uniprot.org/
45.	Pfam	Protein family database for domain analysis and associated information.	https://pfam.xfam.org/
46.	Protein Data Bank	A structural database containing information on the three-dimensional structures of macromolecules that may be used in drug discovery and structural bioinformatics.	https://www.rcsb.org/
47.	AlphaFold	A database for predicting 3D structures of proteins.	https://alphafold.ebi.ac.uk/
48.	PubChem	Small molecule and associated information database, including chemical compound structure.	https://pubchem.ncbi.nlm.nih.gov/
49.	ZINC	Chemical database for use in virtual screening.	https://zinc.docking.org/
50.	GEO	Functional genomics database with gene expression profiles.	https://www.ncbi.nlm.nih.gov/geo/

the merger of various omics tools including genomics, transcriptomics, proteomics, and metabolomics significantly improved our comprehension of the molecular mechanisms of plants (Naik et al., 2023). The analysis of the multi-omics techniques in agricultural science has been carried out using bioinformatics tools employing different software programs (Table 2). There is a great opportunity for scientists/researchers to develop sophisticated, fast algorithms and databases for multi-omics platforms that significantly use omics data in multidisciplinary research in the plant sciences. Following are the important applications of omics technique in crop biofortification.

3.1. Food Safety

Every step of food processing could indeed receive support from the use of omics approaches to maintain quality, improve shelf-life predictions, and detect microbial contamination in real-time (from raw materials to final products).

Table 2: List of computational tools for omics data analysis

S. No.	Database Name	Application	Web Address
1.	AutoDock	Molecular docking.	http://autodock.scripps.ed/
2.	Bowtie 2	Tool for huge reference sequence alignment of sequencing reads.	http://bowtie-bio. sourceforge.net/ bowtie2/ index.shtml
3.	HISAT2	A quick and accurate alignment program for mapping data from next-generation sequencing.	http://daehwankimlab. github.io/hisat2/
4.	STAR	Fast RNA-seq read mapper that supports splice junction and fusion read identification.	https://github.com/ alexdobin/STAR
5.	REVIGO	Revigo can summarize extensive lists of Gene Ontology.	http://revigo.irb.hr/
6.	GROMACS	In particular, proteins, lipids, and nucleic acids are simulated by this molecular dynamics program.	http://www.gromacs.org/
7.	Samtools	Samtools is a collection of programs for editing alignments in the SAM (Sequence Alignment/Map), BAM, and CRAM formats.	http://www.htslib.org/
8.	SwissADME	Calculate physicochemical descriptors as well as estimate ADME parameters, pharmacokinetic features, and chemical molecule drug-like nature.	http://www.swissadme.ch/
9.	Trimmomatic	Trimmomatic handles sequence quality data in both standard (phred+33) and 'legacy' (phred+64) forms.	http://www.usadellab.org/ cms/?page= trimmomatic
10.	BWA	BWA is a tool that allows mapping low-divergence sequences against a large reference genome.	https://bio-bwa. sourceforge.net/
11.	DESeq2	High-dimensional count data can be normalized, visualized, and subjected to differential analysis using the DESeq2 program.	https://bioconductor.org/ packages/release/bioc/ html/DESeq2.html

12.	edgeR	A bioconductor module for examining the expression of different genes in the digital environment.	https://bioconductor.org/packages/release/bioc/html/edgeR.html
13.	BLAST	It is a tool for comparing protein, amino acid, or DNA and/or RNA nucleotide sequences.	https://blast.ncbi.nlm.nih.gov/Blast.cgi
14.	SPAdes	It is a genome assembly system for single and multi-cell bacterial data sets.	https://cab.spbu.ru/software/spades/
15.	MarvinSketch	It is a sophisticated chemical editor that allows you to sketch chemical structures, and reactions.	https://chemaxon.com/products/marvin
16.	Cutadapt	It is a tool for assessing the quality of high-throughput sequencing reads.	https://cutadapt.readthedocs.io/en/ stable/
17.	Cytoscape	It is a tool that combines molecular states, high-throughput expression data, and biomolecular interaction networks into a coherent conceptual framework.	https://cytoscape.org/
18.	AlphaFold 2	It is a complete solution to the protein structure prediction.	https://github.com/deepmind/alphafold
19.	SRA Toolkit	This includes tools for downloading data, converting multiple data types into SRA format, and extracting SRA data in other formats.	https://github.com/ncbi/sra-tools/ wiki/01.-Downloading-SRA-Toolkit
20.	Fastp	Tool for quality testing, adapter trimming, quality filtering, and read pruning.	https://github.com/OpenGene/fastp
21.	FastQ	It is a tool for performing quality control tests on raw sequence data from high-throughput sequencing processes.	https://www.bioinformatics.babraham.ac.uk/projects/fastqc/
22.	Trinity	It is a method for de novo transcriptome assembly using RNA-seq. data.	https://github.com/trinityrnaseq/trinityrnaseq/wiki
23.	WGCNA	It is a bioinformatics program used to investigate the interactions between distinct gene sets.	https://horvath.genetics.ucla.edu/html/CoexpressionNetwork/Rpackages/WGCNA/

(*Contd.*)

Table 2: (*Contd.*)

S. No.	Database Name	Application	Web Address
24.	PyMOL	It is a cross-platform molecular graphic program that has been widely used for 3D macromolecule visualization.	https://pymol.org/2/
25.	Swiss PDB Viewer	It is a program with a user-friendly interface that allows studying many proteins at the same time.	https://spdbv.unil.ch/
26.	STRING	It is a database of known and predicted protein-protein interactions to know direct (physical) and indirect (functional) associations.	https://string-db.org/
27.	DIAMOND	It is a high-throughput alignment tool that compares a file of DNA sequencing reads to a file of reference sequences.	https://github.com/bbuchfink/diamond
28.	Blast2GO	It is a bioinformatics tool for functional genomic data analysis.	https://www.blast2go.com/
29.	Chimera	This is a tool that allows interactively visualizing and analyzing molecular structures and associated data, such as density maps, trajectories, and sequence alignments.	https://www.cgl.ucsf.edu/chimera/
30.	BioModels	It is a database including mathematical models of biological and biomedical processes.	https://www.ebi.ac.uk/biomodels/
31.	Velvet	It is a *de novo* assembly tool that is based on de Bruijn graphs and is ideal for short-read data with good coverage.	https://www.ebi.ac.uk/$\sim$zerbino/ velvet/
32.	KEGG	It is a knowledge repository for systematic investigation of gene functions that connects genomic data with higher level functional data.	https://www.genome.jp/kegg/

33.	MEGA	It is a comprehensive tool for both automatic and manual sequence alignment, phylogenetic tree inference, and other tasks.	https://www.megasoftware.net/
34.	GATK	This program for detecting SNPs and indels in genome data.	Yandell-lab.org/software/maker.html
35.	VCFtools	It can run analysis on the variants in a genome.	https://vcftools.sourceforge.net/
36.	TASSEL	Used for association mapping studies.	https://tassel.bitbucket.io/
37.	GAPIT	Used for association mapping studies.	https://zzlab.net/GAPIT/
38.	STRUCTURE	Tool for population structure analysis.	https://web.stanford.edu/group/pritchardlab/structure.html
39.	RStudio	This connects with R to statistically analyze of data.	https://posit.co/download/rstudio-desktop/
40.	QTL-seq	It combines bulked-segregant analysis (BSA) of a target quantitative traits.	https://github.com/YuSugihara/QTL-seq

Through the use of omics, it is possible to identify protein profiles, biomarkers, and allergy identification (peptide markers) (Cook and Nightingale, 2018).

3.2. Edible Vaccines

An alternative to injecting the vaccine would be to eat fruits or vegetables that are modified to produce the vaccine. The edible vaccinations can offer inexpensive protection, especially in developing nations with low vaccination rates to ensure the integrity of the community and the protection of livestock (Saxena and Rawat, 2014).

3.3. Plantibodies

Therapeutic monoclonal antibodies made from plants are more effective at fighting than antibodies made from animals, according to a comparison with conventionally produced antibodies (Nessa et al., 2020). Using plants as bioreactors for the synthesis and processing of therapeutic proteins has several advantages, including the lack of animal infectious illnesses, low production costs, and simplicity of agricultural scale-up. An additional genetically modified strain of tobacco produces a plantibody that protects against the environmentally harmful chemical microcystin. To get rid of these toxins from the environment, plants send out antibodies from their roots that attach to the toxin compound.

Such plants could reduce the bioavailability of pollutants, purify contaminated water or soils, or even redefine the term 'phytoremediation' (Oluwayelu and Adebiyi, 2016).

3.4. Plants as Synthetic Chemists

Plants are truly excellent synthetic chemists, creating a variety of substances when provided with the proper conditions. With the help of genetically engineered crops, the amount of oleic acid in vegetable oils has increased, while the amount of PUFAs has decreased. The stability of the vegetable oil was subsequently improved by the change in ratios, which ultimately improved the stability of the processed foods (Haslam et al., 2016). Although traditional plant breeding may have increased oleic acid levels, but omics-based technology is crucial to achieving higher levels up to 75–80%. Cottonseed oils with high oleic acid content (78%) and stearic acid content (40%) have been produced using gene silencing. In addition to these synthetic capabilities, plants can also produce industrial products, antibodies, vaccines, and other biological substances (Chen et al., 2021).

3.5. Energy Sustainability

Omics research will progress further as a result of climate change, rising fuel costs, and a growing interest in renewable energy sources. Through modification, sugar cane, maize, and rapeseed can all significant producers of green energy. Approaches for growing another non-domesticated shrub, *Jatropha curcas Linn.* as a sustainable biofuel crop include transcriptome profiling, whole genome sequencing, and molecular markers. Proteomic data from molecular breeding research can help with the later usage plants as sources for sustainable biofuels production without compromising food production.

4. Conclusion and Future Prospects

Climate change, the depletion of natural resources, chronic hunger, and malnutrition few are major problems that threaten global food security. Biofortification is a way to help malnourished people in rural areas that are mostly fed by subsistence agriculture. Recent developments in omics studies have revolutionized crop improvement techniques, enabling researchers to quickly introduce traits that improve grain quality and nutrition. The discovery of superior haplotypes to improve the nutritional value of a variety of crops is made possible by the recent, unprecedented advancements in next-generation DNA sequencing technologies and high-density genotyping-based trait mapping, including genome-wide association studies, fine mapping, QTL discovery, pooled sequencing (QTLseq, BSRseq, MutMap, etc.), and other tools. As a result, association maps appear to be a good solution for plant breeders because they rely on linkage disequilibrium. The NGS approach is used to sequence

GBS data, genomes, exomes, transcriptomes, metagenomes, and epigenomes in addition to sequencing and resequencing normal genomes. Different computational resources are needed for the analysis of bulk of NGS data. The emergence of advanced tools and methods, such as computational algorithms and vast genomic data, offers the chance to biofortify crops which will help to solve the malnutrition issue in the near future.

References

Ali, S., A. Tyagi, and H. Bae. 2021. Ionomic approaches for discovery of novel stress-resilient genes in plants. *Int. J. Mol. Sci.*, 22: 7182.

Artigues, A., O.W. Nadeau, M.A. Rimmer, M.T. Villar, X. Du, A.W. Fenton, and G.M. Carlson. 2016. Protein structural analysis via mass spectrometry-based proteomics. *Adv. Exp. Med. Biol.*, 919: 397–431.

Baxter, I., M. Ouzzani, S. Orcun, B. Kennedy, S.S. Jandhyala, and D.E. Salt. 2007. Purdue ionomics information management system. An integrated functional genomics platform. *Plant Physiol.*, 143: 600–611.

Bayat, A. 2002. Science, medicine, and the future: Bioinformatics. *BMJ.*, 324: 1018–1022.

Behjati, S., and P.S. Tarpey. 2013. What is next generation sequencing? *Arch. Dis. Child. Educ. Pract. Ed.*, 98: 236–238.

Cano-Gamez, E., and G. Trynka. 2020. From GWAS to function: Using functional genomics to identify the mechanisms underlying complex diseases. *Front. Genet.*, 13: 424.

Chaudhary, J., R. Deshmukh, and H.J.P. Sonah. 2019. Mutagenesis approaches and their role in crop improvement. *Plants* (*Basel*), 8: 467.

Chen, Y., M. Fu, H. Li, L. Wang, R. Liu, Z. Liu, X. Zhang, and S. Jin. 2021. High-oleic acid content, nontransgenic allotetraploid cotton (*Gossypium hirsutum* L.) generated by knockout of GhFAD2 genes with CRISPR/Cas9 system. *Plant Biotechnol. J.*, 19: 424–426.

Chong, J., D.S. Wishart, and J. Xia. 2019. Using MetaboAnalyst 4.0 for comprehensive and integrative metabolomics data analysis. *Curr. Protoc. Bioinformatics*, 68: e86.

Cook, P.W., and K.K. Nightingale. 2018. Use of omics methods for the advancement of food quality and food safety. *Anim. Front.*, 8: 33–41.

Coskun, O. 2016. Separation techniques: Chromatography. *North Clin. Istanb.*, 3: 156–160.

Fasoula, D.A., I.M. Ioannides, and M. Omirou. 2020. Phenotyping and plant breeding: Overcoming the barriers. *Front. Plant Sci.*, 10: 1713.

Garg, M., N. Sharma, S. Sharma, P. Kapoor, A. Kumari, V. Chunduri, and P. Arora. 2018. Biofortified crops generated by breeding, agronomy, and transgenic approaches are improving lives of millions of people around the world. *Front. Nutr.*, 5: 12.

Golicz, A.A., P.E. Bayer, G.C. Barker, P.P. Edger, H. Kim, P.A. Martinez, C.K. Chan, A. Severn-Ellis, W.R. McCombie, and I.A. Parkin. 2016. The pangenome of an agronomically important crop plant *Brassica oleracea*. *Nat Commun.*, 7: 13390.

Gonzalez Guzman, M., F. Cellini, V. Fotopoulos, R. Balestrini, and V. Arbona. 2022. New approaches to improve crop tolerance to biotic and abiotic stresses. *Physiol. Plant*, 174: e13547.

Guimaraes, L.C., J. Florczak-Wyspianska, L.B. de Jesus, M.V. Viana, A. Silva, R.T. Ramos, and C. Sde Soares. 2015. Inside the pan-genome-methods and software overview. *Curr. Genomics*, 16: 245–252.

Haslam, R.P., O. Sayanova, H.J. Kim, E.B. Cahoon, and J. Napier. 2016. Synthetic redesign of plant lipid metabolism. *Plant J.*, 87: 76–86.

Huang, X.Y., and D.E. Salt. 2016. Plant ionomics: From elemental profiling to environmental adaptation. *Molecular Plant*, 9: 787–797.

Ishikawa, A. 2017. A strategy for identifying quantitative trait genes using gene expression analysis and causal analysis. *Genes*, 8: 347.

Jain, S., A. Rustagi, D. Kumar, M.A. Yusuf, S. Shekhar, and N.B. Sarin. 2019. Meeting the challenge of developing food crops with improved nutritional quality and food safety: Leveraging proteomics and related omics techniques. *Biotechnol. Lett.*, 41: 471–481.

Kaur, B., K.S. Sandhu, R. Kamal, K. Kaur, J. Singh, M.S. Röder, and Q.H.J.P. Muqaddasi. 2021. Omics for the improvement of abiotic, biotic, and agronomic traits in major cereal crops: Applications, challenges, and prospects. *Plants*, 10: 1989.

Kushanov, F.N., O.S. Turaev, D.K. Ernazarova, B.M. Gapparov, B.B. Oripova, M.K. Kudratova, F.U. Rafieva, K.K. Khalikov, D.S. Erjigitov, and M.T. Khidirov. 2021. Genetic diversity, QTL mapping, and marker-assisted selection technology in cotton (*Gossypium* spp.). *Front. Plant Sci.*, 12: 779386.

Lowe, R., N. Shirley, M. Bleackley, S. Dolan, and T. Shafee. 2017. Transcriptomics technologies. *PLoS Comput. Biol.*, 13: e1005457.

Malleshi, N.G., A. Agarwal, A. Tiwari, and S. Sood. 2021. Nutritional quality and health benefits. *Millets and Pseudo Cereals*, 159–168. Woodhead Publishing.

Maveyraud, L., and L. Mourey. 2020. Protein X-ray crystallography and drug discovery. *Molecules*, 25: 1030.

Miyata, K., M. Naito, T. Miyata, S. Mokuda, and H.J.S.M.D. Asahara. 2017. Bisulfite sequencing for DNA methylation analysis of primary muscle stem cells. *Methods Mol. Biol.*, 1668: 3–13.

Naik, B., V. Kumar, S. Rizwanuddin, M. Chauhan, M. Choudhary, A.K. Gupta, P. Kumar, V. Kumar, P.E.J. Saris, M.A. Rather, S. Bhuyan, R.R. Neog, S. Mishra, and S. Rustagi. 2023. Genomics, proteomics, and metabolomics approaches to improve abiotic stress tolerance in tomato plant. *Int. J. Mol. Sc.*, 24(3): 3025.

Nessa, M.U., M.A. Rahman, and Y. Kabir. 2020. Plant-produced monoclonal antibody as immunotherapy for cancer. *Biomed. Res. Int.*, 2020.

Oluwayelu, D.O., and A.I. Adebiyi. 2016. Plantibodies in human and animal health: A review. *Afr. Health Sci.*, 16: 640–645.

Parent, S.É., L.E. Parent, J.J. Egozcue, D.E. Rozane, A. Hernandes, L. Lapointe, V. Hébert-Gentile, K. Naess, S. Marchand, and J. Lafond. 2013. The plant ionome revisited by the nutrient balance concept. *Front Plant Sci.*, 4: 39.

Pasala, R., and B. Pandey. 2020. Plant phenomics: High-throughput technology for accelerating genomics. *J. Biosci.*, 45: 111.

Rajeevan, M.S., D.G. Ranamukhaarachchi, S.D. Vernon, and E.R. Unger. 2001. Use of real-time quantitative PCR to validate the results of cDNA array and differential display PCR technologies. *Methods*, 25: 443–451.

Raza, A., A. Razzaq, S.S. Mehmood, X. Zou, X. Zhang, Y. Lv, and J. Xu. 2019. Impact of climate change on crops adaptation and strategies to tackle its outcome: A review. *Plants*, 8: 34.

Saxena, J., and S. Rawat. 2014. Edible vaccines. *Advances in Biotechnology*, 22: 207–226.

Shastry, B.S. 2009. SNPs: Impact on gene function and phenotype. *Methods of Mol. Bio.*, 578: 3–22.

Srivastava, R.K., R.B. Singh, V.L. Pujarula, S. Bollam, M. Pusuluri, T.S. Chellapilla, R.S. Yadav, and R. Gupta. 2020. Genome-wide association studies and genomic selection in pearl millet: Advances and prospects. *Front Genet.*, 10: 1389.

Su, W. 2020. Advanced machine learning in point spectroscopy, RGB- and hyperspectral-imaging for automatic discriminations of crops and weeds: A review. *Smart Cities,* 3: 767–792.

Tardy, A.-L., E. Pouteau, D. Marquez, C. Yilmaz, and A. Scholey. 2020. Vitamins and minerals for energy, fatigue, and cognition: A narrative review of the biochemical and clinical evidence. *Nutrients*, 12: 228.

Taylor, S.C., G. Laperriere, and H. Germain. 2017. Droplet Digital PCR versus qPCR for gene expression analysis with low abundant targets: From variable nonsense to publication quality data. *Nutrients*, 7: 2409.

Tiwari, A., D. Sharma, S. Sood, J.P. Jaiswal, S.P. Pachauri, P.W. Ramteke, and A. Kumar. 2020. Genome-wide association mapping for seed protein content in finger millet (*Eleusine coracana*) global collection through genotyping by sequencing. *J. Cereal. Sci.*, 91: 102888.

Wang, Z., M. Gerstein, and M. Snyder. 2009. RNA-Seq: A revolutionary tool for transcriptomics. *Nat. Rev. Genet.*, 10: 57–63.

Zhang, R., C. Zhang, C. Yu, J. Dong, and J. Hu. 2022. Integration of multi-omics technologies for crop improvement: Status and prospects. *Front Bioinform.*, 2: 1027457.

Zhang, Y., K. Massel, I.D. Godwin, and C. Gao. 2018. Applications and potential of genome editing in crop improvement. *Genome Biol.*, 19: 210.

Zhao, C., Y. Zhang, J. Du, X. Guo, W. Wen, S. Gu, J. Wang, and J. Fan. 2019. Crop phenomics: Current status and perspectives. *Front Plant Sci.*, 10: 714.

CHAPTER
19

Meta-Analysis Magic: Unearthing Health Insights in Nutrition

Monika Garg[a*[0000-0002-2297-1137]], Saloni Sharma[a], Vandita Tiwari[a,b] and Apoorv Tiwari[a]

[a] National Agri-Food Biotechnology Institute – Mohali
[b] Panjab University – Chandigarh

1. Introduction

A healthy diet and good health go hand-in-hand; therefore, choosing the right foods is vital for the best life possible. Consequently, it's essential to think carefully about the foods and nutrients that should be prioritized in our diets and how they can improve our health. There are a number of well-documented health benefits associated with diets that are high in fruits, vegetables, whole grains, nuts, and lean meats. These benefits include a reduction in blood pressure, an improvement in diabetes patients' glucose control, the promotion of weight loss, and a reduction in the risk of cardiovascular disease and cancer (Van Duyn and Pivonka, 2000; Lea and Worsley, 2003; Jonnalagadda et al., 2010; Barber et al., 2010; Lund 2013). Countless pre-clinical and clinical research is currently being carried out worldwide to determine the health advantages of known and unknown bioactive components derived from plants or animals.

The key method researchers use to determine whether a new diet or a new food component has an effective nutraceutical potential or to what extent it is bioavailable is to conduct clinical trials. Sometimes clinical trials, such as screening for food allergies, are also used to study any negative effects of long-term usage of any food or diet component for a specific group of people. Before conducting a clinical trial, researchers evaluate the potential through a series of laboratory tests on animals to determine the extent to which it is simultaneously safe and effective. Meta-analysis and systematic reviews have shown themselves

*Corresponding author: mkajgarg@gmail.com

to be the methods of choice when combining the findings of a vast number of smaller studies and large-scale randomized clinical trials to arrive at a conclusive result (Cochran, 1937; Becker 1987).

A meta-analysis combines all the research that has been conducted on a certain subject and method, then subjected to statistical analysis. Sometimes meta-analysis known as a "study of studies", is a type of research that combines the results of several separate studies in order to boost the overall sample size and statistical power of an investigation (Hedges, 1992; Lipsey and Wilson, 2001; Borenstein et al., 2021). The use of meta-analysis can help reconcile inconsistencies in study results that are contradictory. By formally examining the conflicting findings of different studies, it is possible to arrive at new hypotheses and analyze the factors contributing to disagreement. A single study might not provide enough information to address questions that could have a significant impact, but a meta-analysis might. If high-quality studies with similar populations of people are employed, a meta-analysis can, at best, be a helpful overview of a big pool of data.

This book chapter discusses the importance of meta-analysis in evaluating the effects of diverse dietary components on human health. Because researchers in various nations are revealing the health benefits of different dietary components and diets on certain geographical areas, it is vital to combine and analyze the results of these studies. The most challenging part of determining the effects of nutrition is that, in contrast to pharmacological studies, where the interventions are typically identical, nutrition intervention trials frequently use diverse approaches. This presents a significant challenge. In observational studies, it can be difficult to determine how much of a variety of foods and nutrients a community consumes since individuals' eating routines and how they report those routines can vary significantly from one population to the next. Researchers may use various methodologies to examine and document varied nutritional habits. This chapter went through those methodologies and publicly accessible registration and guideline platforms for meta-analysis research.

2. What are Meta-Analyses?

Meta-analysis is an "analysis of analyses" as described by Gene V. Glass (1976). Meta-analyses can also synthesize rare data on adverse events, which require multiple studies to examine (Borenstein et al., 2021).

Meta-analysis is beneficial in many research situations and helps plan new studies. Meta-analyses begin with a well-planned study.

Combining data from earlier studies can provide the needed information, but the research should not be done (Borenstein et al., 2021). There are many techniques to obtain data from research for meta-analyses (Cuijpers, 2016).

Traditional/Narrative reviews: Traditional literature reviews summarize a research subject but do not outline a technique. Data collection and interpretation are unreliable, and findings are subjective. Authors describe and interpret literature theoretically or contextually. This literature review may work if your subject is broad, you are alone, or you are short on time (Grant and Booth, 2009). Since narrative reviews are descriptive, they focus on a subset of research in an area chosen based on availability or author interest. They do not entail a systematic literature search, making them less complete. Literary reviews are informative but often biased towards specific viewpoints. Sometimes they are misleading, especially if a related study's results differ (Uman, 2011).

Systematic reviews and Meta-analysis: Systematic reviews mitigate biasness by identifying, evaluating, and synthesizing significant papers on a particular topic. Systematic reviews sometimes include meta-analysis to aggregate data from several research studies into a single quantitative estimate or summary impact size (Uman, 2011). Effect sizes define the degree of correlation between two variables, revealing the intervention's impact, unlike conventional hypothesis testing, which can reveal statistical significance (i.e., whether the intervention group differed from the control group). Determining the effect size often relies on various factors, including the nature of the outcome and intervention under investigation and the data obtained from the published trials (Ahan and Kang, 2018). An example of a highly regarded non-profit organization in the medical profession is the Cochrane Collaboration (www.cochrane.org). This group supports and disseminates systematic reviews and meta-analyses of medical therapies' efficacy (Uman, 2011). Meta-analysis improves findings' precision and reliability, resulting in better results (Figure 1).

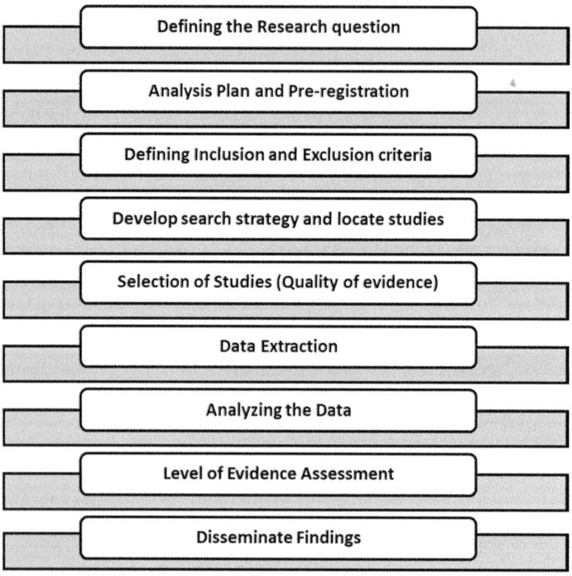

Figure 1: Authenticity and levels of evidence.

***Abbreviations**: SRM: Systematic review and Meta-analysis; RCT: Randomized controlled trial.

3. Stages of Systematic Reviews and Meta-analysis

Setting the study question, locating relevant papers, and categorizing extracted data are crucial. Every effective meta-analysis starts with it. There must be data to conduct a meta-analysis. First, we must define our meta-analysis's study topic and eligibility criteria. We must find and select relevant papers, collect data for our calculations, and code crucial information for presentation (Kang, 2015). Following criteria, standards, and guidelines at each stage can help us develop a high-quality meta-analysis (Piggot and Polanin, 2020). Figure 2 depicts a systematic review and meta-analysis flowchart.

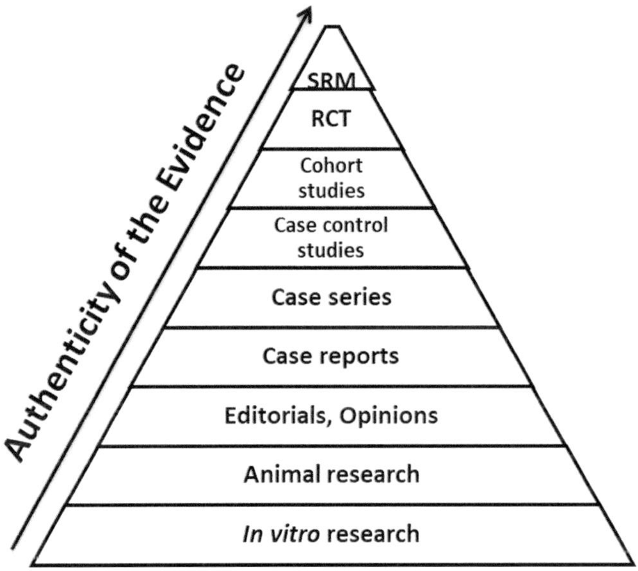

Figure 2: Flowchart illustrating the systematic review and meta-analysis.

3.1. Formulate the Research Question

The initial phase involves defining the review inquiry, developing a hypothesis, and choosing a title. Starting with a problem statement helps create a good research question. A meta-analysis must address a specific issue or challenge for relevance and significance. The Cummings et al. (2013) FINER framework can be used to choose the meta-analysis problem. According to the criteria, research questions should be viable, intriguing, new, ethical, and relevant.

A systematic review uses well-defined and systematic methods to collect all empirical evidence on a topic to answer a question. A meta-analysis is a statistical method used to combine findings from similar studies. When choosing a meta-analysis topic, it is crucial to make sure all the research provides data that can be integrated. A meta-analysis can be conducted with at least two studies if they provide data on the same issue, as Ahan and Kang's (2018) research demonstrated. However, meta-analysis requires a thorough review study selection. A clear explanation of the Population, Intervention, Comparison, and Outcomes (PICO) criteria, which underpin evidence-based research, is crucial. Even without precise data, a research topic should be recognizable to readers (Stewart et al., 2015). Effective search tactics are organized and follow the PICO framework. PICO approaches help researchers to organize and classify search queries. The PICO model helps organize and refine a research problem into a searchable query. Researchers can find relevant literature search phrases and topics by breaking down PICO (Scells et al., 2017).

3.2. Analysis Plan and Pre-registration

A comprehensive research plan must be registered in advance in systematic reviews and meta-analyses. The National Institute of Health Research developed PROSPERO, a global prospective registry for systematic reviews (2011) to consolidate this information and centralize systematic review processes. The PROSPERO database contains future systematic reviews that authors freely register. A protocol that meets eligibility requirements and includes the complete review technique in the registration application will be accepted by PROSPERO. After finalizing the protocol, authors should register in PROSPERO before reviewing publications for eligibility. Primary and secondary outcomes and procedures must be established in advance to maintain study transparency. According to Ahan and Kang (2018), researchers and readers are notified about the modifications made to a particular method, including the details of when, how, and why these changes were implemented. The public can access the PROSPERO database at https://www.crd.york.ac.uk/PROSPERO/. The homepage includes links for registering a review, searching the database, joining PROSPERO, logging into one's account, accessing additional information about PROSPERO, registering a protocol, a concise database description, and a summary of the latest registrations.

3.3. Defining Inclusion and Exclusion Criteria

Once the research issue is specified, the meta-analyst can look for relevant studies for the review. Researchers should use multiple search approaches to find relevant papers that address their research question. According to Swift and Wampold (2018), research that fails the rejection criteria can be excluded later. The Cochrane abbreviation PICO (or PICOC), which stands for population, intervention, comparison, outcomes, and context helps define all necessary components before starting the review. In this context, the authors must determine the community's demographic age range, current conditions, expected outcomes, and intervention and control groups. Only randomized controlled trials (RCTs), a combination of RCTs and almost experimental methods, and qualitative studies must be excluded according to specific criteria.

3.4. Literature Search and Study Selection

Evidence-based research needs a solid foundation, so it is essential to do a comprehensive investigation that includes as many studies as possible that meet the inclusion and exclusion criteria (Ahan and Kang, 2018). Medline, Embase, and the Cochrane Central Register of Controlled Trials (CENTRAL) are used most of the time. In addition, it is important to make a complete list of key terms (called 'MeSH' terms) for each part of PICOC to find all relevant studies in a given area. The ideal search technique combines sensitivity (getting many relevant studies) and specificity (getting few irrelevant studies). Most searches employ numerous relevant electronic databases, but they can also examine article reference lists, hand-search key journals, post requests on listservs, and talk to experts or key researchers (Uman, 2011). The researchers exclude similar studies from the search results, identify studies with abstracts that fulfil inclusion/exclusion criteria, and then choose the final articles with full texts. At least two scientists independently select the studies for clarity and fairness (Kang, 2016). After selecting the final studies for the review, complete the PRISMA flow chart (Page et al., 2021) with all the details. Figure 3 shows all the review steps.

3.5. Assessment of Study Quality

Accurately assessing the quality of the evidence is crucial to determining the strength of the meta-analysis's recommendations. According to Guyatt et al. (2008), when the quality of the included studies could be better, it can adversely affect the overall quality of a meta-analysis, perhaps leading to incorrect results. Many comprehensive guidelines and standards exist in the field, including the Consolidated Standards of Reporting Trials (CONSORT Statement; http://www.consort-statement.org/), which offer suggestions for improving psychological intervention RCTs and meta-analyses (Uman et al., 2010). The Gradeworking Group (http://www.gradeworkinggroup.org/)'s (GRADE) methodology provides a valuable framework for assessing the study's shortcomings, flaws, incomplete outcome data, indirect evidence, and publication bias. The risk of bias in an investigation can be assessed and categorized as low, high, or unclear depending on the existence or absence of six specific processes. Other biases include random sequence creation, allocation concealment, blinding of participants or investigators, inadequate outcome data, and selective reporting (Higgins et al., 2011).

3.6. Extraction of Data

The information collected from each reviewed study (e.g., authors, publication year, number of participants, age range, study design, results, included/excluded) can be organized using a simple data extraction form or table. It is crucial to ensure the presence of a minimum of two reviewers for data extraction to establish inter-rater reliability and mitigate data entry errors. In the event that a

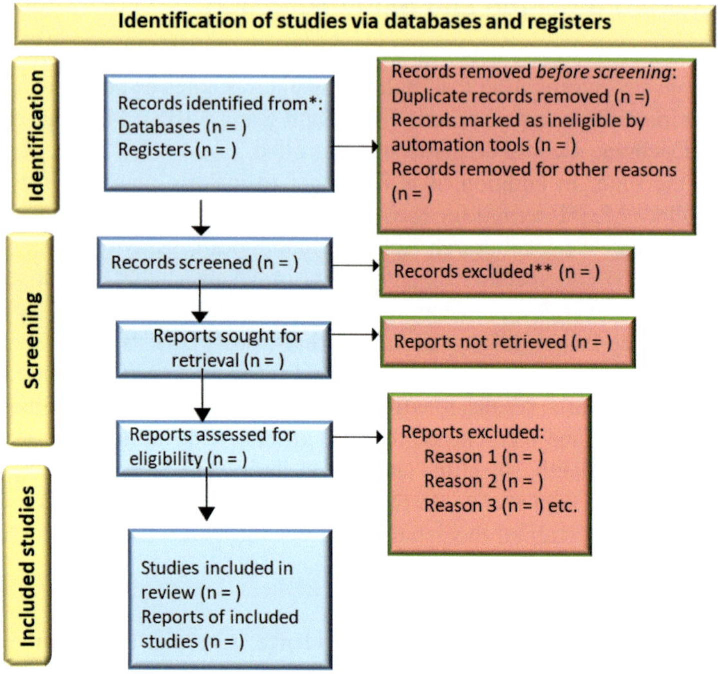

Figure 3: PRISM flowchart for systematic reviews and meta-analysis.
Source: (https://guelphhumber.libguides.com/c.php?g=213266&p=1406923)

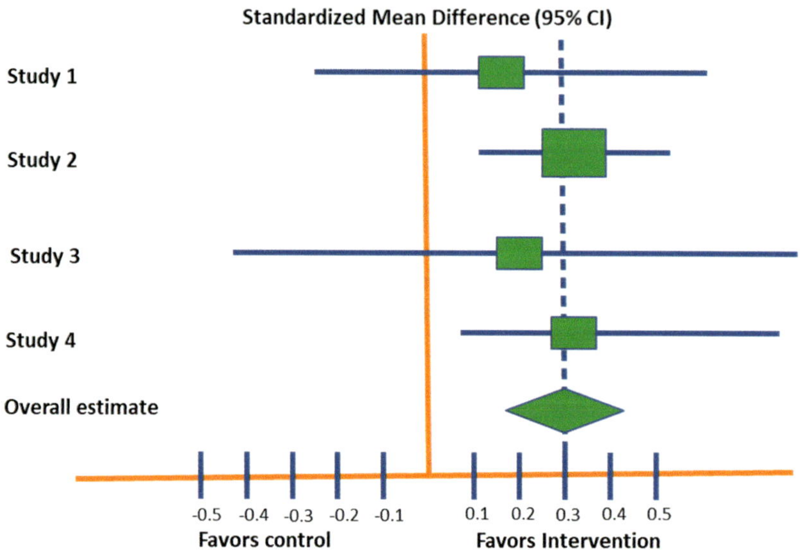

Figure 4: Example of a Forest Plot.

consensus cannot be reached between the two reviewers, a third reviewer should be consulted as suggested by Ahn and Kang (2018). Due to each variable's varying sizes and formats, the resulting outcomes likewise exhibit differences in size and format. Consequently, when merging the data, it may be necessary to make small modifications, as mentioned by Egger (1998).

4. Data Analysis

Meta-analyses aim to produce more accurate and reliable results than individual studies. Meta-analyses use statistical methods like Review Manager (RevMan) and Cochrane Collaboration R packages to assess effect sizes (Deeks et al., 2019). An effect size and 95% confidence interval (CI) range are shown in numerical and graphical formats like forest plots. Forest plots are visual representations that display each trial as a horizontal diamond shape. The centre of the diamond represents the effect size, such as the standardized mean difference (SMD). In contrast, the endpoints of the diamond represent the upper and lower bounds of the confidence interval (CI) (Figure 4). These diamonds are depicted on a graph with a central line denoting the zero point. The left side of the graph (zero) represents the treatment condition, whereas the right side (>zero) represents the control condition. At the bottom of the graph is a summary effect size or diamond indicating the combined results of all separate research. The intervention should be more effective than the control if this full diamond (effect size and both CI anchors) goes below zero. Most systems also generate a heterogeneity score to determine if studies are comparable (Dettori et al., 2021).

5. Statistical Analysis

Researchers employ various methodologies to integrate and summarize the outcomes of investigations incorporated into their research. Before conducting analysis, it is imperative to do a crucial first step known as data extraction in the extraction sheet. During this process, the analyst arranges the data within the extraction sheet in a format compatible with analytical tools, facilitating further analysis (Sivarajah et al., 2017). Subgroup analysis, sensitivity analysis, cumulative analyses, and meta-regression are:

- Suitable methodologies for assessing the consistency of results
- Exploring the impact of specific findings on the outcome
- Identifying the most effective models
- Assessment of publication bias is crucial to examine the potential impact of missing research on the overall summary.

Significance tests to detect publication bias are:

a. Begg and Mazumdar test (1994).
b. Egger's test (1997).

Egger's test is an analysis of the Y-intercept equalling zero in a linear regression model that analyses the relationship between the normalized effect estimate (the estimate divided by its standard error) and precision (the reciprocal of the standard error of the estimate) (Egger et al., 1997). Begg's test examines whether standardized effect sizes and variances are correlated. Publication bias is present when the two-sided P-value is low for both tests. With a smaller number of studies, each method will have limited power of identification of bias power (Begg et al., 1994).

Is it possible to correct the effect of publication bias using the available data? Various methodologies, such as trim and fill, selection models, and meta-regression, can be employed in this context. The trim and fill deal with two distinct phases. The initial step is data trimming, which entails the removal of studies in a sequential manner, beginning with those that possess minor statistical strength until a state of symmetry is achieved in the funnel plot. A novel aggregated estimation is derived from the remaining studies. Additionally, the gaps found are addressed by including them in the pooled estimate line and introducing additional research to compensate for the excluded ones (Shi et al., 2019).

6. Analyzing Data in Meta-analyses

Meta-analysis refers to the statistical analysis of findings derived from several independent investigations. The occurrence of a significant finding, referred to as a "diamond at the bottom of a plot", is often a highly expected instance for researchers. However, it is essential to recognize that the findings of meta-analyses can only be accurate if proper attention has been devoted to various essential aspects. These include formulating a well-defined review question, establishing clear eligibility criteria, meticulously identifying and selecting relevant studies, critically appraising their quality, collecting appropriate data, and determining the meaningful aspects to be analyzed (Jahan et al., 2016). These steps are necessary to maintain the validity and reliability of the meta-analysis results. Having a comprehensive understanding of the various types of data, such as dichotomous or continuous, that arise from measuring an outcome in particular research is necessary.

Additionally, it is essential to select appropriate impact measures when comparing different intervention groups carefully. Most meta-analysis techniques include modifications to a weighted mean of the impact estimates derived from many studies (Haidich et al., 2010). The presence of heterogeneity, which refers to variation between studies, is an essential factor to examine in research. The concept of "random effects" refers to a statistical modeling approach that considers the Meta-analyses to include heterogeneity by assuming that the underlying effects conform to a normal distribution. The process of preparing a Cochrane review or meta-analysis requires a series of several decisions. Sensitivity analyses are vital to assess the robustness of overall findings in light of possibly influencing decisions. The phenomenon that arises from the comparison of outcomes between two groups subjected to distinct treatments is sometimes referred to as the 'effect', "treatment effect", or "intervention effect" (Nagendrababu et al., 2021).

Cohen's *d* is a statistical measure employed to quantify the standardized difference between two means. Furthermore, it is extensively utilized in the field of meta-analysis. Cohen's *d* is a suitable effect size measure for evaluating the difference between two means. Cohen's *d* can be computed by dividing the difference between the means by the pooled standard deviation (Sullivan and Feinn, 2016). When assessing an effect, it is commonly recommended to consider the effect size as small (0.2), medium (0.5), and large (0.8), according to general recommendations.

$$\text{Cohen's } d = (M_2 - M_1)/SD_{\text{pooled}}$$

$$SD_{\text{pooled}} = \sqrt{((SD_1^{\,2} + SD_2^{\,2})/2)}$$

where M_1, M_2 are sample means and SD_1 and SD_2 are standard deviations of two different independent studies.

The RevMan program necessitates that researchers deliberately choose a preferred strategy for each outcome. If explicit specifications are not provided, the program will employ the fixed-effect Mantel-Haenszel odds ratio for dichotomous outcomes, the fixed-effect mean difference for continuous outcomes, and the fixed-effect model for generic inverse-variance outcomes (Chang et al., 2017).

6.1. Heterogeneity

Heterogeneity is a term used to describe any form of variability observed in research included in a systematic review. Differentiating between various forms of heterogeneity can be advantageous. Clinical variety, or clinical heterogeneity, refers to the variability observed in the participants, interventions, and outcomes examined within a study. On the other hand, methodological diversity, sometimes referred to as methodological heterogeneity, relates to the variability in study design and the potential for bias. Statistical heterogeneity refers to the variability observed in the intervention effects being assessed across many studies, which can be attributed to differences in clinical or methodological characteristics, or a combination of both, among the studies. Statistical heterogeneity is evident when the observed effects of interventions exhibit more significant dissimilarity than what would be anticipated only based on random error (chance). The term "statistical heterogeneity" will be referred to as 'heterogeneity' by convention (Higgins et al., 2002).

When the confidence intervals for individual research outcomes, usually displayed graphically with horizontal lines, exhibit limited overlap, it generally suggests the existence of statistical heterogeneity. There exists a statistical test for heterogeneity that is characterized by a more formal approach. The chi-squared ($\chi 2$, or Chi2) test is incorporated inside the forest plots featured in Cochrane reviews. A small P value (or a high chi-squared statistic compared to its degree of freedom) indicates the presence of heterogeneity in intervention effects, suggesting that there is variance in the estimated effects that cannot be attributed to chance alone (Thorlund et al., 2012).

It is imperative to use caution when interpreting the chi-squared test, as it exhibits limited statistical power in the prevalent scenario of a meta-analysis where studies possess small sample sizes or are limited in number. This implies that whereas a statistically significant outcome may suggest the presence of heterogeneity, a lack of statistical significance should not be interpreted as proof of the absence of heterogeneity. An additional issue pertaining to the test, which is infrequently observed in Cochrane reviews, is its propensity to exhibit great statistical power in detecting even minimal levels of heterogeneity in meta-analyses, despite the potential lack of clinical significance (Higgins et al., 2002). A useful statistic for quantifying inconsistency is

$$I^2 = \left(\frac{Q - df}{Q}\right) \times 100\%$$

The chi-squared statistic, denoted as Q, and its degrees of freedom (df) are referenced by Higgins (2002). This pertains to the percentage of the variability in effect estimates that may be attributed to heterogeneity rather than random sampling error.

The interpretation of I^2 can be potentially deceptive since the significance of inconsistency is dependent upon multiple conditions. A fundamental framework for interpretation can be outlined as follows:

- The observed data may not have significant importance in the range of 0%–40%.
- When the observed data falls within the 30%– 60% range, it is likely to indicate moderate heterogeneity.
- If the observed data falls within the 50%–90% range, it may suggest substantial heterogeneity.
- When the observed data falls within the range of 75%–100%, it indicates considerable heterogeneity.

The significance of the observed value of I^2 is contingent upon two factors: (1) the magnitude and direction of effects and (2) the degree of evidence for heterogeneity, which can be determined by the P value from the chi-squared test or a confidence interval for I^2.

Role of R Package in Meta-Analysis

Meta-analysis is a statistical methodology that enables researchers to combine effect sizes derived from several primary studies for a comprehensive synthesis. The open-source statistical environment R is increasingly favored as a popular tool for estimating meta-analysis models. The meta-analytic community has facilitated the expansion in the field of meta-analysis, which has developed a wide range of specialized software programs for conducting meta-analyses.

There is a comprehensive list of meta-analytic R programs that are publicly accessible. The R software environment, accessed at https://www.r-project.org, and the commercially available Stata5 software offer the most extensive range of statistical tools for meta-analyses. RStudio, available at https://www.rstudio.com/, is widely recognized and endorsed as an integrated development environment (IDE) for the R programming language. It offers a user-friendly interface with various features, including graphical tools for generating plots, a command history for referencing past R commands, capabilities for managing data, and the ability to install and update R packages conveniently. Before conducting a meta-analysis, installing the R packages 'meta' and 'metasens' is necessary. These packages encompass all the required functions for performing the analyses and generating the figures as provided in this paper (Lortie et al., 2020). To install these packages, the following command can be used: install.packages ('meta', 'metasns').

7. Efficacious Implication of Meta-analysis on Nutrition and Health Evaluation

The field of nutrition science poses specific challenges for meta-analyses. Nutrition intervention clinical trials have substantial methodological heterogeneity, undermining the case for pooling findings compared to drug trials. There is considerable variation in how various research address these concerns, like food intakes, and these data may have been generated using different dietary assessment methods. In observational studies, it can be challenging to reliably quantify the consumption of most common foods and nutrients due to the considerable variation in dietary practises across populations, even though some dietary characteristics, like caffeine intake, are constant among people from day-to-day life and are reliably reported. So, there should be some points that need to be considered when combining studies for meta-analysis, described as follows with examples.

7.1. Differences in Population

In meta-analysis, pooling data from multiple studies can boost statistical power; however, when there are substantial differences across studies in participant demographics or research methodology, the pooled data may show more variation in results, making it more difficult to discern true effects. For this reason, a null result in a meta-analysis may indicate heterogeneity among studies that dilute real results. Kapoor et al. (2023) showed excluding studies with wider cumulative interval values impacted the meta-analysis outcome. Hence, sometimes it's helpful to conduct a sensitivity analysis which involves methodically excluding certain studies from the analysis to get the most relevant data. A meta-analysis's ability to explain why various studies have yielded inconsistent findings is perhaps more valuable than a single statistical summary of effect size. Unexplained heterogeneity should be recognized, and subgroup analyses may shed light on the causes of observed differences (Zhang et al., 2022; Tiwari et al., 2023). When there is a great deal of heterogeneity between studies regarding population size and research design, a meta-analysis may be less informative than a single or small number of carefully executed studies.

7.2. Comparison Strategy

Any conclusions on the impact of a change in the diet must be drawn compared to some baseline. Like Tiwari et al. (2023), meta-analysis compared the effect of dietary anthocyanins from food sources either in extract form or as whole anthocyanin-rich food (fruit, vegetables, grains etc.). To compare all studies, they converted their anthocyanins-rich food and extracts into anthocyanin doses per kg of body weight of animals and humans accordingly to get same baseline. On the other hand, O'Connor et al. (2017), meta-analysis included 39 studies that linked red meat consumption with blood lipid concentrations. Their meta-analysis combined the 34 studies that combined red meat with other meats and showed a lower decrease in LDL-cholesterol and 5 more studies that compared red meat with plant-based foods and showed no significant association with LDL-cholesterol. However, researchers pooled the data from all 39 studies together and concluded that red meat has no effect on CVD risk variables, resulting in a false outcome of meta-analysis. Thus, a more effective strategy would involve isolating a particular comparison and checking if enough studies had employed the method of interest.

7.3. Quality of Comparing Studies

The quality of studies varies substantially. It depends upon the kind of trials because some trials are randomized, and others are not. It can be cleared by looking into two independent meta-analysis studies. One was conducted in 2014, showing the impact of palm oil consumption on blood lipids (Fattore et al., 2014), inconsistent with study quality and inconsistent results. A similar kind of study was conducted in 2015, but with high-quality randomized trials that concluded significant rise in LDL cholesterol by palm oils (Sun et al., 2015). Thus, all meta-analyses must adhere to reporting guidelines like PRISMA (Preferred Reporting Items for Systematic Reviews and Meta-Analyses) and incorporate a comprehensive risk of bias evaluation in individual research due to the significance of trial quality.

8. Conclusion

Meta-analyses peer reviewers and authors need additional checks and balances in the peer-review process to ensure that no harm will come from the findings and unbiased analysis needs to be performed. Methods to improve the quality of meta-analyses include but are not limited to (1) giving more weight to meta-analyses based on pooling original primary data than those based on secondary data; (2) needing to be edited by editors who are both familiar in meta-analysis and the specific field being discussed; and (3) requiring writers to provide enough information for others to replicate their work, including summary data and details about their methodology. Meta-analyses need extensive research on the studies included and any potential conflicts of interest. Meta-analyses of nutritional research or other subjects may benefit from these procedures but will not eliminate disputed results.

References

Ahn, E., and H. Kang. 2018. Introduction to systematic review and meta-analysis. *Korean J. Anesthesiol.*, 71: 103–112.

Barber, T.M., S. Kabisch, A.F. Pfeiffer, and M.O. Weickert. 2020. The health benefits of dietary fibee. *Nutrients*, 12: 3209.

Becker, B.J. 1987. Applying tests of combined significance in meta-analysis. *Psychol. Bull.*, 102: 164.

Begg, C.B., and M. Mazumdar. 1994. Operating characteristics of a rank correlation test for publication bias. *Biometrics*, 1088–1101.

Begg, C.B., and J.A. Berlin. 1988. Publication bias: A problem in interpreting medical data. *J.R. Stat. Soc. Ser. A (Stat Soc.)*, 151: 419–445.

Borenstein, M., L.V. Hedges, J.P. Higgins, and H.R. Rothstein. 2021. *Introduction to Meta-analysis*. John Wiley & Sons.

Chang, B.H., and D.C. Hoaglin. 2017. Meta-analysis of odds ratios: Current good practices. *Med. Care*, 55: 328.

Cheng, H.M., G. Koutsidis, J.K. Lodge, A.W. Ashor, M. Siervo, and J. Lara. 2019. Lycopene and tomato and risk of cardiovascular diseases: A systematic review and meta-analysis of epidemiological evidence. *Crit. Rev. Food Sci. Nutr.*, 59: 141–158.

Cochran, W.G. 1937. Problems arising in the analysis of a series of similar experiments. *Suppl. J.R. Stat. Soc.*, 4: 102–118.

Cuijpers, P. 2016. *Meta-analyses in Mental Health Research: A Practical Guide*, Vol. 15. Amsterdam: Vrije Universiteit.

Cummings, S.R., W.S. Browner, and S.B. Hulley. 2013. Conceiving the research question and developing the study plan. *Designing Clinical Research*, 4: 14–22.

Deeks, J.J., J.P. Higgins, D.G. Altman, and Cochrane. 2019. Statistical Methods Group. 2019. Analysing data and undertaking meta-analyses. *Cochrane Handbook for Systematic Reviews of Interventions*, 241–284.

Dettori, J.R., D.C. Norvell, and J.R. Chapman. 2021. Seeing the forest by looking at the trees: How to interpret a meta-analysis forest plot. *Global Spine J.*, 11: 614–616.

Egger, M., M. Schneider, and G.D. Smith. 1998. Meta-analysis Spurious precision? Meta-analysis of observational studies. *BMJ*, 316: 140–144.

Egger, M., G.D. Smith, M. Schneider, and C. Minder. 1997. Bias in meta-analysis detected by a simple, graphical test. *BMJ*, 315: 629–634.

Fattore, E., C. Bosetti, F. Brighenti, C. Agostoni, and G. Fattore. 2014. Palm oil and blood lipid–related markers of cardiovascular disease: A systematic review and meta-analysis of dietary intervention trials. *Am. J. Clin. Nutr.*, 99: 1331–1350.

Glass, G.V. 1976. Primary, secondary, and meta-analysis of research. *Educ. Res.*, 5: 3–8.

Grant, M.J., and A. Booth. 2009. A typology of reviews: An analysis of 14 review types and associated methodologies. *Health Inf. Libr. J.*, 26: 91–108.

Haidich, A.B. 2010. Meta-analysis in medical research. *Hippokratia*, 14: 29.

Harrer, M., P. Cuijpers, T. Furukawa, and D. Ebert. 2021. *Doing Meta-analysis with R: A Hands-on Guide*. Chapman and Hall/CRC.

Hedges, L.V. 1992. Meta-analysis. *J. Educ. Stat.*, 17: 279–296.

Higgins, J.P., J. Thomas, J. Chandler, M. Cumpston, T. Li, M.J. Page, and V.A. Welch. 2019. *Cochrane Handbook for Systematic Reviews of Interventions*. John Wiley & Sons.

Higgins, J.P., D.G. Altman, P.C. Gøtzsche, P. Jüni, D. Moher, A.D. Oxman, and J.A. Sterne. 2011. The Cochrane Collaboration's tool for assessing risk of bias in randomized trials. *BMJ*, 343: D5928.

Higgins, J.P., and S.G. Thompson. 2002. Quantifying heterogeneity in a meta-analysis. *Stat Med.*, 21: 1539–1558.

Jahan, N., S. Naveed, M. Zeshan, and M.A. Tahir. 2016. How to conduct a systematic review: A narrative literature review. *Cureus*, 4 Nov. 8(11): e864.

Jonnalagadda, S.S., L. Harnack, R. Hai Liu, N. McKeown, C. Seal, S. Liu, and G.C. Fahey. 2011. Putting the whole grain puzzle together: Health benefits associated with whole grains—Summary of American Society for Nutrition 2010 Satellite Symposium. *J. Nutr.*, 141: 1011S–1022S.

Kang, H. 2016. How to understand and conduct evidence-based medicine. *Korean J. Anesthesiol.*, 69: 435–445.

Kang, H. 2015. Statistical considerations in meta-analysis. *Hanyang Med. Rev.*, 35: 23–32.

Kapoor, P., A. Tiwari, S. Sharma, V. Tiwari, B. Sheoran, U. Ali, and M. Garg. 2023. Effect of anthocyanins on gut health markers, Firmicutes-Bacteroidetes ratio and short-chain fatty acids: A systematic review via meta-analysis. *Sci. Rep.*, 13: 1729.

Lea, E., and T. Worsley. 2003. The factors associated with the belief that vegetarian diets provide health benefits. *Asia Pac. J. Clin. Nut.*, 12(3): 296–303.

Lipsey, M.W., and D.B. Wilson. 2001. *Practical Meta-analysis*. Sage Publications, Inc.

Lortie, C.J., and A. Filazzola. 2020. A contrast of meta and metafor packages for meta-analyses in R. *Ecol. Evol.*, 10: 10916.

Lund, E.K. 2013. Health benefits of seafood; is it just the fatty acids? *Food Chem.*, 140: 413–420.

Nagendrababu, V., H.F. Duncan, S.J. Pulikkotil, and P.M.H. Dummer. 2021. Glossary for randomized clinical trials. *Int. Endod. J.*, 54: 354–365.

O'Connor, L.E., J.E. Kim, and W.W. Campbell. 2017. Total red meat intake of ≥0.5 servings/d does not negatively influence cardiovascular disease risk factors: A systemically searched meta-analysis of randomized controlled trials. *Am. J. Clin. Nutr.*, 105: 57–69.

Page, M.J., J.E. McKenzie, P.M. Bossuyt, I. Boutron, T.C. Hoffmann, C.D. Mulrow, and D. Moher. 2021. The PRISMA 2020 statement: An updated guideline for reporting systematic reviews. *Int. J. Surg.*, 88: 105906.

Pigott, T.D., and J.R. Polanin. 2020. Methodological guidance paper: High-quality meta-analysis in a systematic review. *Rev. Educ. Res.*, 90: 24–46.

Scells, H., G. Zuccon, B. Koopman, A. Deacon, L. Azzopardi, and S. Geva. 2017. Integrating the framing of clinical questions via PICO into the retrieval of medical literature for systematic reviews. *In*: *Proceedings of the 2017 ACM on Conference on Information and Knowledge Management*, 2291–2294.

Schiavo, J.H. 2019. PROSPERO: An international register of systematic review protocols. *Med. Ref. Serv. Q.*, 38: 171–180.

Shi, L., and L. Lin. 2019. The trim-and-fill method for publication bias: Practical guidelines and recommendations based on a large database of meta-analyses. *Medicine*, 98(23): e15987.

Sivarajah, U., M.M. Kamal, Z. Irani, and V. Weerakkody. 2017. Critical analysis of Big Data challenges and analytical methods. *J. Bus. Res.*, 70: 263–286.

Stewart, L.A., M. Clarke, M. Rovers, R.D. Riley, M. Simmonds, G. Stewart, and J.F. Tierney. 2015. Preferred reporting items for a systematic review and meta-analysis of individual participant data: The PRISMA-IPD statement. *Jama.*, 313: 1657–1665.

Sullivan, G.M., and R. Feinn. 2012. Using effect size: Or why the P value is not enough. *J. Grad. Med. Educ.*, 4: 279–282.

Sun, Y., N. Neelakantan, Y. Wu, R. Lote-Oke, A. Pan, and R.M. van Dam. 2015. Palm oil consumption increases LDL cholesterol compared with vegetable oils low in saturated fat in a meta-analysis of clinical trials. *J. Nutr.*, 145: 1549–1558.

Swift, J.K., and B.E. Wampold. 2018. Inclusion and exclusion strategies for conducting meta-analyses. *Psychother. Res.*, 28: 356–366.

Thorlund, K., G. Imberger, B.C. Johnston, M. Walsh, T. Awad, L. Thabane, and J. Wetterslev. 2012. Evolution of heterogeneity (I2) estimates and their 95% confidence intervals in large meta-analyses. *PloS One*, 7: e39471.

Tiwari, V., S. Sharma, A. Tiwari, B. Sheoran, S. Kaur, A. Sharma, and M. Garg. 2023. Effect of dietary anthocyanins on biomarkers of type 2 diabetes and related obesity: A systematic review and meta-analysis. *Crit. Rev. Food Sci. Nutr.* (March): 1–18.

Uman, L.S. 2011. Systematic reviews and meta-analyses. *J. Can. Acad. Child Adolesc. Psychiatry*, 20: 57.

Uman, L.S., C.T. Chambers, P.J. McGrath, S. Kisely, D. Matthews, and K. Hayton. 2010. Assessing the quality of randomized controlled trials examining psychological interventions for pediatric procedural pain: Recommendations for quality improvement. *J. Pediatr. Psychol.*, 35: 693–703.

Van Duyn, M.A.S., and E. Pivonka. 2000. Overview of the health benefits of fruit and vegetable consumption for the dietetics professional: Selected literature. *J. Am. Diet. Assoc.*, 100: 1511–1521.

Zhang, Y., J. Ding, and J. Liang. 2022. Associations of dietary vitamin A and beta-carotene intake with depression: A meta-analysis of observational studies. *Front. Nutr.*, 9: 881139.

Index

For Product Safety Concerns and Information please contact our EU representative GPSR@taylorandfrancis.com Taylor & Francis Verlag GmbH, Kaufingerstraße 24, 80331 München, Germany